Universitext

Editorial Board
(North America):

S. Axler

K.A. Ribet

Universitext

Editors (North America): S. Axler and K.A. Ribet

(continued after index)

Klaus-Jochen Engel
Rainer Nagel

A Short Course on
Operator Semigroups

Klaus-Jochen Engel
Faculty of Engineering
Division of Mathematics
Piazzale E. Pontieri, 2
I-67040 Monteluco di Roio (AQ)
Italy

Rainer Nagel
Mathematics Institute
Tübingen University
Auf der Morgenstelle 10
D-72076 Tübingen
Germany

Mathematics Subject Classification (2000): 47Dxx

ISBN   978-1-4419-2174-1   e-ISBN   978-0-387-36619-7

Printed on acid-free paper.

10 9 8 7 6 5 4 3 2 1

springer.com

Preface

The theory of strongly continuous semigroups of linear operators on Banach spaces, *operator semigroups* for short, has become an indispensable tool in a great number of areas of modern mathematical analysis. In our Springer Graduate Text [EN00] we presented this beautiful theory, together with many applications, and tried to show the progress made since the publication in 1957 of the now classical monograph [HP57] by E. Hille and R. Phillips. However, the wealth of results exhibited in our Graduate Text seems to have discouraged some of the potentially interested readers. With the present text we offer a streamlined version that strictly sticks to the essentials. We have skipped certain parts, avoided the use of sophisticated arguments, and, occasionally, weakened the formulation of results and modified the proofs. However, to a large extent this book consists of excerpts taken from our Graduate Text, with some new material on positive semigroups added in Chapter VI.

We hope that the present text will help students take their first step into this interesting and lively research field. On the other side, it should provide useful tools for the working mathematician.

Acknowledgments

This book is dedicated to our students. Without them we would not be able to do and to enjoy mathematics. Many of them read previous versions when it served as the text of our Seventh Internet Seminar 2003/04. Here

 Genni Fragnelli, Marc Preunkert and Mark C. Veraar

were among the most active readers. Particular thanks go to

 Tanja Eisner, Vera Keicher, Agnes Radl

for proposing considerable improvements in the final version.

How much we owe to our colleague and friend Ulf Schlotterbeck cannot be seen from the pages of this book.

Klaus-Jochen Engel
Rainer Nagel
October 2005

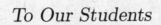

To Our Students

Contents

Chapter I

Introduction

Generally speaking, a *dynamical system* is a family $(T(t))_{t \geq 0}$ of mappings on a set X satisfying

$$\begin{cases} T(t+s) = T(t)T(s) & \text{for all } t, s \geq 0, \\ T(0) = id. \end{cases}$$

Here X is viewed as the set of all states of a system, $t \in \mathbb{R}_+ := [0, \infty)$ as time and $T(t)$ as the map describing the change of a state $x \in X$ at time 0 into the state $T(t)x$ at time t. In the linear context, the state space X is a vector space, each $T(t)$ is a linear operator on X, and $(T(t))_{t \geq 0}$ is called a (one-parameter) *operator semigroup*.

The standard situation in which such operator semigroups naturally appear are so-called *Abstract Cauchy Problems*

(ACP) $\qquad \begin{cases} \dot{u}(t) = Au(t) & \text{for } t \geq 0, \\ u(0) = x, \end{cases}$

where A is a linear operator on a Banach space X. Here, the problem consists in finding a differentiable function u on $\mathbb{R}_+$ such that (ACP) holds. If for each initial value $x \in X$ a unique solution $u(\cdot, x)$ exists, then

$$T(t)x := u(t, x), \ t \geq 0, \ x \in X,$$

defines an operator semigroup. For the "working mathematician," (ACP) is the problem, and $(T(t))_{t \geq 0}$ the solution to be found. The opposite point of view also makes sense: given an operator semigroup (i.e., a dynamical system) $(T(t))_{t \geq 0}$, under what conditions can it be "described" by a differential equation (ACP), and how can the operator A be found?

1

In some simple and concrete situations (see Section 2 below) the relation between $(T(t))_{t \geq 0}$ and A is given by the formulas

$$T(t) = e^{tA} \qquad \text{and} \qquad A = \tfrac{d}{dt} T(t)|_{t=0}.$$

In general, a comparably simple relation seems to be out of reach. However, miraculously as it may seem, a simple continuity assumption on the semigroup (see Definition 1.1) produces, in the usual Banach space setting, a rich and beautiful theory with a broad and almost universal field of applications. It is the aim of this course to develop this theory.

1. Strongly Continuous Semigroups

The following is our basic definition.

1.1 Definition. *A family $(T(t))_{t \geq 0}$ of bounded linear operators on a Banach space X is called a strongly continuous (one-parameter) semigroup (or C_0-semigroup[1]) if it satisfies the functional equation*

(FE) $\begin{cases} T(t+s) = T(t)T(s) & \text{for all } t, s \geq 0, \\ T(0) = I \end{cases}$

and is strongly continuous in the following sense. For every $x \in X$ the orbit maps

(SC) $\xi_x : t \mapsto \xi_x(t) := T(t)x$

are continuous from $\mathbb{R}_+$ into X for every $x \in X$.

The property (SC) can also be expressed by saying that the map

$$t \mapsto T(t)$$

is continuous from $\mathbb{R}_+$ into the space $\mathcal{L}_s(X)$ of all bounded operators on X endowed with the strong operator topology (see Appendix A, (A.2)).

Finally, if these properties hold for $\mathbb{R}$ instead of $\mathbb{R}_+$, we call $(T(t))_{t \in \mathbb{R}}$ a *strongly continuous (one-parameter) group* (or C_0-*group*) on X.

a. Basic Properties

Our first goal is to facilitate the verification of the strong continuity (SC) required in Definition 1.1. This is possible thanks to the uniform boundedness principle, which implies the following frequently used equivalence. (See also Exercise 1.8.(1) and the more abstract version in Proposition A.3.)

---

[1] Although we prefer the terminology "strongly continuous," we point out that the symbol C_0 abbreviates "Cesàro summable of order 0."

1.2 Lemma. *Let X be a Banach space and let F be a function from a compact set $K \subset \mathbb{R}$ into $\mathcal{L}(X)$. Then the following assertions are equivalent.*

(a) *F is continuous for the strong operator topology; i.e., the mappings $K \ni t \mapsto F(t)x \in X$ are continuous for every $x \in X$.*

(b) *F is uniformly bounded on K, and the mappings $K \ni t \mapsto F(t)x \in X$ are continuous for all x in some dense subset D of X.*

(c) *F is continuous for the topology of uniform convergence on compact subsets of X; i.e., the map*

$$K \times C \ni (t, x) \mapsto F(t)x \in X$$

is uniformly continuous for every compact set C in X.

PROOF. The implication (c) $\Rightarrow$ (a) is trivial, whereas (a) $\Rightarrow$ (b) follows from the uniform boundedness principle, because the mappings $t \mapsto F(t)x$ are continuous, hence bounded, on the compact set K.

To show (b) $\Rightarrow$ (c), we assume $\|F(t)\| \leq M$ for all $t \in K$ and fix some $\varepsilon > 0$ and a compact set C in X. Then there exist finitely many $x_1, \ldots, x_m \in D$ such that $C \subset \bigcup_{i=1}^{m} (x_i + \varepsilon/M \, U)$, where U denotes the unit ball of X. Now choose $\delta > 0$ such that $\|F(t)x_i - F(s)x_i\| \leq \varepsilon$ for all $i = 1, \ldots, m$, and for all $t, s \in K$, such that $|t - s| \leq \delta$. For arbitrary x, $y \in C$ and $t, s \in K$ with $\|x - y\| \leq \varepsilon/M$ $|t - s| \leq \delta$, this yields

$$\|F(t)x - F(s)y\| \leq \|F(t)(x - x_j)\| + \|(F(t) - F(s))x_j\|$$
$$+ \|F(s)(x_j - x)\| + \|F(s)(x - y)\| \leq 4\varepsilon,$$

where we choose $j \in \{1, \ldots, m\}$ such that $\|x - x_j\| \leq \varepsilon/M$. This estimate proves that $(t, x) \mapsto F(t)x$ is uniformly continuous with respect to $t \in K$ and $x \in C$. □

As an easy consequence of this lemma, in combination with the functional equation (FE), we obtain that the continuity of the orbit maps

$$\xi_x : t \mapsto T(t)x$$

at each $t \geq 0$ and for each $x \in X$ is already implied by much weaker properties.

1.3 Proposition. *For a semigroup $(T(t))_{t \geq 0}$ on a Banach space X, the following assertions are equivalent.*

(a) *$(T(t))_{t \geq 0}$ is strongly continuous.*

(b) *$\lim_{t \downarrow 0} T(t)x = x$ for all $x \in X$.*

(c) *There exist $\delta > 0$, $M \geq 1$, and a dense subset $D \subset X$ such that*

    (i) *$\|T(t)\| \leq M$ for all $t \in [0, \delta]$,*

    (ii) *$\lim_{t \downarrow 0} T(t)x = x$ for all $x \in D$.*

PROOF. The implication (a) $\Rightarrow$ (c.ii) is trivial. In order to prove that (a) $\Rightarrow$ (c.i), we assume, by contradiction, that there exists a sequence $(\delta_n)_{n\in\mathbb{N}} \subset \mathbb{R}_+$ converging to zero such that $\|T(\delta_n)\| \to \infty$ as $n \to \infty$. Then, by the uniform boundedness principle, there exists $x \in X$ such that $(\|T(\delta_n)x\|)_{n\in\mathbb{N}}$ is unbounded, contradicting the fact that $T(\cdot)x$ is continuous at $t = 0$.

In order to verify that (c) $\Rightarrow$ (b), we put $K := \{t_n : n \in \mathbb{N}\} \cup \{0\}$ for an arbitrary sequence $(t_n)_{n\in\mathbb{N}} \subset [0,\infty)$ converging to $t = 0$. Then $K \subset [0,\infty)$ is compact, $T(\cdot)_{|K}$ is bounded, and $T(\cdot)_{|K}x$ is continuous for all $x \in D$. Hence, we can apply Lemma 1.2.(b) to obtain

$$\lim_{n\to\infty} T(t_n)x = x$$

for all $x \in X$. Because $(t_n)_{n\in\mathbb{N}}$ was chosen arbitrarily, this proves (b).

To show that (b) $\Rightarrow$ (a), let $t_0 > 0$ and let $x \in X$. Then

$$\lim_{h\downarrow 0} \|T(t_0 + h)x - T(t_0)x\| \leq \|T(t_0)\| \cdot \lim_{h\downarrow 0} \|T(h)x - x\| = 0,$$

which proves right continuity. If $h < 0$, the estimate

$$\|T(t_0 + h)x - T(t_0)x\| \leq \|T(t_0 + h)\| \cdot \|x - T(-h)x\|$$

implies left continuity whenever $\|T(t)\|$ remains uniformly bounded for $t \in [0, t_0]$. This, however, follows as above first for some small interval $[0, \delta]$ by the uniform boundedness principle and then on each compact interval using (FE). □

Because in many cases the uniform boundedness of the operators $T(t)$ for $t \in [0, t_0]$ is obvious, one obtains strong continuity by checking (right) continuity of the orbit maps ξ_x at $t = 0$ for a dense set of "nice" elements $x \in X$ only.

We demonstrate the advantage of this procedure in the examples discussed below (e.g., in Paragraph 3.15).

We repeat that for a strongly continuous semigroup $(T(t))_{t\geq 0}$ the finite orbits

$$\{T(t)x : t \in [0, t_0]\}$$

are continuous images of a compact interval, hence compact and therefore bounded for each $x \in X$. So by the uniform boundedness principle, each strongly continuous semigroup is uniformly bounded on each compact interval, a fact that implies exponential boundedness on $\mathbb{R}_+$.

1.4 Proposition. *For every strongly continuous semigroup* $(T(t))_{t\geq 0}$*, there exist constants* $w \in \mathbb{R}$ *and* $M \geq 1$ *such that*

$$(1.1) \qquad\qquad \|T(t)\| \leq Me^{wt}$$

for all $t \geq 0$.

PROOF. Choose $M \geq 1$ such that $\|T(s)\| \leq M$ for all $0 \leq s \leq 1$ and write $t \geq 0$ as $t = s + n$ for $n \in \mathbb{N}$ and $0 \leq s < 1$. Then

$$\|T(t)\| \leq \|T(s)\| \cdot \|T(1)\|^n \leq M^{n+1}$$
$$= Me^{n \log M} \leq Me^{wt}$$

holds for $w := \log M$ and each $t \geq 0$. $\square$

The infimum of all exponents w for which an estimate of the form (1.1) holds for a given strongly continuous semigroup plays an important role in the sequel. We therefore reserve a name for it.

1.5 Definition. *For a strongly continuous semigroup* $\mathcal{T} = \big(T(t)\big)_{t \geq 0}$, *we call*

$$\omega_0 := \omega_0(\mathcal{T}) := \inf \left\{ w \in \mathbb{R} : \begin{array}{l} \text{there exists } M_w \geq 1 \text{ such that} \\ \|T(t)\| \leq M_w e^{wt} \text{ for all } t \geq 0 \end{array} \right\}$$

its growth bound (or type). Moreover, a semigroup is called bounded if we can take $w = 0$ in (1.1), and contractive if $w = 0$ and $M = 1$ is possible. Finally, the semigroup $\big(T(t)\big)_{t \geq 0}$ is called isometric if $\|T(t)x\| = \|x\|$ for all $t \geq 0$ and $x \in X$.

It becomes clear in the discussion below, but is presently left as a challenge to the reader that

- $\omega_0 = -\infty$ may occur,
- The infimum in (1.1) may not be attained; i.e, it might happen that no constant M exists such that $\|T(t)\| \leq Me^{\omega_0 t}$ for all $t \geq 0$, and
- Constants $M > 1$ may be necessary; i.e., no matter how large $w \geq \omega_0$ is chosen, $\|T(t)\|$ will not be dominated by e^{wt} for all $t \geq 0$.

We close this subsection by showing that using the weak operator topology instead of the strong operator topology in Definition 1.1 will not change our class of semigroups.

This is a surprising result, and its proof needs more sophisticated tools from functional analysis than we have used up to this point. So the beginner may just skip the proof.

1.6 Theorem. *A semigroup $\big(T(t)\big)_{t \geq 0}$ on a Banach space X is strongly continuous if and only if it is weakly continuous, i.e., if the mappings*

$$\mathbb{R}_+ \ni t \mapsto \langle T(t)x, x' \rangle \in \mathbb{C}$$

are continuous for each $x \in X$, $x' \in X'$.

PROOF. We have only to show that weak implies strong continuity. As a first step, we use the principle of uniform boundedness twice to conclude that on compact intervals, $(T(t))_{t\geq 0}$ is pointwise and then uniformly bounded. Therefore (use Proposition 1.3.(c)), it suffices to show that

$$E := \left\{ x \in X : \lim_{t\downarrow 0} \|T(t)x - x\| = 0 \right\}$$

is a (strongly) dense subspace of X.

To this end, we define for $x \in X$ and $r > 0$ a linear form x_r on X' by

$$\langle x_r, x' \rangle := \frac{1}{r} \int_0^r \langle T(s)x, x' \rangle \, ds \qquad \text{for } x' \in X'.$$

Then x_r is bounded and hence $x_r \in X''$. On the other hand, the set

$$\{ T(s)x : s \in [0, r] \}$$

is the continuous image of $[0, r]$ in the space X endowed with the weak topology, hence is weakly compact in X. Kreĭn's theorem (see Proposition A.1.(ii)) implies that its closed convex hull

$$\overline{\text{co}} \left\{ T(s)x : s \in [0, r] \right\}$$

is still weakly compact in X. Because x_r is a $\sigma(X'', X')$-limit of Riemann sums, it follows that

$$x_r \in \overline{\text{co}} \left\{ T(s)x : s \in [0, r] \right\},$$

whence $x_r \in X$. (See also [Rud73, Thm. 3.27].)

It is clear from the definition that the set

$$D := \left\{ x_r : r > 0, x \in X \right\}$$

is weakly dense in X. On the other hand, for $x_r \in D$ we obtain

$$
\begin{aligned}
\|T(t)x_r - x_r\| &= \sup_{\|x'\|\leq 1} \left| \frac{1}{r} \int_t^{t+r} \langle T(s)x, x' \rangle \, ds - \frac{1}{r} \int_0^r \langle T(s)x, x' \rangle \, ds \right| \\
&\leq \sup_{\|x'\|\leq 1} \left(\left| \frac{1}{r} \int_r^{r+t} \langle T(s)x, x' \rangle \, ds \right| + \left| \frac{1}{r} \int_0^t \langle T(s)x, x' \rangle \, ds \right| \right) \\
&\leq \frac{2t}{r} \|x\| \sup_{0\leq s\leq r+t} \|T(s)\| \to 0
\end{aligned}
$$

as $t \downarrow 0$; i.e., $D \subset E$. We conclude that E is weakly, hence by Proposition A.1.(i) strongly, dense in X. $\square$

1.7 Final Comment. In the next subsection we start a relatively long discussion of elementary constructions and concrete examples. The reader impatient to see the general theory can immediately proceed to Chapter II.

1.8 Exercises. (1) Let X be a Banach space and let $(T_n)_{n\in\mathbb{N}}$ be a sequence in $\mathcal{L}(X)$. Then the following assertions are equivalent.

(a) $(T_n x)_{n\in\mathbb{N}}$ converges for all $x \in X$.

(b) $(T_n)_{n\in\mathbb{N}} \subset \mathcal{L}(X)$ is bounded and $(T_n x)_{n\in\mathbb{N}}$ converges for all x in some dense subset D of X.

(c) $(T_n x)_{n\in\mathbb{N}}$ converges uniformly for all $x \in C$ and every compact set C in X.

(2) Show that the left translation semigroup $(T_l(t))_{t\geq 0}$ defined by

$$(T_l(t)f)(s) := f(s+t), \quad s, t \geq 0,$$

is strongly continuous on each of the Banach spaces

(a) $\mathrm{C}_0(\mathbb{R}_+) := \{f \in \mathrm{C}(\mathbb{R}_+) : \lim_{s\to\infty} f(s) = 0\}$ endowed with the sup-norm,

(b) $\mathrm{C}_{ub}(\mathbb{R}_+) := \{f \in \mathrm{C}(\mathbb{R}_+) : f \text{ is bounded and uniformly continuous}\}$ endowed with the sup-norm,

(c) $\mathrm{C}_0^1(\mathbb{R}_+) := \{f \in \mathrm{C}^1(\mathbb{R}_+) : \lim_{s\to\infty} f(s) = \lim_{s\to\infty} f'(s) = 0\}$ endowed with the norm $\|f\| := \sup_{s\geq 0} |f(s)| + \sup_{s\geq 0} |f'(s)|$.

(3) Define

$$(T(t)f)(s) := f(se^t), \quad s, t \geq 0,$$

and show that $(T(t))_{t\geq 0}$ yields strongly continuous semigroups on

$$X_\infty := \mathrm{C}_0[1, \infty) := \{f \in \mathrm{C}[1, \infty) : \lim_{s\to\infty} f(s) = 0\}$$

and $X_p := \mathrm{L}^p[1, \infty)$ for $1 \leq p < \infty$.

(4) Show that for a group $(T(t))_{t\in\mathbb{R}}$ on a Banach space X the following conditions are equivalent.

(a) The group $(T(t))_{t\geq 0}$ is strongly continuous; i.e., the map $\mathbb{R} \ni t \mapsto T(t)x \in X$ is continuous for all $x \in X$.

(b) $\lim_{t\to t_0} T(t)x = T(t_0)x$ for some $t_0 \in \mathbb{R}$ and all $x \in X$.

(c) There exist constants $t_0 \in \mathbb{R}$, $\delta > 0$, $M \geq 0$ and a dense subset $D \subset X$ such that

　(i) $\|T(t)\| \leq M$ for all $t \in [t_0, t_0 + \delta]$,

　(ii) $\lim_{t\downarrow t_0} T(t)x = T(t_0)x$ for all $x \in D$.

(5) Show that a strongly continuous semigroup $(T(t))_{t\geq 0}$ containing an invertible operator $T(t_0)$ for some $t_0 > 0$ can be extended to a strongly continuous group $(T(t))_{t\in\mathbb{R}}$.

(6) On $X := \mathrm{C}[0,1]$, define bounded operators $T(t)$, $t > 0$, by

$$(T(t)f)(s) := \begin{cases} \mathrm{e}^{t \log s}[f(s) - f(0) \log s] & \text{if } s \in (0,1], \\ 0 & \text{if } s = 0 \end{cases}$$

for $f \in X$ and put $T(0) := I$. Prove the following assertions.

(i) $(T(t))_{t \geq 0}$ is a semigroup that is strongly continuous only on $(0, \infty)$.

(ii) $\lim_{t \downarrow 0} \|T(t)\| = \infty$.

(7) Construct a strongly continuous semigroup that is not nilpotent (hence satisfies $T(t) \neq 0$ for all $t \geq 0$), but has growth bound $\omega_0 = -\infty$. (Hint: Consider $(T(t)f)(s) := \mathrm{e}^{-t^2 + 2st} f(s - t)$ on the Banach space $\mathrm{C}_0(-\infty, 0] = \{f \in \mathrm{C}[-\infty, 0) : \lim_{s \to -\infty} f(s) = 0\}$.)

b. Standard Constructions

We now explain how one can construct in various ways new strongly continuous semigroups from a given one. This might seem trivial and/or boring, but there will be many occasions to appreciate the toolbox provided by these constructions. In any case, it is a series of instructive exercises.

In the following, we always assume $\mathcal{T} = (T(t))_{t \geq 0}$ to be a strongly continuous semigroup on a Banach space X.

1.9 Similar Semigroups. Given another Banach space Y and an isomorphism V from Y onto X, we obtain (as in Lemma 2.4) a new strongly continuous semigroup $\mathcal{S} = (S(t))_{t \geq 0}$ on Y by defining

$$S(t) := V^{-1} T(t) V \qquad \text{for } t \geq 0.$$

Without explicit reference to the isomorphism V, we call the two semigroups $\mathcal{T}$ and $\mathcal{S}$ *similar* or *isomorphic*. Two such semigroups have the same topological properties; e.g., $\omega_0(\mathcal{T}) = \omega_0(\mathcal{S})$.

1.10 Rescaled Semigroups. For any numbers $\mu \in \mathbb{C}$ and $\alpha > 0$, we define the *rescaled semigroup* $(S(t))_{t \geq 0}$ by

$$S(t) := \mathrm{e}^{\mu t} T(\alpha t)$$

for $t \geq 0$.

For example, taking $\mu = -\omega_0$ (or $\mu < -\omega_0$) and $\alpha = 1$ the rescaled semigroup will have growth bound equal to (or less than) zero. This is an assumption we make without loss of generality in many situations.

1.11 Subspace Semigroups. If Y is a closed subspace of X such that $T(t)Y \subseteq Y$ for all $t \geq 0$, i.e., if Y is $(T(t))_{t \geq 0}$-*invariant*, then the restrictions

$$T(t)_| := T(t)_{|Y}$$

form a strongly continuous semigroup $(T(t)_|)_{t \geq 0}$, called the *subspace semigroup*, on the Banach space Y.

1.12 Quotient Semigroups. For a closed $(T(t))_{t \geq 0}$-invariant subspace Y of X, we consider the quotient Banach space $X_/ := {}^X/_Y$ with canonical quotient map $q : X \to X_/$. The quotient operators $T(t)_/$ given by

$$T(t)_/ q(x) := q(T(t)x) \qquad \text{for } x \in X \text{ and } t \geq 0$$

are well-defined and form a strongly continuous semigroup, called the *quotient semigroup* $(T(t)_/)_{t \geq 0}$ on $X_/$.

1.13 Adjoint Semigroups. The *adjoint semigroup* $(T(t)')_{t \geq 0}$ consisting of all adjoint operators $T(t)'$ on the dual space X' is, in general, *not* strongly continuous.

An example is provided by the (left) translation group on $L^1(\mathbb{R})$ (see Section 3.c). Its adjoint is the (right) translation group on $L^\infty(\mathbb{R})$, which is not strongly continuous (see the proposition in Paragraph 3.15). However, it is easy to see that $(T(t)')_{t \geq 0}$ is always *weak*-continuous* in the sense that the maps

$$t \mapsto \xi_{x,x'}(t) := \langle x, T(t)'x' \rangle = \langle T(t)x, x' \rangle$$

are continuous for all $x \in X$ and $x' \in X'$.

Because on the dual of a reflexive Banach space weak and weak* topology coincide, the adjoint semigroup on such spaces is weakly, and hence by Theorem 1.6 strongly, continuous.

Proposition. *The adjoint semigroup of a strongly continuous semigroup on a reflexive Banach space is again strongly continuous.*

1.14 Product Semigroups. Let $(S(t))_{t \geq 0}$ be another strongly continuous semigroup commuting with $(T(t))_{t \geq 0}$; i.e., $S(t)T(t) = T(t)S(t)$ for all $t \geq 0$. Then the operators

$$U(t) := S(t)T(t)$$

form a strongly continuous semigroup $(U(t))_{t \geq 0}$, called the *product semigroup* of $(T(t))_{t \geq 0}$ and $(S(t))_{t \geq 0}$.

PROOF. Clearly, $U(0) = I$. In order to show the semigroup property for $(U(t))_{t \geq 0}$, we first show that $T(s)$ and $S(r)$ commute for all $s, r \geq 0$. To this end, we first take $r = p_1/q$ and $s = p_2/q \in \mathbb{Q}_+$. Then

$$S(r)T(s) = S\left(1/q\right)^{p_1} \cdot T\left(1/q\right)^{p_2}$$
$$= T\left(1/q\right)^{p_2} \cdot S\left(1/q\right)^{p_1} = T(s)S(r);$$

i.e., $F(r, s) = G(r, s)$ for all $r, s \in \mathbb{Q}_+$, where

$$F : [0, \infty) \times [0, \infty) \to \mathcal{L}(X), \qquad F(r, s) := S(r)T(s),$$
and
$$G : [0, \infty) \times [0, \infty) \to \mathcal{L}(X), \qquad G(r, s) := T(s)S(r).$$

Now, for fixed $x \in X$, the functions $F(\cdot, \cdot)x$ and $G(\cdot, \cdot)x$ are continuous in each coordinate and coincide on $\mathbb{Q}_+ \times \mathbb{Q}_+$; hence we conclude that $F = G$. This shows that

$$S(r)T(s) = T(s)S(r)$$

for all $s, r \geq 0$, and the semigroup property $U(r + s) = U(r)U(s)$ for $s, r \geq 0$ follows immediately. Finally, the strong continuity of $(U(t))_{t \geq 0}$ follows from Lemma A.18. □

1.15 Exercises. (1) Let $(T_l(t))_{t \geq 0}$ be the left translation semigroup (cf. Exercise 1.8.(2)) on $X := C_0(\mathbb{R}_+)$ or $C_{ub}(\mathbb{R}_+)$ and take a nonvanishing, continuous function q on $\mathbb{R}_+$ such that q and $1/q$ are bounded. Then the multiplication operator M_q defined by $(M_q f)(s) := q(s) \cdot f(s)$ yields a similarity transformation. Determine the semigroup $(S(t))_{t \geq 0}$ defined by

$$S(t) := M_q T_l(t) M_{1/q}, \quad t \geq 0.$$

(2) On $X := C_0(\mathbb{R}^2) = \{f \in C(\mathbb{R}^2) : \lim_{\|(x,y)\| \to \infty} f(x, y) = 0\}$ or

$$C_{ub}(\mathbb{R}^2) := \{f \in C(\mathbb{R}^2) : f \text{ is bounded and uniformly continuous}\}$$

endowed with the sup-norm, consider the two semigroups $(T(t))_{t \geq 0}$ and $(S(t))_{t \geq 0}$ defined by

$$(S(t)f)(x, y) := f(x + t, y) \qquad \text{and} \qquad (T(t)f)(x, y) := f(x, y + t)$$

for $f \in X$, $t \geq 0$. Show that both are strongly continuous and determine their product semigroup.

(3) Consider the function space

$$Y := \{f : [0, 1] \to \mathbb{C} : |f(s)| \leq ns \text{ for all } s \in [0, 1] \text{ and some } n \in \mathbb{N}\},$$

which becomes a Banach space for the norm

$$\|f\| := \inf\{c \geq 0 : |f(s)| \leq cs \text{ for all } s \in [0, 1]\}.$$

On $X := \mathbb{C} \oplus Y$, we define a "translation" semigroup $(T(t))_{t \geq 0}$ by $T(0) := I$ and

$$T(t)\begin{pmatrix} \alpha \\ f \end{pmatrix} := \begin{pmatrix} 0 \\ g \end{pmatrix} \qquad \text{for } t > 0,$$

where

$$g(s) := \begin{cases} 0 & \text{for } s < t, \\ \alpha & \text{for } s = t, \\ f(s-t) & \text{for } s > t. \end{cases}$$

(i) Show that $\|T(t)\| = t^{-1}$ for $t \in (0,1)$, and hence $(T(t))_{t \geq 0}$ is not exponentially bounded.

(ii) Find the largest $(T(t))_{t \geq 0}$-invariant closed subspace of X on which the restriction of $(T(t))_{t \geq 0}$ becomes strongly continuous for $t > 0$ ($t \geq 0$, respectively).

2. Examples

In order to create a feeling for the concepts introduced so far, we discuss first the case in which the semigroup $(T(t))_{t \geq 0}$ can be represented as an operator-valued exponential function $(e^{tA})_{t \geq 0}$. Due to this representation, we later consider this case as rather trivial.

a. Finite-Dimensional Systems: Matrix Semigroups

We start with a reasonably detailed discussion of the finite-dimensional situation; i.e., $X = \mathbb{C}^n$. Here, $\mathcal{L}(X)$ is identified with the space $M_n(\mathbb{C})$ of all complex $n \times n$ matrices.

Because on $M_n(\mathbb{C})$ all Hausdorff vector space topologies coincide, we simply speak of continuity of a semigroup $(T(t))_{t \geq 0}$ on X. We want to determine all continuous semigroups on $X = \mathbb{C}^n$ and start by looking at the natural examples in the form of matrix exponentials.

2.1 Proposition. *For any $A \in M_n(\mathbb{C})$ and $t \geq 0$, the series*

$$(2.1) \qquad e^{tA} := \sum_{k=0}^{\infty} \frac{t^k A^k}{k!}$$

converges absolutely. Moreover, the mapping

$$\mathbb{R}_+ \ni t \mapsto e^{tA} \in M_n(\mathbb{C})$$

is continuous and satisfies

$$(FE) \qquad \begin{cases} e^{(t+s)A} = e^{tA}e^{sA} & \text{for } t, s \geq 0, \\ e^{0A} = I. \end{cases}$$

PROOF. Because the series $\sum_{k=0}^{\infty} t^k \|A\|^k / k!$ converges, one can show, as for the Cauchy product of scalar series, that

$$\sum_{k=0}^{\infty} \frac{t^k A^k}{k!} \cdot \sum_{k=0}^{\infty} \frac{s^k A^k}{k!} = \sum_{n=0}^{\infty} \sum_{k=0}^{n} \frac{t^{n-k} A^{n-k}}{(n-k)!} \cdot \frac{s^k A^k}{k!}$$

$$= \sum_{n=0}^{\infty} \frac{(t+s)^n A^n}{n!}.$$

This proves (FE). In order to show that $t \mapsto e^{tA}$ is continuous, we first observe that by (FE) one has

$$e^{(t+h)A} - e^{tA} = e^{tA}\left(e^{hA} - I\right)$$

for all $t, h \in \mathbb{R}$. Therefore, it suffices to show that $\lim_{h \to 0} e^{hA} = I$. This follows from the estimate

$$\left\| e^{hA} - I \right\| = \left\| \sum_{k=1}^{\infty} \frac{h^k A^k}{k!} \right\|$$

$$\leq \sum_{k=1}^{\infty} \frac{|h|^k \cdot \|A\|^k}{k!} = e^{|h| \cdot \|A\|} - 1.$$

$\square$

2.2 Definition. *We call $\left(e^{tA}\right)_{t \geq 0}$ the (one-parameter) semigroup generated by the matrix $A \in M_n(\mathbb{C})$.*

As the reader may have already realized, there is no need in Proposition 2.1 (and in Definition 2.2) to restrict the (time) parameter t to $\mathbb{R}_+$. The definition, the continuity, and the functional equation (FE) hold for any real and even complex t. Then the map

$$T(\cdot) : t \mapsto e^{tA}$$

extends to a continuous (even analytic) homomorphism from the additive group $(\mathbb{R}, +)$ (or, $(\mathbb{C}, +)$) into the multiplicative group $GL(n, \mathbb{C})$ of all invertible, complex $n \times n$ matrices. We call $\left(e^{tA}\right)_{t \in \mathbb{R}}$ the (one-parameter) *group* generated by A.

Before proceeding with the abstract theory, the reader might appreciate some examples of matrix semigroups.

2.3 Examples. (i) The (semi) group generated by a diagonal matrix $A = \text{diag}(a_1, \ldots, a_n)$ is given by

$$e^{tA} = \text{diag}\left(e^{ta_1}, \ldots, e^{ta_n}\right).$$

(ii) Less trivial is the case of a $k \times k$ Jordan block

$$
A = \begin{pmatrix} \lambda & 1 & 0 & \cdots & 0 \\ 0 & \lambda & 1 & \ddots & \vdots \\ \vdots & \ddots & \ddots & \ddots & 0 \\ \vdots & & \ddots & \ddots & 1 \\ 0 & \cdots & \cdots & 0 & \lambda \end{pmatrix}_{k \times k}
$$

with eigenvalue $\lambda \in \mathbb{C}$. Decompose A into a sum $A = D + N$ where $D = \lambda I$. Then the kth power of N is zero, and the power series (2.1) (with A replaced by N) becomes

$$
(2.2) \qquad e^{tN} = \begin{pmatrix} 1 & t & \frac{t^2}{2} & \cdots & \frac{t^{k-1}}{(k-1)!} \\ 0 & 1 & t & \cdots & \frac{t^{k-2}}{(k-2)!} \\ \vdots & \ddots & \ddots & \ddots & \vdots \\ \vdots & & \ddots & \ddots & t \\ 0 & \cdots & \cdots & 0 & 1 \end{pmatrix}_{k \times k}.
$$

Because D and N commute, we obtain

$$
(2.3) \qquad e^{tA} = e^{t\lambda} e^{tN}
$$

(see Exercise 2.9.(1)).

For arbitrary matrices A, the direct computation of e^{tA} (using the above definition) is very difficult if not impossible. Fortunately, thanks to the existence of the Jordan normal form, the following lemma shows that in a certain sense the Examples 2.3.(i) and (ii) suffice.

2.4 Lemma. *Let $B \in M_n(\mathbb{C})$ and take an invertible matrix $S \in M_n(\mathbb{C})$. Then the (semi) group generated by the matrix $A := S^{-1}BS$ is given by*

$$
e^{tA} = S^{-1} e^{tB} S.
$$

PROOF. Because $A^k = S^{-1} B^k S$ for all $k \in \mathbb{N}$ and because S, S^{-1} are continuous operators, we obtain

$$
e^{tA} = \sum_{k=0}^{\infty} \frac{t^k A^k}{k!} = \sum_{k=0}^{\infty} \frac{t^k S^{-1} B^k S}{k!}
$$

$$
= S^{-1} \left(\sum_{k=0}^{\infty} \frac{t^k B^k}{k!} \right) S = S^{-1} e^{tB} S.
$$

$\square$

The content of this lemma is that *similar* matrices (for the definition of similarity see Paragraph 1.9) generate *similar* (semi) groups. Because we know that any complex $n \times n$ matrix is similar to a direct sum of Jordan blocks, we conclude that any matrix (semi) group is similar to a direct sum of (semi) groups as in Example 2.3.(ii). Already in the case of 2×2 matrices, the necessary computations are lengthy; however, they yield explicit formulas for the matrix exponential function.

2.5 More Examples. (iii) Take an arbitrary 2×2 matrix $A = \begin{pmatrix} a & b \\ c & d \end{pmatrix}$, define $\delta := ad - bc$, $\tau := a + d$, and take $\gamma \in \mathbb{C}$ such that $\gamma^2 = 1/4(\tau^2 - 4\delta)$. Then the (semi) group generated by A is given by the matrices

(2.4)
$$
\mathrm{e}^{tA} = \begin{cases} \mathrm{e}^{t\tau/2}\left(1/\gamma \sinh(t\gamma)A + \big(\cosh(t\gamma) - \tau/2\gamma \sinh(t\gamma)\big)I \right) & \text{if } \gamma \neq 0, \\[2mm] \mathrm{e}^{t\tau/2}\left(tA + (1 - t\tau/2)I \right) & \text{if } \gamma = 0. \end{cases}
$$

We list some special cases yielding simpler formulas:

$$
A = \begin{pmatrix} 0 & 1 \\ -1 & 0 \end{pmatrix}, \qquad \mathrm{e}^{tA} = \begin{pmatrix} \cos(t) & \sin(t) \\ -\sin(t) & \cos(t) \end{pmatrix},
$$

$$
A = \begin{pmatrix} 0 & 1 \\ 1 & 0 \end{pmatrix}, \qquad \mathrm{e}^{tA} = \begin{pmatrix} \cosh(t) & \sinh(t) \\ \sinh(t) & \cosh(t) \end{pmatrix},
$$

$$
A = \begin{pmatrix} 1 & 1 \\ -1 & -1 \end{pmatrix}, \qquad \mathrm{e}^{tA} = \begin{pmatrix} 1+t & t \\ -t & 1-t \end{pmatrix}.
$$

In the case where the spectral projections of a general $n \times n$ matrix are known, the corresponding (semi)group can be calculated explicitly by the following formula. We recall that the *minimal polynomial* m_A of a matrix A is the polynomial of lowest degree with leading coefficient 1 satisfying $m_A(A) = 0$. Moreover, the set of zeros of m_A coincides with the spectrum $\sigma(A)$ of A.

2.6 Proposition. *Let $A \in \mathrm{M}_n(\mathbb{C})$ with eigenvalues $\lambda_1, \ldots, \lambda_m$ and respective multiplicities $k_1, \ldots, k_m$ as zeros of the minimal polynomial m_A of A. If P_i denotes the spectral projection associated with $\{\lambda_i\}$, $1 \leq i \leq m$, (cf. (1.7) in Chapter V), then*

$$
\mathrm{e}^{tA} = \sum_{i=1}^{m} \sum_{j=0}^{k_i-1} \mathrm{e}^{t\lambda_i} \frac{t^j}{j!} (A - \lambda_i)^j P_i \qquad \text{for } t \in \mathbb{R}.
$$

PROOF. Because the spectral projections P_i, $1 \leq i \leq m$, sum up to the identity, we have

$$
\mathrm{e}^{tA} = \sum_{i=1}^{m} \mathrm{e}^{tA} P_i = \sum_{i=1}^{m} \mathrm{e}^{t\lambda_i} \mathrm{e}^{t(A - \lambda_i)} P_i.
$$

Recall that $\left((A - \lambda_i)|_{\text{rg}\,P_i}\right)^{k_i} = 0$ by Exercise V.1.20.(1). Therefore, we obtain

$$e^{tA} = \sum_{i=1}^{m} \sum_{j=0}^{k_i-1} e^{t\lambda_i} \frac{t^j}{j!} (A - \lambda_i)^j P_i$$

as claimed.   □

Returning to one of the questions posed at the very beginning of this text, namely if a given semigroup can be described by a differential equation, we now proceed in two more steps. First, we show that in the case $T(t) = e^{tA}$ we even have differentiability of the map $t \mapsto T(t)$ (from $\mathbb{R}$ to $M_n(\mathbb{C})$), and that $U(t) := e^{tA}$ solves the differential equation

(DE) $\begin{cases} \frac{d}{dt} U(t) = A U(t) & \text{for } t \geq 0, \\ U(0) = I. \end{cases}$

In a second step, we show that a general continuous operator semigroup on $X = \mathbb{C}^n$ is even differentiable in $t = 0$ and is the exponential of its derivative at $t = 0$.

2.7 Proposition. *Let $T(t) := e^{tA}$ for some $A \in M_n(\mathbb{C})$. Then the function $T(\cdot) : \mathbb{R}_+ \to M_n(\mathbb{C})$ is differentiable and satisfies the differential equation (DE). Conversely, every differentiable function $T(\cdot) : \mathbb{R}_+ \to M_n(\mathbb{C})$ satisfying (DE) is already of the form $T(t) = e^{tA}$ for[2] $A := \dot{T}(0) \in M_n(\mathbb{C})$.*

PROOF. We only show that $T(\cdot)$ satisfies (DE). Because the functional equation (FE) in Proposition 2.1 implies

$$\frac{T(t+h) - T(t)}{h} = \frac{T(h) - I}{h} \cdot T(t)$$

for all $t, h \in \mathbb{R}$, (DE) is proved if $\lim_{h \to 0} \frac{T(h) - I}{h} = A$. This, however, follows, because

$$\left\| \frac{T(h) - I}{h} - A \right\| \leq \sum_{k=2}^{\infty} \frac{|h|^{k-1} \cdot \|A\|^k}{k!}$$

$$= \frac{e^{|h| \cdot \|A\|} - 1}{|h|} - \|A\| \to 0 \quad \text{as } h \to 0.$$

The proof of remaining assertions is left to the reader; cf. Exercise 2.9.(5).
                                                                                      □

---

[2] Here $\dot{T}(0) := \frac{d}{dt} T(t)_{|t=0}$.

2.8 Theorem. *Let $T(\cdot) : \mathbb{R}_+ \to M_n(\mathbb{C})$ be a continuous function satisfying* (FE). *Then there exists $A \in M_n(\mathbb{C})$ such that*

$$T(t) = e^{tA} \quad \text{for all } t \geq 0.$$

PROOF. Because $T(\cdot)$ is continuous and $T(0) = I$ is invertible, the matrices

$$V(t_0) := \int_0^{t_0} T(s)\, ds$$

are invertible for sufficiently small $t_0 > 0$ (use that $\lim_{t \downarrow 0} 1/t\, V(t) = T(0) = I$). The functional equation (FE) now yields

$$T(t) = V(t_0)^{-1} V(t_0) T(t) = V(t_0)^{-1} \int_0^{t_0} T(t+s)\, ds$$
$$= V(t_0)^{-1} \int_t^{t+t_0} T(s)\, ds = V(t_0)^{-1}\bigl(V(t+t_0) - V(t)\bigr)$$

for all $t \geq 0$. Hence, $T(\cdot)$ is differentiable with derivative

$$\tfrac{d}{dt} T(t) = \lim_{h \downarrow 0} \frac{T(t+h) - T(t)}{h}$$
$$= \lim_{h \downarrow 0} \frac{T(h) - T(0)}{h} T(t) = \dot{T}(0) T(t) \quad \text{for all } t \geq 0.$$

This shows that $T(\cdot)$ satisfies (DE) with $A = \dot{T}(0)$. $\qquad\qquad\square$

     With this theorem we have characterized all continuous one-parameter (semi) groups on $\mathbb{C}^n$ as matrix-valued exponential functions $\bigl(e^{tA}\bigr)_{t \geq 0}$.

2.9 Exercises. (1) If $A, B \in M_n(\mathbb{C})$ commute, then $e^{A+B} = e^A e^B$.

(2) Let $A \in M_n(\mathbb{C})$ be an $n \times n$ matrix and denote by m_A its minimal polynomial. If p is a polynomial such that $p \equiv \exp (\mathrm{mod}\, m_A)$; i.e., if the function $(p - \exp)/m_A$ can be analytically extended to $\mathbb{C}$, then $p(A) = \exp(A)$. Use this fact in order to verify Formula (2.4).

(3) Show that $A \in M_n(\mathbb{C})$ generates a bounded group, i.e., $\|e^{tA}\| \leq M$ for all $t \in \mathbb{R}$ and some $M \geq 1$, if and only if A is similar to a diagonal matrix with purely imaginary entries.

(4) Characterize semigroups $\bigl(e^{tA}\bigr)_{t \geq 0}$ satisfying $e^A = I$ in terms of the eigenvalues of the matrix $A \in M_n(\mathbb{C})$.

(5) Show that every differentiable function $T(\cdot) : \mathbb{R}_+ \to M_n(\mathbb{C})$ satisfying (DE) is already of the form $T(t) = e^{tA}$ for $A := \dot{T}(0) \in M_n(\mathbb{C})$.

(6) For $A \in M_n(\mathbb{C})$, we call $\lambda \in \sigma(A) \cap \mathbb{R}$ a *dominant eigenvalue* if

$$\operatorname{Re}\mu < \lambda \qquad \text{for all } \mu \in \sigma(A) \setminus \{\lambda\}$$

and if the Jordan blocks corresponding to λ are all 1×1. Show that the following properties are equivalent.

(a) The eigenvalue $0 \in \sigma(A)$ is dominant.

(b) There exist $P = P^2 \in M_n(\mathbb{C})$ and $M \geq 1$, $\varepsilon > 0$ such that

$$\|e^{tA} - P\| \leq Me^{-\varepsilon t} \qquad \text{for all } t \geq 0.$$

b. Uniformly Continuous Operator Semigroups

We now desire to extend the above results to semigroups $(T(t))_{t \geq 0}$ on an infinite-dimensional Banach space X. To this purpose, it suffices to assume continuity of the map $t \mapsto T(t) \in \mathcal{L}(X)$ in the operator norm. Then we can replace the matrix $A \in M_n(\mathbb{C})$ by a bounded operator $A \in \mathcal{L}(X)$ and argue as in Section 2.a.

2.10 Definition. *For $A \in \mathcal{L}(X)$ we define*

$$(2.5) \qquad e^{tA} := \sum_{n=0}^{\infty} \frac{t^n A^n}{n!}$$

for each $t \geq 0$.

It follows from the completeness of X that e^{tA} is a well-defined bounded operator on X.

2.11 Proposition. *For $A \in \mathcal{L}(X)$ define $\left(e^{tA}\right)_{t \geq 0}$ by (2.5). Then the following properties hold.*

(i) *$\left(e^{tA}\right)_{t \geq 0}$ is a semigroup on X such that*

$$\mathbb{R}_+ \ni t \mapsto e^{tA} \in (\mathcal{L}(X), \|\cdot\|)$$

is continuous.

(ii) *The map $\mathbb{R}_+ \ni t \mapsto T(t) := e^{tA} \in (\mathcal{L}(X), \|\cdot\|)$ is differentiable and satisfies the differential equation*

$$\text{(DE)} \qquad \begin{aligned} \tfrac{d}{dt}T(t) &= AT(t) \quad \text{for } t \geq 0, \\ T(0) &= I. \end{aligned}$$

Conversely, every differentiable function $T(\cdot) : \mathbb{R}_+ \to (\mathcal{L}(X), \|\cdot\|)$ satisfying (DE) is already of the form $T(t) = e^{tA}$ for $A = \dot{T}(0) \in \mathcal{L}(X)$.

The proof of this result can be adapted from Section 2.a and is left to the reader.

Semigroups having the continuity property stated in Proposition 2.11.(i) are called *uniformly continuous* (or *norm-continuous*) .

2.12 Theorem. *Every uniformly continuous semigroup* $(T(t))_{t \geq 0}$ *on a Banach space X is of the form*

$$T(t) = e^{tA}, \quad t \geq 0,$$

for some bounded operator $A \in \mathcal{L}(X)$.

PROOF. Because the following arguments were already used in the matrix-valued cases (see Section 2.a), a brief outline of the proof should be sufficient.

For a uniformly continuous semigroup $(T(t))_{t \geq 0}$ the operators

$$V(t) := \int_0^t T(s) \, ds, \quad t \geq 0$$

are well-defined, and $1/t V(t)$ converges (in norm!) to $T(0) = I$ as $t \downarrow 0$. Hence, for $t > 0$ sufficiently small, the operator $V(t)$ becomes invertible. Repeat now the computations from the proof of Theorem 2.8 in order to obtain that $t \mapsto T(t)$ is differentiable and satisfies (DE). Then Proposition 2.11 yields the assertion. □

Before adding some comments on and further properties of uniformly continuous semigroups we state the following question leading directly to the main objects of this text.

2.13 Problem. *Do there exist "natural" one-parameter semigroups of linear operators on Banach spaces that are not uniformly continuous, hence not of the form* $(e^{tA})_{t \geq 0}$ *for some bounded operator A?*

2.14 Comments. (i) The operator A in Theorem 2.12 is determined uniquely as the derivative of $T(\cdot)$ at zero; i.e., $A = \dot{T}(0)$. We call it the *generator* of $(T(t))_{t \geq 0}$.

(ii) Because Definition 2.10 for e^{tA} works also for $t \in \mathbb{R}$ and even for $t \in \mathbb{C}$, it follows that each uniformly continuous semigroup can be extended to a uniformly continuous group $(e^{tA})_{t \in \mathbb{R}}$, or to $(e^{tA})_{t \in \mathbb{C}}$, respectively.

(iii) From the differentiability of $t \mapsto T(t)$ it follows that for each $x \in X$ the orbit map $\mathbb{R}_+ \ni t \mapsto T(t)x \in X$ is differentiable as well. Therefore, the map $x(t) := T(t)x$ is the unique solution of the X-valued initial value problem (or *abstract Cauchy problem*)

(ACP)
$$\begin{cases} \dot{x}(t) = Ax(t) & \text{for } t \geq 0, \\ x(0) = x. \end{cases}$$

2.15 Exercises. (1) On $X := C_0(\mathbb{R}) := \{f \in C(\mathbb{R}) : \lim_{|s| \to \infty} f(s) = 0\}$ and for a fixed constant $\alpha > 0$, we define an operator A_α by the difference quotients

$$A_\alpha f(s) := \frac{1}{\alpha}\big(f(s+\alpha) - f(s)\big), \quad f \in X, \ s \in \mathbb{R}.$$

Show that $A_\alpha \in \mathcal{L}(X)$ with $\|A_\alpha\| = 2/\alpha$, and hence one has the estimate

$$\left\|e^{tA_\alpha}\right\| \le e^{2t/\alpha} \qquad \text{for all } t \ge 0.$$

However, e^{tA_α} can be computed explicitly as

$$e^{tA_\alpha} f(s) = e^{-t/\alpha} \sum_{k=0}^{\infty} \frac{(t/\alpha)^k}{k!} f(s + k\alpha), \quad f \in X, \ s \in \mathbb{R},$$

hence it satisfies

$$\left\|e^{tA_\alpha}\right\| = 1 \qquad \text{for all } t \ge 0.$$

(2) Let X be a Banach space. For which operators $T \in \mathcal{L}(X)$ can we find $A \in \mathcal{L}(X)$ such that $T = e^A$; i.e, which bounded T can be embedded into a uniformly continuous semigroup? (Hint: Find (sufficient) conditions on T such that $A := \log T$ can be defined in analogy to Definition 2.10.) Show that such operators T are *infinitely divisible*; i.e., for each $n \in \mathbb{N}$ there exists $S \in \mathcal{L}(X)$ such that $S^n = T$.

(3) Show that for $A, B \in \mathcal{L}(X)$, X a Banach space, the following assertions are equivalent.

  (a) $AB = BA$.

  (b) $e^{t(A+B)} = e^{tA} \cdot e^{tB}$ for all $t \in \mathbb{R}$.

(Hint: To show that (a) implies (b) proceed as in the proof of Lemma 2.4. For the converse implication, compute the second derivative of the functions appearing in (b).)

(4) The reader familiar with Banach algebras should reformulate the notion of "uniformly continuous semigroup" and Theorem 2.12 by replacing the operator algebra $\mathcal{L}(X)$ by an arbitrary Banach algebra.

3. More Semigroups

In order to convince the reader that new and interesting phenomena appear for semigroups on infinite-dimensional Banach spaces, we first discuss several classes of one-parameter semigroups on concrete spaces. These semigroups are not uniformly continuous anymore and hence, unlike those in Section 2.b, not of the form $(e^{tA})_{t\ge 0}$ for some bounded operator A. On the other hand, they are not "pathological" in the sense of being completely unrelated to any analytic structure as the semigroup constructed in Exercise II.2.13.(4). In addition, these semigroups accompany us through the further development of the theory and provide a source of illuminating examples and counterexamples.

a. Multiplication Semigroups on $C_0(\Omega)$

Multiplication operators can be considered as an infinite-dimensional generalization of diagonal matrices. They are extremely simple to construct, and most of their properties are evident. Nevertheless, their value should not be underestimated. They appear, for example, naturally in the context of Fourier analysis or when one applies the spectral theorem for self adjoint operators on Hilbert spaces (see Theorem 3.9). We therefore strongly recommend that any first attempt to illustrate a result or disprove a conjecture on semigroups should be made using multiplication semigroups.

We start from a locally compact space Ω and define the Banach space (endowed with the sup-norm $\|f\|_\infty := \sup_{s \in \Omega} |f(s)|$)

$$C_0(\Omega) := \left\{ f \in C(\Omega) : \begin{array}{l} \text{for all } \varepsilon > 0 \text{ there exists a compact } K_\varepsilon \subset \Omega \\ \text{such that } |f(s)| < \varepsilon \text{ for all } s \in \Omega \setminus K_\varepsilon \end{array} \right\}$$

of all continuous, complex-valued functions on Ω that vanish at infinity. As a typical example the reader might always take Ω to be a bounded or unbounded interval in $\mathbb{R}$.

With any continuous function $q : \Omega \to \mathbb{C}$ we associate a linear operator M_q on $C_0(\Omega)$ defined on its "maximal domain" $D(M_q)$ in $C_0(\Omega)$.

3.1 Definition. *The* multiplication operator M_q *induced on* $C_0(\Omega)$ *by some continuous function* $q : \Omega \to \mathbb{C}$ *is defined by*

$$M_q f := q \cdot f \quad \text{for all } f \text{ in the domain}$$

$$D(M_q) := \{ f \in C_0(\Omega) : q \cdot f \in C_0(\Omega) \}.$$

The main feature of these multiplication operators is that most operator-theoretic properties of M_q can be characterized by analogous properties of the function q. In the following proposition we give some examples for this correspondence.

3.2 Proposition. *Let* M_q *with domain* $D(M_q)$ *be the multiplication operator induced on* $C_0(\Omega)$ *by some continuous function* q. *Then the following assertions hold.*

(i) *The operator* $(M_q, D(M_q))$ *is closed and densely defined.*

(ii) *The operator* M_q *is bounded (with* $D(M_q) = C_0(\Omega)$*) if and only if the function* q *is bounded. In that case, one has*

$$\|M_q\| = \|q\| := \sup_{s \in \Omega} |q(s)|.$$

(iii) *The operator* M_q *has a bounded inverse if and only if the function* q *has a bounded inverse* $1/q$; *i.e.,* $0 \notin \overline{q(\Omega)}$. *In that case, one has*

$$M_q^{-1} = M_{1/q}.$$

(iv) *The spectrum of* M_q *is the closed range of* q; *i.e.,*

$$\sigma(M_q) = \overline{q(\Omega)}.$$

PROOF. (i) The domain $D(M_q)$ always contains the space

$$C_c(\Omega) := \{f \in C(\Omega) : \operatorname{supp} f \text{ is compact}\}$$

of all continuous functions having compact support

$$\operatorname{supp} f := \overline{\{s \in \Omega : f(s) \neq 0\}}.$$

In order to show that these functions form a dense subspace, we first observe that the one-point compactification of Ω is a normal topological space (cf. [Dug66, Chap. XI, Thm. 8.4 and Thm. 1.2] or [Kel75, Chap. 5, Thm. 21 and Thm. 10]). Hence, by Urysohn's lemma (cf. [Dug66, Chap. VII, Thm. 4.1] or [Kel75, Chap. 4, Lem. 4]), for every compact subset $K \subseteq \Omega$ we can find a function $h_K \in C(\Omega)$ still having compact support satisfying[3]

$$0 \leq h_K \leq \mathbb{1} \qquad \text{and} \qquad h_K(s) = 1 \text{ for all } s \in K.$$

Then, for each $f \in C_0(\Omega)$, the function $f \cdot h_K$ has compact support, and

$$\|f - f \cdot h_K\| = \sup_{s \in \Omega \setminus K} |f(s)(1 - h_K(s))|$$
$$\leq 2 \sup_{s \in \Omega \setminus K} |f(s)|.$$

This implies that the continuous functions with compact support are dense in $C_0(\Omega)$; hence M_q is densely defined.

To show the closedness of M_q, we take a sequence $(f_n)_{n \in \mathbb{N}} \subset D(M_q)$ converging to $f \in C_0(\Omega)$ such that $\lim_{n \to \infty} q f_n =: g \in C_0(\Omega)$ exists. Clearly, this implies $g = qf$ and hence $f \in D(M_q)$ and $M_q f = g$.

(ii) If q is bounded, we have

$$\|M_q f\| = \sup_{s \in \Omega} |q(s)f(s)| \leq \|q\| \cdot \|f\|$$

for any $f \in C_0(\Omega)$; hence M_q is bounded with $\|M_q\| \leq \|q\|$. On the other hand, if M_q is bounded, for every $s \in \Omega$ we choose, again using Urysohn's lemma, a continuous function f_s with compact support satisfying $\|f_s\| = 1 = f_s(s)$. This implies

$$\|M_q\| \geq \|M_q f_s\| \geq |q(s)f_s(s)| = |q(s)| \qquad \text{for all } s \in \Omega;$$

hence q is bounded with $\|M_q\| \geq \|q\|$.

---

[3] Here $\mathbb{1}$ denotes the constant function with $\mathbb{1}(s) = 1$ for all $s \in \Omega$.

(iii) If $0 \notin \overline{q(\Omega)}$, then $1/q$ is a bounded continuous function and $M_{1/q}$ is the bounded inverse of M_q. Conversely, assume M_q to have a bounded inverse M_q^{-1}. Then we obtain

$$\|f\| \leq \|M_q^{-1}\| \cdot \|M_q f\| \qquad \text{for all } f \in D(M_q),$$

whence

$$(3.1) \qquad \delta := \frac{1}{\|M_q^{-1}\|} \leq \sup_{s \in \Omega} |q(s)f(s)| \qquad \text{for all } f \in D(M_q), \ \|f\| = 1.$$

Now assume $\inf_{s \in \Omega} |q(s)| < \delta/2$. Then there exists an open set $\mathcal{O} \subset \Omega$ such that $|q(s)| < \delta/2$ for all $s \in \mathcal{O}$. On the other hand, by Urysohn's lemma we find a function $f_0 \in C_0(\Omega)$ such that $\|f_0\| = 1$ and $f_0(s) = 0$ for all $s \in \Omega \setminus \mathcal{O}$. This implies $\sup_{s \in \Omega} |q(s)f_0(s)| \leq \delta/2$, contradicting (3.1). Hence $0 < \delta/2 \leq |q(s)|$ for all $s \in \Omega$; i.e., $M_{1/q}$ is bounded, and one easily verifies that it yields the inverse of the operator M_q.

(iv) By definition, one has $\lambda \in \sigma(M_q)$ if and only if $\lambda - M_q = M_{\lambda - q}$ is not invertible. Thus (iii) applied to the function $\lambda - q$ yields the assertion. $\square$

With any continuous function $q : \Omega \to \mathbb{C}$ we now associate the exponential function

$$e^{tq} : s \mapsto e^{tq(s)} \qquad \text{for } s \in \Omega, \ t \geq 0.$$

It is then immediate that the corresponding multiplication operators

$$T_q(t)f := e^{tq} f, \ f \in C_0(\Omega),$$

formally satisfy the semigroup law (FE) from Definition 1.1. So, in order to obtain a one-parameter semigroup on $C_0(\Omega)$, we have only to make sure that these multiplication operators $T_q(t)$ are bounded operators on $C_0(\Omega)$. Using Proposition 3.2.(ii), we see that this is the case if and only if

$$\sup_{s \in \Omega} |e^{tq(s)}| = \sup_{s \in \Omega} e^{t \operatorname{Re} q(s)}$$
$$= e^{t \sup_{s \in \Omega} \operatorname{Re} q(s)} < \infty.$$

This observation leads to the following definition.

3.3 Definition. *Let* $q : \Omega \to \mathbb{C}$ *be a continuous function such that*

$$\sup_{s \in \Omega} \operatorname{Re} q(s) < \infty.$$

Then the semigroup $\big(T_q(t)\big)_{t \geq 0}$ *defined by*

$$T_q(t)f := e^{tq} f$$

for $t \geq 0$ *and* $f \in C_0(\Omega)$ *is called the* multiplication semigroup *generated by the multiplication operator* M_q *(or, the function* q*) on* $C_0(\Omega)$.

By Proposition 2.11.(i) and Theorem 2.12 the semigroup $\big(T_q(t)\big)_{t\geq 0}$ is uniformly continuous if and only if it is of the form $\big(e^{tA}\big)_{t\geq 0}$ for some bounded operator A. As predicted, this can already be read off from the function q.

3.4 Proposition. *The multiplication semigroup $\big(T_q(t)\big)_{t\geq 0}$ generated by $q : \Omega \to \mathbb{C}$ is uniformly continuous if and only if q is bounded.*

PROOF. If q and hence M_q are bounded, it is easy to see that $T_q(t)$ coincides with the exponential e^{tM_q}, hence is uniformly continuous by Proposition 2.11.(i).

Now let q be unbounded and choose $(s_n)_{n\in\mathbb{N}} \subset \Omega$ such that $|q(s_n)| \to \infty$ for $n \to \infty$. Then we take $t_n := 1/|q(s_n)| \to 0$. Because $e^z \neq 1$ for all $|z| = 1$, there exists $\delta > 0$ such that

$$\left| 1 - e^{t_n q(s_n)} \right| \geq \delta$$

for all $n \in \mathbb{N}$. With functions $f_n \in C_0(\Omega)$ satisfying $\|f_n\| = 1 = f_n(s_n)$, we finally obtain

$$\|T_q(0) - T_q(t_n)\| \geq \left\| f_n - e^{t_n q} f_n \right\|$$
$$\geq \left| 1 - e^{t_n q(s_n)} \right| \geq \delta > 0$$

for all $n \in \mathbb{N}$; i.e., $\big(T_q(t)\big)_{t\geq 0}$ is not uniformly continuous. $\square$

This means that for every unbounded continuous function $q : \Omega \to \mathbb{C}$ satisfying

$$\sup_{s\in\Omega} \operatorname{Re} q(s) < \infty,$$

we obtain a one-parameter semigroup that is *not* uniformly continuous, hence to which Theorem 2.12 does not apply. In order to prepare for later developments, we now show that these multiplication semigroups are strongly continuous.

3.5 Proposition. *Let $\big(T_q(t)\big)_{t\geq 0}$ be the multiplication semigroup generated by a continuous function $q : \Omega \to \mathbb{C}$ satisfying*

$$w := \sup_{s\in\Omega} \operatorname{Re} q(s) < \infty.$$

Then the mappings

$$\mathbb{R}_+ \ni t \mapsto T_q(t)f = e^{tq} f \in C_0(\Omega)$$

are continuous for every $f \in C_0(\Omega)$.

PROOF. Let $f \in C_0(\Omega)$ with $\|f\| \le 1$. For $\varepsilon > 0$ take a compact subset K of Ω such that $|f(s)| \le \varepsilon/(e^{|w|}+1)$ for all $s \in \Omega \setminus K$. Because the exponential function is uniformly continuous on compact sets, there exists $t_0 \in (0,1]$ such that

$$\left|e^{tq(s)} - 1\right| \le \varepsilon$$

for all $s \in K$ and $0 \le t \le t_0$. Hence, we obtain

$$\|e^{tq}f - f\| \le \sup_{s \in K}\left(\left|e^{tq(s)} - 1\right| \cdot |f(s)|\right) + \left(e^{|w|} + 1\right) \cdot \sup_{s \in \Omega \setminus K} |f(s)|$$
$$\le 2\varepsilon$$

for all $0 \le t \le t_0$. $\square$

Finally, we show that each strongly continuous semigroup consisting of multiplication operators on $C_0(\Omega)$ is a *multiplication semigroup* in the sense of Definition 3.3.

3.6 Proposition. *For $t \ge 0$, let $m_t : \Omega \to \mathbb{C}$ be bounded continuous functions and assume that the corresponding multiplication operators*

$$T(t)f := m_t \cdot f$$

form a strongly continuous semigroup $(T(t))_{t \ge 0}$ of bounded operators on $C_0(\Omega)$. Then there exists a continuous function $q : \Omega \to \mathbb{C}$ satisfying

$$\sup_{s \in \Omega} \operatorname{Re} q(s) < \infty$$

such that $m_t(s) = e^{tq(s)}$ for every $s \in \Omega$, $t \ge 0$.

PROOF. For fixed $s \in \Omega$ choose $f \in C_0(\Omega)$ such that $f \equiv 1$ in some neighborhood of s. Then, by assumption,

$$\mathbb{R}_+ \ni t \mapsto \left(T(t)f\right)(s) = m_t(s) \in \mathbb{C}$$

is a continuous function satisfying the functional equation (FE) from Definition 1.1. Therefore, by Proposition 2.11 (for $X := \mathbb{C}$), there exists a unique $q(s) \in \mathbb{C}$ such that $m_t(s) = e^{tq(s)}$ for all $t \ge 0$. Because the map $s \mapsto m_t(s)$ in a neighborhood of s coincides with $s \mapsto \left(T(t)f\right)(s) \in C_0(\Omega)$, the functions $\Omega \ni s \mapsto e^{tq(s)} \in \mathbb{C}$ are continuous for all $t \ge 0$. In order to show that q is continuous, we first observe that q is bounded on compact subsets of Ω. In fact, if $K \subset \Omega$ is compact, then $(T(t))_{t \ge 0}$ induces a uniformly continuous semigroup $(T_K(t))_{t \ge 0}$ on $C(K)$ given by

$$\left(T_K(t)f\right)(s) = e^{tq(s)}f(s), \qquad f \in C(K), \ s \in K,$$

and the same arguments as in the second part of the proof of Proposition 3.4 show that q is bounded on K. This implies that the convergence in

$$\lim_{t \downarrow 0} \frac{e^{tq(s)} - 1}{t} = q(s)$$

is uniform on compact sets in Ω. Because every point in Ω possesses a compact neighborhood, we conclude that q, being the uniform limit (on compact subsets) of the continuous functions $s \mapsto (e^{tq(s)}-1)/t$, is continuous as well.

Finally, the multiplication operators $T(t)f = e^{tq} \cdot f$ are supposed to be bounded; hence the real part of q must be bounded from above.   □

We conclude this subsection with some simple observations and concrete examples.

3.7 Examples. (i) On a compact space, every multiplication operator given by a continuous function is already bounded, and hence every multiplication semigroup is uniformly continuous.

(ii) We can choose Ω and q in such a way that the closed range of q is a given closed subset of $\mathbb{C}$. When q generates a multiplication semigroup $(T_q(t))_{t \geq 0}$, this has obvious consequences for the operators $T_q(t)$. Choose any closed subset Ω of $\mathbb{C}$ and define

$$q(s) := s$$

for $s \in \Omega$. Then $\sigma(M_q) = \Omega$ and $\sigma(T_q(t)) = \overline{e^{t\Omega}} := \overline{\{e^{ts} : s \in \Omega\}}$ for all $t \geq 0$. In particular, if $\Omega \subseteq \{\lambda \in \mathbb{C} : \mathrm{Re}\,\lambda \leq 0\}$ (or $\Omega \subseteq i\mathbb{R}$), we conclude that $(T_q(t))_{t \geq 0}$ consists of contractions (or isometries, respectively) on $\mathrm{C}_0(\Omega)$.

(iii) For $\Omega := \mathbb{N}$ each complex sequence $(q_n)_{n \in \mathbb{N}} \subset \mathbb{C}$ defines a multiplication operator

$$(x_n)_{n \in \mathbb{N}} \mapsto (q_n \cdot x_n)_{n \in \mathbb{N}}$$

on the space $\mathrm{C}_0(\Omega) = \mathrm{c}_0$. For $q_n := in$ we obtain a group of isometries

$$T(t)(x_n)_{n \in \mathbb{N}} = (e^{int} x_n)_{n \in \mathbb{N}}, \quad t \in \mathbb{R},$$

and for $q_n := -n^2$ we obtain a semigroup of contractions

$$T(t)(x_n)_{n \in \mathbb{N}} = (e^{-n^2 t} x_n)_{n \in \mathbb{N}}, \quad t \geq 0.$$

(iv) This simple example serves just to explain the first sentence in this subsection. Take $\Omega = \{1, 2, \ldots, m\}$ to be a finite set. Then $\mathrm{C}_0(\Omega)$ is simply $\mathbb{C}^m$, and the multiplication operator $(x_n) \mapsto (q_n \cdot x_n)$ corresponds to the diagonal matrix $A = \mathrm{diag}(q_1, \ldots, q_m)$. The corresponding multiplication semigroup is given by $e^{tA} = \mathrm{diag}(e^{tq_1}, \ldots, e^{tq_m})$ as in Example 2.3.(i).

3.8 Exercises. (1) For a sequence $q = (q_n)_{n \in \mathbb{N}} \subset \mathbb{C}$ define the corresponding multiplication operator M_q on $X := c_0$ or $X := \ell^p$, $1 \leq p \leq \infty$. Show that its point spectrum is given by $P\sigma(M_q) = \{q_n : n \in \mathbb{N}\}$ and that $\sigma(M_q) = \overline{P\sigma(M_q)}$.

(2) Many properties of the multiplication semigroup $(T_q(t))_{t \geq 0}$ generated by a multiplication operator M_q on $X := C_0(\Omega)$ can be characterized by properties of the function $q : \Omega \to \mathbb{C}$.

(i) $(T_q(t))_{t \geq 0}$ is bounded (contractive) if and only if

$$\operatorname{Re} q(s) \leq 0 \qquad \text{for all } s \in \Omega.$$

(ii) $(T_q(t))_{t \geq 0}$ satisfies $T_q(2\pi) = I$ if and only if

$$q(\Omega) \subseteq i\mathbb{Z}.$$

(3) Take $X := C_0(\mathbb{R})$ and $q(s) := \frac{-1}{1+|s|} + is$, $s \in \mathbb{R}$. Show that the growth bound of the corresponding multiplication semigroup $\mathcal{T} = (T_q(t))_{t \geq 0}$ does not satisfy $\omega_0(\mathcal{T}) < 0$, whereas

$$\lim_{t \to \infty} \|T_q(t)f\| = 0$$

for each $f \in X$.

b. Multiplication Semigroups on $L^p(\Omega, \mu)$

As claimed at the beginning of the previous subsection, multiplication operators arise in a natural way in various instances. For example, if one applies the Fourier transform to a linear differential operator on $L^2(\mathbb{R}^n)$, this operator becomes a multiplication operator on $L^2(\mathbb{R}^n)$. Moreover, the classical "spectral theorem" asserts that each self-adjoint or, more generally, normal operator[4] on a Hilbert space is (isomorphic to) a multiplication operator on some L^2-space. This viewpoint is emphasized in Halmos's article [Hal63] and motivates our systematic analysis of multiplication operators. We therefore formulate this version of the spectral theorem explicitly (see also [Con85, Chap. 10, Thm. 4.19] or [Wei80, Chap. 7, Thm. 7.33]).

---

[4] We recall that an operator A on a Hilbert space H is called *normal* if $D(A^*A) = D(AA^*) =: D$ and $A^*Ax = AA^*x$ for all $x \in D$; i.e., $A^*A = AA^*$.

3.9 Spectral Theorem. *If A is a normal operator on a separable Hilbert space H, then there is a σ-finite measure space (Σ, Ω, μ) and a measurable function $q : \Omega \to \mathbb{C}$ such that A is unitarily equivalent to the multiplication operator M_q on $L^2(\Omega, \mu)$; i.e., there exists a unitary operator $U \in \mathcal{L}(H, L^2(\Omega, \mu))$ such that the diagram*

$$
\begin{array}{ccc}
H \;\supseteq\; D(A) & \xrightarrow{\ \ A\ \ } & H \\
\Big\downarrow{\scriptstyle U} \qquad \Big\downarrow{\scriptstyle U} & & \Big\uparrow{\scriptstyle U^*=U^{-1}} \\
L^2(\Omega, \mu) \supseteq D(M_q) & \xrightarrow{\ \ M_q\ \ } & L^2(\Omega, \mu)
\end{array}
$$

commutes.

In order to define what we mean by a multiplication operator, we take some σ-finite measure space (Ω, Σ, μ); see, e.g., [Hal74, Chap. II] or [Rao87, Chap. 2]. Then, for fixed $1 \le p < \infty$, we consider the Banach space

$$X := L^p(\Omega, \mu)$$

of all (equivalence classes of) p-integrable complex functions on Ω endowed with the p-norm

$$\|f\|_p := \left(\int_\Omega |f(s)|^p \, d\mu(s) \right)^{1/p}.$$

Next, for a measurable function

$$q : \Omega \to \mathbb{C},$$

we call the set

$$q_{\mathrm{ess}}(\Omega) := \Big\{ \lambda \in \mathbb{C} : \mu\big(\{ s \in \Omega : |q(s) - \lambda| < \varepsilon \} \big) \ne 0 \text{ for all } \varepsilon > 0 \Big\},$$

its *essential range* and define the associated *multiplication operator* M_q by

$$(3.2) \qquad \begin{aligned} M_q f &:= q \cdot f \quad \text{for all } f \text{ in the domain} \\ D(M_q) &:= \{ f \in L^p(\Omega, \mu) : q \cdot f \in L^p(\Omega, \mu) \}. \end{aligned}$$

In analogy to Proposition 3.2, we now have the following result.

3.10 Proposition. *Let M_q with domain $D(M_q)$ be the multiplication operator induced on $L^p(\Omega, \mu)$ by some measurable function q. Then the following assertions hold.*

(i) *The operator $(M_q, D(M_q))$ is closed and densely defined.*

(ii) *The operator M_q is bounded (with $D(M_q) = L^p(\Omega, \mu)$) if and only if the function q is essentially bounded; i.e., the set $q_{\mathrm{ess}}(\Omega)$ is bounded in $\mathbb{C}$. In this case, one has*

$$\|M_q\| = \|q\|_\infty := \sup\left\{|\lambda| : \lambda \in q_{\mathrm{ess}}(\Omega)\right\}.$$

(iii) *The operator M_q has a bounded inverse if and only if $0 \notin q_{\mathrm{ess}}(\Omega)$. In that case, one has*
$$M_q^{-1} = M_r$$

for $r : \Omega \to \mathbb{C}$ defined by

$$r(s) := \begin{cases} 1/q(s) & \text{if } q(s) \neq 0, \\ 0 & \text{if } q(s) = 0. \end{cases}$$

(iv) *The spectrum of M_q is the essential range of q; i.e.,*

$$\sigma(M_q) = q_{\mathrm{ess}}(\Omega).$$

The proof uses measure theory and is left as Exercise 3.13.(2).

Also, the other results of Section 3.a, after the appropriate changes, remain valid in the L^p-case. For the convenience of the reader and due to their importance for the applications, we state them explicitly. The proofs, however, are left as Exercises 3.13.(3) and (4).

3.11 Proposition. *Let $(T_q(t))_{t \geq 0}$ be the multiplication semigroup generated by a measurable function $q : \Omega \to \mathbb{C}$ satisfying*

$$\operatorname*{ess\,sup}_{s \in \Omega} \operatorname{Re} q(s) := \sup_{\lambda \in q_{\mathrm{ess}}(\Omega)} \operatorname{Re} \lambda < \infty;$$

i.e.,
$$T_q(t)f := e^{tq}f \qquad \text{for every } f \in L^p(\Omega, \mu),\ t \geq 0.$$

Then the mappings

$$\mathbb{R}_+ \ni t \mapsto T_q(t)f = e^{tq}f \in L^p(\Omega, \mu)$$

are continuous for every $f \in L^p(\Omega, \mu)$. Moreover, the semigroup $(T_q(t))_{t \geq 0}$ is uniformly continuous if and only if q is essentially bounded.

3.12 Proposition. *For $t \geq 0$, let $m_t : \Omega \to \mathbb{C}$ be bounded measurable functions and assume that*

(i) *The corresponding multiplication operators*

$$T(t)f := m_t \cdot f$$

form a semigroup $(T(t))_{t\geq 0}$ of bounded operators on $L^p(\Omega, \mu)$, and

(ii) *The mappings*

$$\mathbb{R}_+ \ni t \mapsto T(t)f \in L^p(\Omega, \mu)$$

are continuous for every $f \in L^p(\Omega, \mu)$; i.e. $(T(t))_{t\geq 0}$ is strongly continuous.

Then there exists a measurable function $q : \Omega \to \mathbb{C}$ satisfying

$$\operatorname{ess\,sup}_{s\in\Omega} \operatorname{Re} q(s) := \sup_{\lambda \in q_{\mathrm{ess}}(\Omega)} \operatorname{Re} \lambda < \infty$$

such that $m_t = e^{tq}$ almost everywhere for every $t \geq 0$.

3.13 Exercises. (1) On the spaces $X := c_0$ and $X := \ell^p, 1 \leq p < \infty$, there exist multiplication semigroups $(T_q(t))_{t\geq 0}$ such that each $T_q(t)$, $t > 0$, is a compact operator. Construct concrete examples. Observe that this is not possible if

(i) The function spaces are $X := C_0(\mathbb{R})$ or $X := L^p(\mathbb{R})$, or if

(ii) The function q is bounded.

(2) Prove Proposition 3.10. (Hints: To prove that M_q is closed, use the fact that every convergent sequence in $L^p(\Omega, \mu)$ has a μ-almost everywhere convergent subsequence; see, e.g., [Rud86, Chap. 3, Thm. 3.12]. In order to show that M_q is densely defined, combine the fact that Ω is σ-finite with Lebesgue's convergence theorem (cf. [Rud86, Chap. 1, 1.34]). For the "only if" part of (ii), assume q not to be essentially bounded and choose suitable characteristic functions to conclude that M_q is unbounded. In the "only if" part of (iii), show first that M_q^{-1} is given by a multiplication operator and then apply (ii).)

(3) Prove Proposition 3.11. (Hint: Use Lebesgue's convergence theorem.)

(4) Prove Proposition 3.12.

(5) For every measurable function $q : \Omega \to \mathbb{C}$ we can define the multiplication operator M_q on $L^\infty(\Omega, \mu)$ as we did for $L^p(\Omega, \mu)$, $1 \leq p < \infty$. Show that M_q is densely defined if and only if q is essentially bounded.

(6) Let $A := M_q$ be a multiplication operator on $L^p(\Omega, \mu)$, $1 \leq p < \infty$. Show that $\lambda \in \mathbb{C}$ is an eigenvalue of A if and only if $\mu(\{s \in \Omega : q(s) = \lambda\}) > 0$.

(7) A bounded linear operator $T : L^p(\Omega, \mu) \to L^p(\Omega, \mu)$, $1 \leq p \leq \infty$, is called *local* if for every measurable subset $S \subset \Omega$ one has $Tf = Tg$ almost everywhere on S if $f = g$ almost everywhere on S. Show that every local operator is a multiplication operator M_q for some $q \in L^\infty(\Omega, \mu)$. Extend this characterization to unbounded multiplication operators. (Hint: See [Nag86, C-II, Thm. 5.13].)

c. Translation Semigroups

Another important class of examples is obtained by "translating," to the left or to the right, complex-valued functions defined on (subsets of) $\mathbb{R}$. We first define these "translation operators" and only then specify the appropriate spaces.

3.14 Definition. *For a function $f : \mathbb{R} \to \mathbb{C}$ and $t \geq 0$, we call*

$$\big(T_l(t)f\big)(s) := f(s + t), \quad s \in \mathbb{R},$$

the left translation (of f by t), and

$$\big(T_r(t)f\big)(s) := f(s - t), \quad s \in \mathbb{R},$$

is the right translation (of f by t).

It is immediately clear that the operators $T_l(t)$ (and $T_r(t)$) satisfy the semigroup law (FE). We have only to choose appropriate function spaces to produce one-parameter operator semigroups. For that purpose, we start with spaces of continuous or integrable functions and the translation on *all* of $\mathbb{R}$.

3.15 Translations on $\mathbb{R}$. As Banach space X we take one of the spaces

- $X_\infty := L^\infty(\mathbb{R})$ of all bounded measurable functions on $\mathbb{R}$,
- $X_b := C_b(\mathbb{R})$ of all bounded continuous functions on $\mathbb{R}$,
- $X_{ub} := C_{ub}(\mathbb{R})$ of all bounded, uniformly continuous functions on $\mathbb{R}$,
- $X_0 := C_0(\mathbb{R})$ of all continuous functions on $\mathbb{R}$ vanishing at infinity,
- $X_{2\pi} := C_{2\pi}(\mathbb{R})$ of all 2π-periodic continuous functions on $\mathbb{R}$,

all endowed with the sup-norm $\| \cdot \|_\infty$, or we take the spaces

- $X_p := L^p(\mathbb{R})$, $1 \leq p < \infty$, of all p-integrable functions on $\mathbb{R}$

endowed with the corresponding p-norm $\| \cdot \|_p$.

Then each left translation operator $T_l(t)$ is an isometry on each of these spaces, having as inverse the right translation operator $T_r(t)$. This means that $\big(T_l(t)\big)_{t \in \mathbb{R}}$ and $\big(T_r(t)\big)_{t \in \mathbb{R}}$ form one-parameter groups on X, called the *(left or right) translation group*.

For our purposes, the following continuity properties of these translation groups on the various function spaces are fundamental.

Proposition. *The (left) translation group* $(T_l(t))_{t\in\mathbb{R}}$

(a) *Is not uniformly continuous on any of the above spaces;*

(b) *Is strongly continuous on* X_{ub}, X_0 *and on* X_p *for all* $1 \le p < \infty$.

PROOF. The proof of (a) is left as Exercise 3.19.(4).

(b) The strong continuity of $(T_l(t))_{t\in\mathbb{R}}$ on X_{ub} and X_0 is a direct consequence of the uniform continuity of any f in X_{ub} and X_0. So it remains to show strong continuity on $X_p = \mathrm{L}^p(\mathbb{R})$.

It is evident that each $T(t)$ is a contraction, so $(T(t))_{t\ge 0}$ is uniformly bounded on $\mathbb{R}$. Now take a continuous function f on $\mathbb{R}$ with compact support and observe that it is uniformly continuous. Therefore,

$$\lim_{t\downarrow 0} \|T(t)f - f\|_\infty = \lim_{t\downarrow 0} \sup_{s\in\mathbb{R}} |f(t+s) - f(s)| = 0,$$

and because the p-norm (for functions on bounded intervals) is weaker,

$$\lim_{t\downarrow 0} \|T(t)f - f\|_p = 0.$$

Because the continuous functions with compact support are dense in $\mathrm{L}^p(\mathbb{R})$ for all $1 \le p < \infty$, the assertion now follows from the adaptation of Proposition 1.3 to groups (see Exercise 1.8.(4)). $\qquad\square$

We now modify the spaces on which translation takes place. As a first case, we consider functions defined on $\mathbb{R}_+$ only.

3.16 Translations on $\mathbb{R}_+$. In analogy to Paragraph 3.15, let X denote one of the spaces

- $X_\infty := \mathrm{L}^\infty(\mathbb{R}_+)$ of all bounded measurable functions on $\mathbb{R}_+$,
- $X_{\mathrm{b}} := \mathrm{C}_{\mathrm{b}}(\mathbb{R}_+)$ of all bounded continuous functions on $\mathbb{R}_+$,
- $X_{\mathrm{ub}} := \mathrm{C}_{\mathrm{ub}}(\mathbb{R}_+)$ of all bounded, uniformly continuous functions on $\mathbb{R}_+$,
- $X_0 := \mathrm{C}_0(\mathbb{R}_+)$ of all continuous functions on $\mathbb{R}_+$ vanishing at infinity,
- $X_p := \mathrm{L}^p(\mathbb{R}_+)$, $1 \le p < \infty$, of all p-integrable functions on $\mathbb{R}_+$,

and observe that the left translations $T_l(t)$ are well-defined contractions on these spaces, but now yield a *semi*group only, called the *left translation semigroup* $(T_l(t))_{t\ge 0}$ on $\mathbb{R}_+$.

For the right translations $T_r(t)$, however, the value $(T_r(t)f)(s) = f(s-t)$ is not defined if $s - t < 0$. To overcome this obstacle, we put

$$(T_r(t)f)(s) := \begin{cases} f(s-t) & \text{for } s - t \ge 0, \\ f(0) & \text{for } s - t < 0 \end{cases}$$

for $f \in X = X_{\mathrm{b}}, X_{\mathrm{ub}}, X_0$, and

$$(T_r(t)f)(s) := \begin{cases} f(s-t) & \text{for } s - t \ge 0, \\ 0 & \text{for } s - t < 0 \end{cases}$$

for $f \in X_p$. In this way, we again obtain semigroups of contractions on X called the *right translation semigroups* $(T_r(t))_{t\geq 0}$ on $\mathbb{R}_+$. Clearly, the continuity properties stated in the proposition in Paragraph 3.15 prevail. Moreover, it is not difficult to see that on X_p, for $1 < p < \infty$, the semigroups $(T_l(t))_{t\geq 0}$ and $(T_r(t))_{t\geq 0}$ are adjoint; i.e., $T_l(t)'$ on X_p' coincides with $T_r(t)$ on $X_{p'}$ where $1/p + 1/p' = 1$.

Even on function spaces on finite intervals, we can define translation semigroups.

3.17 Translations on finite intervals. If we take the Banach space $C[a, b]$ and look at the left translations, we have to specify the values $(T_l(t)f)(s)$ for $s + t > b$. Imitating the idea above, we put

$$(T_l(t)f)(s) := \begin{cases} f(s+t) & \text{for } s+t \leq b, \\ f(b) & \text{for } s+t > b. \end{cases}$$

We note that this choice is not the only one to extend the translations to a semigroup on $C[a, b]$ (see, e.g., Paragraph II.3.29). In any case, we still call $(T_l(t))_{t\geq 0}$ a *left translation semigroup* on $C[a, b]$. By a similar definition, fixing the value at the left endpoint, we obtain a *right translation semigroup* $(T_r(t))_{t\geq 0}$ on the space $C[a, b]$.

On the Banach spaces $L^p[a, b]$, $1 \leq p \leq \infty$, we can modify this definition by taking

$$(T_l(t)f)(s) := \begin{cases} f(s+t) & \text{for } s+t \leq b, \\ 0 & \text{for } s+t > b, \end{cases}$$

and again this yields a semigroup. However, now a completely new phenomenon appears: this semigroup, i.e., this "exponential function," vanishes for $t > b - a$.

Proposition. *The left translation semigroup* $(T_l(t))_{t\geq 0}$ *is nilpotent on* $L^p[a, b]$; *that is,*

$$T_l(t) = 0$$

for all $t \geq b - a$.

3.18 Rotations on the torus. Take $\Gamma := \{z \in \mathbb{C} : |z| = 1\}$ and $X := C(\Gamma)$. Then the operators $T(t)$, $t \in \mathbb{R}$, defined by

$$(T(t)f)(s) := f(e^{it} \cdot s) \qquad \text{for } f \in C(\Gamma) \text{ and } s \in \Gamma$$

form the so-called *rotation group*. It enjoys the same continuity properties as the translation group on X_{ub} in Paragraph 3.16. This can be seen by identifying $C(\Gamma)$ with the Banach space $C_{2\pi}(\mathbb{R}) \subset X_{ub}$ of all 2π-periodic continuous functions on $\mathbb{R}$. After this identification, the above rotation group becomes the translation group $(T_l(t))_{t\in\mathbb{R}}$ on $C_{2\pi}(\mathbb{R})$ satisfying

$$T(2\pi) = I.$$

We call such a group *periodic* (of period 2π).

Because the operators $T(t)$ are isometries for the p-norm and because $C(\Gamma)$ is dense in $L^p(\Gamma, \mu)$, $1 \le p < \infty$, and μ the Lebesgue measure on Γ, the above definition can be extended to $f \in L^p(\Gamma, \mu)$, and we obtain a periodic rotation group on each L^p-space for $1 \le p < \infty$.

3.19 Exercises. (1) Show that the space $C_{ub}(\mathbb{R})$ of all bounded, uniformly continuous functions on $\mathbb{R}$ is the maximal subspace X of $C_b(\mathbb{R})$ such that the orbits of the left translation group $(T_l(t))_{t \in \mathbb{R}}$; i.e., the mappings

$$\mathbb{R} \ni t \mapsto T_l(t)f \in C_b(\mathbb{R}),$$

become continuous for each $f \in X$.

(2) Show that in the context of Paragraphs 3.15 and 3.16 and on the corresponding L^p-spaces, the right translation semigroups are the adjoints of the left translation semigroups; i.e.,

$$T_l(t)' = T_r(t) \qquad \text{for } t \ge 0.$$

(3) Construct more (left) translation semigroups on $L^p[a, b]$ by defining $(T_l(t)f)(s)$ for $s + t > b$ in an appropriate way. For example, take $\alpha \in \mathbb{C}$ and put

$$(T_l(t)f)(s) := \alpha^k f(s + t - k(b - a))$$

for $s + t - a \in [k(b - a), (k + 1)(b - a)]$, $k = 0, 1, 2, \ldots$. This semigroup becomes nilpotent for $\alpha = 0$, whereas it is periodic for $\alpha = 1$. For which α is this semigroup contractive?

(4) Prove part (a) of the proposition in Paragraph 3.15.

(5) Take $X := C_0(\mathbb{R}^n)$ or $C_{ub}(\mathbb{R}^n)$ and choose some $B \in M_n(\mathbb{R})$. Show that $(T(t)f)(s) := f(e^{tB}s)$ for $t \in \mathbb{R}$, $f \in X$, defines a group $(T(t))_{t \in \mathbb{R}}$ on X which is strongly continuous on $C_0(\mathbb{R}^n)$ but not on $C_{ub}(\mathbb{R}^n)$.

Chapter II

Semigroups, Generators, and Resolvents

In this chapter it is our aim to achieve what we obtained, without too much effort, for uniformly continuous semigroups in Section I.2.b. There, we characterized every uniformly continuous semigroup $(T(t))_{t \geq 0}$ on a Banach space X as an operator-valued exponential function; i.e., we found an operator $A \in \mathcal{L}(X)$ such that

$$T(t) = \mathrm{e}^{tA}$$

for all $t \geq 0$ (see Theorem I.2.12). For *strongly continuous semigroups*, we succeed in defining an analogue of A, called the *generator* of the semigroup. It is a linear, but generally unbounded, operator defined only on a dense subspace $D(A)$ of the Banach space X. In order to retrieve the semigroup $(T(t))_{t \geq 0}$ from its generator $(A, D(A))$, we need a third object, namely the *resolvent*

$$\lambda \mapsto R(\lambda, A) := (\lambda - A)^{-1} \in \mathcal{L}(X)$$

of A, which is defined for all complex numbers in the *resolvent set* $\rho(A)$. For the basic concepts from spectral theory we refer to Section V.1.a.

To find and discuss the various relations between these objects is the theme of this chapter, which can be illustrated by the following triangle.

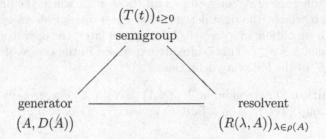

$(T(t))_{t\geq 0}$

semigroup

generator ———————— resolvent

$(A, D(A))$ $(R(\lambda, A))_{\lambda\in\rho(A)}$

1. Generators of Semigroups and Their Resolvents

We recall that for a one-parameter semigroup $(T(t))_{t\geq 0}$ on a Banach space X *uniform* continuity implies differentiability of the map $t \mapsto T(t) \in \mathcal{L}(X)$. The right derivative of $T(\cdot)$ at $t = 0$ then yields a bounded operator A for which $T(t) = e^{tA}$ for all $t \geq 0$.

We now hope that *strong* continuity of a semigroup $(T(t))_{t\geq 0}$ still implies some differentiability of the *orbit maps*

$$\xi_x : t \mapsto T(t)x \in X.$$

In order to pursue this idea we first show, in analogy to Proposition I.1.3, that differentiability of ξ_x is already implied by right differentiability at $t = 0$.

1.1 Lemma. *Take a strongly continuous semigroup $(T(t))_{t\geq 0}$ and an element $x \in X$. For the orbit map $\xi_x : t \mapsto T(t)x$, the following properties are equivalent.*

(a) $\xi_x(\cdot)$ *is differentiable on $\mathbb{R}_+$.*

(b) $\xi_x(\cdot)$ *is right differentiable at $t = 0$.*

PROOF. We have only to show that (b) implies (a). For $h > 0$, one has

$$\lim_{h\downarrow 0} \tfrac{1}{h}\big(T(t+h)x - T(t)x\big) = T(t) \lim_{h\downarrow 0} \tfrac{1}{h}\big(T(h)x - x\big)$$

$$= T(t)\,\dot{\xi}_x(0),$$

and hence $\xi_x(\cdot)$ is right differentiable on $\mathbb{R}_+$.

On the other hand, for $-t \leq h < 0$, we write

$$\tfrac{1}{h}\big(T(t+h)x - T(t)x\big) - T(t)\dot{\xi}_x(0) = T(t+h)\left(\tfrac{1}{h}\big(x - T(-h)x\big) - \dot{\xi}_x(0)\right)$$

$$+ T(t+h)\dot{\xi}_x(0) - T(t)\dot{\xi}_x(0).$$

As $h \uparrow 0$, the first term on the right-hand side converges to zero, because $\|T(t+h)\|$ remains bounded. The remaining part converges to zero by the strong continuity of $(T(t))_{t\geq 0}$. Hence, ξ_x is also left differentiable, and its derivative is

(1.1) $$\dot{\xi}_x(t) = T(t)\,\dot{\xi}_x(0)$$

for each $t \geq 0$. $\qquad\qquad\qquad\qquad\qquad\qquad\qquad\qquad\qquad\qquad\square$

On the subspace of X consisting of all those x for which the orbit maps ξ_x are differentiable, the right derivative at $t = 0$ then yields an operator A from which we obtain, in a sense to be specified later, the operators $T(t)$ as the "exponentials e^{tA}." This is already expressed in the choice of the term "generator" in the following definition.

1.2 Definition. *The* generator $A : D(A) \subseteq X \to X$ *of a strongly continuous semigroup* $(T(t))_{t \geq 0}$ *on a Banach space* X *is the operator*

$$(1.2) \qquad Ax := \dot{\xi}_x(0) = \lim_{h \downarrow 0} \tfrac{1}{h}\big(T(h)x - x\big)$$

defined for every x *in its domain*

$$(1.3) \qquad D(A) := \{x \in X : \xi_x \text{ is differentiable in } \mathbb{R}_+\}.$$

We observe from Lemma 1.1 that the domain $D(A)$ is also given as the set of all elements $x \in X$ for which $\xi_x(\cdot)$ is right differentiable in $t = 0$; i.e.,

$$(1.4) \qquad D(A) = \Big\{x \in X : \lim_{h \downarrow 0} \tfrac{1}{h}\big(T(h)x - x\big) \text{ exists}\Big\}.$$

The domain $D(A)$, which is a linear subspace, is an essential part of the definition of the generator A. Accordingly, we should always denote it by the *pair* $\big(A, D(A)\big)$, but for convenience, we often only write A and assume implicitly that its domain is given by (1.4).

To ensure that the operator $\big(A, D(A)\big)$ has reasonable properties, we proceed as in Chapter I. There we used the "smoothing operators" $V(t) := \int_0^t T(s)\, ds$ to prove differentiability of the semigroup $(T(t))_{t \geq 0}$ (see the proof of Theorem I.2.12). Because we now assume that the orbit maps ξ_x are only continuous, we need to look at "smoothed" elements of the form

$$y_t := \frac{1}{t} \int_0^t \xi_x(s)\, ds = \frac{1}{t} \int_0^t T(s)x\, ds \qquad \text{for } x \in X, t > 0.$$

It is a simple consequence of the definition of the integral as a limit of Riemann sums that the vectors y_t converge to x as $t \downarrow 0$. In addition, they always belong to the domain $D(A)$. This and other elementary facts are collected in the following result.

1.3 Lemma. *For the generator* $\big(A, D(A)\big)$ *of a strongly continuous semigroup* $(T(t))_{t \geq 0}$, *the following properties hold.*

(i) $A : D(A) \subseteq X \to X$ *is a linear operator.*

(ii) *If* $x \in D(A)$, *then* $T(t)x \in D(A)$ *and*

$$(1.5) \qquad \tfrac{d}{dt}T(t)x = T(t)Ax = AT(t)x \quad \text{for all } t \geq 0.$$

(iii) *For every $t \geq 0$ and $x \in X$, one has*

$$\int_0^t T(s)x\, ds \in D(A).$$

(iv) *For every $t \geq 0$, one has*

$$(1.6) \qquad T(t)x - x = A \int_0^t T(s)x\, ds \quad \text{if } x \in X,$$

$$(1.7) \qquad\qquad = \int_0^t T(s)Ax\, ds \quad \text{if } x \in D(A).$$

PROOF. Assertion (i) is trivial. To prove (ii) take $x \in D(A)$. Then it follows from (1.1) that $1/h\big(T(t+h)x - T(t)x\big)$ converges to $T(t)Ax$ as $h \downarrow 0$. Therefore,

$$\lim_{h \downarrow 0} \frac{1}{h}\big(T(h)T(t)x - T(t)x\big)$$

exists, and hence $T(t)x \in D(A)$ by (1.4) with $AT(t)x = T(t)Ax$.

The proof of assertion (iii) is included in the following proof of (iv). For $x \in X$ and $t \geq 0$, one has

$$\frac{1}{h}\left(T(h)\int_0^t T(s)x\, ds - \int_0^t T(s)x\, ds\right)$$

$$= \frac{1}{h}\int_0^t T(s+h)x\, ds - \frac{1}{h}\int_0^t T(s)x\, ds$$

$$= \frac{1}{h}\int_h^{t+h} T(s)x\, ds - \frac{1}{h}\int_0^t T(s)x\, ds$$

$$= \frac{1}{h}\int_t^{t+h} T(s)x\, ds - \frac{1}{h}\int_0^h T(s)x\, ds,$$

which converges to $T(t)x - x$ as $h \downarrow 0$. Hence (1.6) holds.

If $x \in D(A)$, then the functions $s \mapsto T(s)\,{}^{(T(h)x-x)}/_h$ converge uniformly on $[0,t]$ to the function $s \mapsto T(s)Ax$ as $h \downarrow 0$. Therefore,

$$\lim_{h \downarrow 0} \frac{1}{h}\big(T(h) - I\big)\int_0^t T(s)x\, ds = \lim_{h \downarrow 0} \int_0^t T(s)\frac{1}{h}\big(T(h) - I\big)x\, ds$$

$$= \int_0^t T(s)Ax\, ds.$$

$\square$

With the help of this lemma we now show that the generator introduced in Definition 1.2, although unbounded in general, has nice properties.

1.4 Theorem. *The generator of a strongly continuous semigroup is a closed and densely defined linear operator that determines the semigroup uniquely.*

PROOF. Let $\bigl(T(t)\bigr)_{t\geq 0}$ be a strongly continuous semigroup on a Banach space X. As already noted, its generator $\bigl(A, D(A)\bigr)$ is a linear operator. To show that A is closed, consider a sequence $(x_n)_{n\in\mathbb{N}} \subset D(A)$ for which $\lim_{n\to\infty} x_n = x$ and $\lim_{n\to\infty} Ax_n = y$ exist. By (1.7) in the previous lemma, we have

$$T(t)x_n - x_n = \int_0^t T(s)Ax_n \, ds$$

for $t > 0$. The uniform convergence of $T(\cdot)Ax_n$ on $[0, t]$ for $n \to \infty$ implies that

$$T(t)x - x = \int_0^t T(s)y \, ds.$$

Multiplying both sides by $1/t$ and taking the limit as $t \downarrow 0$, we see that $x \in D(A)$ and $Ax = y$; i.e., A is closed.

By Lemma 1.3.(iii) the elements $1/t \int_0^t T(s)x \, ds$ always belong to $D(A)$. Because the strong continuity of $\bigl(T(t)\bigr)_{t\geq 0}$ implies

$$\lim_{t\downarrow 0} \frac{1}{t} \int_0^t T(s)x \, ds = x$$

for every $x \in X$, we conclude that $D(A)$ is dense in X.

Finally, let $\bigl(S(t)\bigr)_{t\geq 0}$ be another strongly continuous semigroup having the same generator $\bigl(A, D(A)\bigr)$. For $x \in D(A)$ and $t > 0$, we consider the map

$$s \mapsto \eta_x(s) := T(t - s)S(s)x$$

for $0 \leq s \leq t$. Because for fixed s the set

$$\left\{ \frac{S(s + h)x - S(s)x}{h} : h \in (0, 1] \right\} \cup \{AS(s)x\}$$

is compact, the difference quotients

$$\frac{1}{h}\bigl(\eta_x(s + h) - \eta_x(s)\bigr) = T(t - s - h)\frac{1}{h}\bigl(S(s + h)x - S(s)x\bigr)$$

$$+ \frac{1}{h}\bigl(T(t - s - h) - T(t - s)\bigr)S(s)x$$

converge by Lemma I.1.2 and Lemma 1.3.(ii) to

$$\tfrac{d}{ds}\eta_x(s) = T(t - s)AS(s)x - AT(t - s)S(s)x = 0.$$

From $\eta_x(0) = T(t)x$ and $\eta_x(t) = S(t)x$ we obtain

$$T(t)x = S(t)x$$

for all x in the dense domain $D(A)$. Hence, $T(t) = S(t)$ for each $t \geq 0$. $\square$

Combining these properties of the generator with the closed graph theorem gives a new characterization of uniformly continuous semigroups, thus complementing Theorem I.2.12.

1.5 Corollary. *For a strongly continuous semigroup $(T(t))_{t\geq0}$ on a Banach space X with generator $(A, D(A))$, the following assertions are equivalent.*

(a) *The generator A is bounded; i.e., there exists $M > 0$ such that $\|Ax\| \leq M\,\|x\|$ for all $x \in D(A)$.*

(b) *The domain $D(A)$ is all of X.*

(c) *The domain $D(A)$ is closed in X.*

(d) *The semigroup $(T(t))_{t\geq0}$ is uniformly continuous.*

In each case, the semigroup is given by

$$T(t) = e^{tA} := \sum_{n=0}^{\infty} \frac{t^n A^n}{n!}, \quad t \geq 0.$$

The proof of this corollary and of some more equivalences is left as Exercise 1.15.(1).

Property (b) indicates that the domain of the generator contains important information about the semigroup and therefore has to be taken into account carefully. However, in many examples (see, e.g., Paragraph 2.6 and Example 4.11 below) it is often routine to compute the expression Ax for some or even many elements in the domain $D(A)$, although it is difficult to identify $D(A)$ precisely. In these situations, the following concept helps to distinguish between "small" and "large" subspaces of $D(A)$.

1.6 Definition. *A subspace D of the domain $D(A)$ of a linear operator $A : D(A) \subseteq X \to X$ is called a* core *for A if D is dense in $D(A)$ for the* graph norm

$$\|x\|_A := \|x\| + \|Ax\|.$$

We now state a useful criterion for subspaces to be a core for the generator.

1.7 Proposition. *Let $(A, D(A))$ be the generator of a strongly continuous semigroup $(T(t))_{t\geq0}$ on a Banach space X. A subspace D of $D(A)$ that is $\|\cdot\|$-dense in X and invariant under the semigroup $(T(t))_{t\geq0}$ is always a* core *for A.*

PROOF. For every $x \in D(A)$ we can find a sequence $(x_n)_{n\in\mathbb{N}} \subset D$ such that $\lim_{n\to\infty} x_n = x$. Because for each n the map $s \mapsto T(s)x_n \in D$ is continuous for the graph norm $\|\cdot\|_A$ (use (1.5)), it follows that

$$\int_0^t T(s)x_n\,ds,$$

being a Riemann integral, belongs to the $\|\cdot\|_A$-closure of D. Similarly, the $\|\cdot\|_A$-continuity of $s \mapsto T(s)x$ for $x \in D(A)$ implies that

$$\left\| \frac{1}{t} \int_0^t T(s)x\,ds - x \right\|_A \to 0 \qquad \text{as } t \downarrow 0 \text{ and}$$

$$\left\| \frac{1}{t} \int_0^t T(s)x_n\,ds - \frac{1}{t} \int_0^t T(s)x\,ds \right\|_A \to 0 \qquad \text{as } n \to \infty \text{ and for each } t > 0.$$

This proves that for every $\varepsilon > 0$ we can find $t > 0$ and $n \in \mathbb{N}$ such that

$$\left\| \frac{1}{t} \int_0^t T(s)x_n\,ds - x \right\|_A < \varepsilon.$$

Hence, $x \in \overline{D}^{\,\|\cdot\|_A}$. $\qquad\qquad\qquad\qquad\qquad\qquad\qquad\qquad\qquad\qquad$ $\square$

Important examples of cores are given by the domains $D(A^n)$ of the powers A^n of a generator A.

1.8 Proposition. *For the generator* $(A, D(A))$ *of a strongly continuous semigroup* $(T(t))_{t \geq 0}$ *the space*

$$D(A^\infty) := \bigcap_{n \in \mathbb{N}} D(A^n),$$

hence each $D(A^n) := \{x \in D(A^{n-1}) : A^{n-1}x \in D(A)\}$, *is a core for* A.

PROOF. Because the space $D(A^\infty)$ is a $(T(t))_{t \geq 0}$-invariant subspace of $D(A)$, it remains to show that it is dense in X. To that purpose, we prove that for each function $\varphi \in C^\infty(-\infty, \infty)$ with compact support in $(0, \infty)$ and each $x \in X$ the element

$$x_\varphi := \int_0^\infty \varphi(s)T(s)x\,ds$$

belongs to $D(A^\infty)$. In fact, if we set

$$\mathcal{D} := \{\varphi \in C^\infty(-\infty, \infty) : \operatorname{supp}\varphi \text{ is compact in } (0, \infty)\},$$

then for $x \in X$, $\varphi \in \mathcal{D}$, and $h > 0$ sufficiently small we have

$$\frac{T(h) - I}{h}\,x_\varphi = \frac{1}{h} \int_0^\infty \varphi(s)\big(T(s+h) - T(s)\big)x\,ds$$

$$= \frac{1}{h} \int_h^\infty \big(\varphi(s-h) - \varphi(s)\big)T(s)x\,ds - \frac{1}{h} \int_0^h \varphi(s)T(s)x\,ds$$

$$(1.8) \qquad = \int_0^\infty \frac{1}{h}\big(\varphi(s-h) - \varphi(s)\big)T(s)x\,ds.$$

The integrand in (1.8) converges uniformly on $[0, \infty)$ to $-\varphi'(s)T(s)x$ as $h \downarrow 0$. This shows that $x_\varphi \in D(A)$ and

$$Ax_\varphi = -\int_0^\infty \varphi'(s)T(s)x \, ds.$$

Because $\varphi^{(n)} \in \mathcal{D}$ for all $n \in \mathbb{N}$, we conclude by induction that $x_\varphi \in D(A^n)$ for all $n \in \mathbb{N}$; i.e., $x_\varphi \in D(A^\infty)$. Assume that the linear span

$$D := \lim \{x_\varphi : x \in X, \ \varphi \in \mathcal{D}\}$$

is not dense in X. By the Hahn–Banach theorem there is a linear functional $0 \neq x' \in X'$ such that $\langle y, x' \rangle = 0$ for all $y \in D$; i.e.,

$$(1.9) \qquad \int_0^\infty \varphi(s) \langle T(s)x, x' \rangle \, ds = \left\langle \int_0^\infty \varphi(s) \, T(s)x \, ds, x' \right\rangle = 0$$

for all $x \in X$ and $\varphi \in \mathcal{D}$. This implies that the continuous functions $s \mapsto \langle T(s)x, x' \rangle$ vanish on $[0, \infty)$ for all $x \in X$. Otherwise there would exist $\varphi \in \mathcal{D}$ such that the left-hand side of (1.9) does not vanish. Choosing $s = 0$, we obtain $\langle x, x' \rangle = 0$ for all $x \in X$; hence $x' = 0$. This contradicts the choice of $x' \neq 0$, and therefore $D \subset X$ is dense.

Because we have seen in the first step that $D \subset D(A^\infty)$, and because $D(A^\infty)$ is invariant under $(T(t))_{t \geq 0}$, the assertion follows from Proposition 1.7. $\qquad\qquad\square$

In the remaining part of this section we introduce some basic spectral properties for generators of strongly continuous semigroups. Such properties are studied in more detail in Section V.1.a. We start by introducing the relevant notions (see also Definition V.1.1)

spectrum $\sigma(A) := \{\lambda \in \mathbb{C} : \lambda - A \text{ is not bijective}\}$,

resolvent set $\rho(A) := \mathbb{C} \setminus \sigma(A)$, and

resolvent $R(\lambda, A) := (\lambda - A)^{-1}$ at $\lambda \in \rho(A)$

for a closed operator $(A, D(A))$ on a Banach space X.

Our starting points are the following two identities, which are easily derived from their predecessors in Lemma 1.3.(iv). We stress that these identities will be used very frequently throughout these notes.

1.9 Lemma. *Let $(A, D(A))$ be the generator of a strongly continuous semigroup $(T(t))_{t \geq 0}$. Then, for every $\lambda \in \mathbb{C}$ and $t > 0$, the following identities hold.*

$$(1.10) \quad e^{-\lambda t}T(t)x - x = (A - \lambda) \int_0^t e^{-\lambda s}T(s)x \, ds \qquad \text{if } x \in X,$$

$$(1.11) \qquad\qquad = \int_0^t e^{-\lambda s}T(s)(A - \lambda)x \, ds \qquad \text{if } x \in D(A).$$

PROOF. It suffices to apply Lemma 1.3.(iv) to the rescaled semigroup

$$S(t) := e^{-\lambda t}T(t), \quad t \geq 0,$$

whose generator is $B := A - \lambda$ with domain $D(B) = D(A)$. $\qquad\qquad\square$

Next, we give an important formula relating the semigroup to the resolvent of its generator.

1.10 Theorem. *Let* $(T(t))_{t \geq 0}$ *be a strongly continuous semigroup on the Banach space X and take constants $w \in \mathbb{R}$, $M \geq 1$ (see Proposition I.1.4) such that*

$$(1.12) \qquad \qquad \|T(t)\| \leq Me^{wt}$$

for $t \geq 0$. For the generator $(A, D(A))$ of $(T(t))_{t \geq 0}$ the following properties hold.

- (i) *If $\lambda \in \mathbb{C}$ such that $R(\lambda)x := \int_0^\infty e^{-\lambda s} T(s)x \, ds$ exists for all $x \in X$, then $\lambda \in \rho(A)$ and $R(\lambda, A) = R(\lambda)$.*
- (ii) *If $\operatorname{Re} \lambda > w$, then $\lambda \in \rho(A)$, and the resolvent is given by the integral expression in (i).*
- (iii) *$\|R(\lambda, A)\| \leq \frac{M}{\operatorname{Re} \lambda - w}$ for all $\operatorname{Re} \lambda > w$.*

The formula for $R(\lambda, A)$ in (i) is called the *integral representation of the resolvent*. Of course, the integral has to be understood as an improper Riemann integral; i.e.,

$$(1.13) \qquad \qquad R(\lambda, A)x = \lim_{t \to \infty} \int_0^t e^{-\lambda s} T(s)x \, ds$$

for all $x \in X$.

Having in mind this interpretation, we frequently write

$$(1.14) \qquad \qquad R(\lambda, A) = \int_0^\infty e^{-\lambda s} T(s) \, ds.$$

PROOF OF THEOREM 1.10. (i) By a simple rescaling argument (cf. Paragraph I.1.10) we may assume that $\lambda = 0$. Then, for arbitrary $x \in X$ and $h > 0$, we have

$$
\begin{aligned}
\frac{T(h) - I}{h} R(0)x &= \frac{T(h) - I}{h} \int_0^\infty T(s)x \, ds \\
&= \frac{1}{h} \int_0^\infty T(s+h)x \, ds - \frac{1}{h} \int_0^\infty T(s)x \, ds \\
&= \frac{1}{h} \int_h^\infty T(s)x \, ds - \frac{1}{h} \int_0^\infty T(s)x \, ds \\
&= -\frac{1}{h} \int_0^h T(s)x \, ds.
\end{aligned}
$$

By taking the limit as $h \downarrow 0$, we conclude that[1] $\operatorname{rg} R(0) \subseteq D(A)$ and $AR(0) = -I$. On the other hand, for $x \in D(A)$ we have

$$\lim_{t \to \infty} \int_0^t T(s)x\,ds = R(0)x,$$

and

$$\lim_{t \to \infty} A \int_0^t T(s)x\,ds = \lim_{t \to \infty} \int_0^t T(s)Ax\,ds = R(0)Ax,$$

where we have used Lemma 1.3.(iv) for the second equality. Because by Theorem 1.4 the operator A is closed, this implies $R(0)Ax = AR(0)x = -x$ and therefore $R(0) = (-A)^{-1}$ as claimed.

Parts (ii) and (iii) then follow easily from (i) and the estimate

$$\left\| \int_0^t e^{-\lambda s} T(s)\,ds \right\| \leq M \int_0^t e^{(w - \operatorname{Re}\lambda)s}\,ds,$$

because for $\operatorname{Re}\lambda > w$ the right-hand side converges to $M/(\operatorname{Re}\lambda - w)$ as $t \to \infty$. $\qquad\square$

The above integral representation can now be used to represent and estimate the powers of $R(\lambda, A)$.

1.11 Corollary. *For the generator* $(A, D(A))$ *of a strongly continuous semigroup* $(T(t))_{t \geq 0}$ *satisfying*

$$\|T(t)\| \leq Me^{wt} \qquad \text{for all } t \geq 0,$$

one has, for $\operatorname{Re}\lambda > w$ *and* $n \in \mathbb{N}$, *that*

$$(1.15) \qquad R(\lambda, A)^n x = \frac{(-1)^{n-1}}{(n-1)!} \cdot \frac{d^{n-1}}{d\lambda^{n-1}} R(\lambda, A)x$$

$$(1.16) \qquad = \frac{1}{(n-1)!} \int_0^\infty s^{n-1} e^{-\lambda s} T(s)x\,ds$$

for all $x \in X$. *In particular, the estimates*

$$(1.17) \qquad \|R(\lambda, A)^n\| \leq \frac{M}{(\operatorname{Re}\lambda - w)^n}$$

hold for all $n \in \mathbb{N}$ *and* $\operatorname{Re}\lambda > w$.

---

[1] Here $\operatorname{rg} T := TX$ indicates the *range* of an operator $T : X \to Y$.

PROOF. Equation (1.15) is actually valid for every operator with nonempty resolvent set; see Proposition V.1.3.(ii). On the other hand, Theorem 1.10.(i) implies

$$\frac{d}{d\lambda}R(\lambda, A)x = \frac{d}{d\lambda}\int_0^\infty e^{-\lambda s}T(s)x\,ds$$

$$= -\int_0^\infty se^{-\lambda s}T(s)x\,ds$$

for $\operatorname{Re}\lambda > w$ and all $x \in X$. Proceeding by induction, we deduce (1.16). Finally, the estimate (1.17) follows from

$$\|R(\lambda, A)^n x\| = \frac{1}{(n-1)!} \cdot \left\|\int_0^\infty s^{n-1}e^{-\lambda s}T(s)x\,ds\right\|$$

$$\leq \frac{M}{(n-1)!} \cdot \int_0^\infty s^{n-1}e^{(w-\operatorname{Re}\lambda)s}\,ds \cdot \|x\|$$

$$= \frac{M}{(\operatorname{Re}\lambda - w)^n} \cdot \|x\|$$

for all $x \in X$. $\qquad\square$

Property (ii) in Theorem 1.10 says that the spectrum of a semigroup generator is always contained in a left half-plane. The number determining the smallest such half-plane is an important characteristic of any linear operator and is now defined explicitly.

1.12 Definition. *With any linear operator A we associate its spectral bound defined by*

$$s(A) := \sup\{\operatorname{Re}\lambda : \lambda \in \sigma(A)\}.$$

As an immediate consequence of Theorem 1.10.(ii) the following relation holds between the growth bound of a strongly continuous semigroup (see Definition I.1.5) and the spectral bound of its generator.

1.13 Corollary. *For a strongly continuous semigroup $(T(t))_{t\geq 0}$ with generator A, one has*

$$-\infty \leq s(A) \leq \omega_0 < +\infty.$$

1.14 Diagram. To conclude this section, we collect in a diagram the information obtained so far on the relations between a semigroup, its generator, and its resolvent.

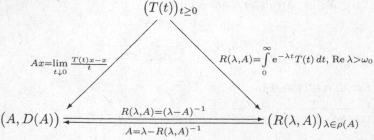

By developing our theory further, we are able to add one of the missing links in this diagram (see Diagram IV.2.6 below).

1.15 Exercises. (1) Prove that the statements (a)–(d) in Corollary 1.5 are equivalent to each of the following conditions.

(e) $\|T(t) - I\| \le ct$ for $0 \le t \le 1$ and some $c > 0$.

(f) $\overline{\lim}_{\lambda \to \infty} \|\lambda AR(\lambda, A)\| < \infty$.

(2) Show that for a closed linear operator $(A, D(A))$ on a Banach space X and a linear subspace $Y \subset D(A)$ the following assertions are equivalent.

(a) Y is a core for $(A, D(A))$.

(b) $\overline{A_{|Y}} = A$.

If, in addition, $\rho(A) \ne \emptyset$, then these assertions are equivalent to

(c) $(\lambda - A)Y$ is dense in X for one/all $\lambda \in \rho(A)$.

(3) Show that the space of all continuous functions with compact support forms a core for each multiplication operator M_q on $C_0(\Omega)$.

(4) Decide whether $\mathcal{D} := \{ f \in C^\infty(\mathbb{R}_+) : f'(0) = 0 \text{ and supp } f \text{ is compact} \}$ is a core for

(i) The generator of the left translation semigroup on $C_0(\mathbb{R}_+)$, and

(ii) The generator of the right translation semigroup on $C_0(\mathbb{R}_+)$,

as defined in Paragraph I.3.16. (Hint: Compare the hint in (Exercise 6.iii).)

(5) Consider the Banach space $X := C_0(\Omega)$ for some locally compact space Ω. Show that for a strongly continuous semigroup $(T(t))_{t \ge 0}$ with generator $(A, D(A))$ on X the following statements are equivalent.

(a) $(T(t))_{t \ge 0}$ is a semigroup of *algebra homomorphisms* on X; i.e., $T(t)(f \cdot g) = T(t)f \cdot T(t)g$ for $f, g \in X$ and $t \ge 0$.

(b) $(A, D(A))$ is a *derivation*; i.e., $D(A)$ is a subalgebra of X and

$$A(f \cdot g) = (Af) \cdot g + f \cdot Ag$$

for $f, g \in D(A)$.

(Hint: For the implication (b) $\Rightarrow$ (a) consider the maps $s \mapsto T(t-s)[T(s)f \cdot T(s)g]$ for each $0 \le s \le t$ and $f, g \in D(A)$. For more information see [Nag86, B-II, Sect. 3])

(6) Let $(A, D(A))$ be the generator of a contraction semigroup $(T(t))_{t \ge 0}$ on some Banach space X. Establish the following assertions.

(i) The *Landau–Kolmogorov inequality*, which states that

$$\|Ax\|^2 \le 4 \|A^2 x\| \cdot \|x\|$$

for each $x \in D(A^2)$. (Hint: As a first step, verify Taylor's formula

$$T(t)x = x + tAx + \int_0^t (t - s)T(s)A^2 x \, ds$$

for $x \in D(A^2)$.)

(ii) If $(T(t))_{t \geq 0}$ is a group of isometries, then (i) can be improved to

$$\|Ax\|^2 \leq 2\|A^2x\| \cdot \|x\|$$

for $x \in D(A^2)$.

(iii) Apply (i) and (ii) to the various translation semigroups of Section I.3.c, in particular to the left translation semigroup on $L^p(\mathbb{R}_+)$. (Hint: The generator of the (left) translation semigroup is the differentiation operator with appropriate domain; see Paragraph 2.9. For details see [Gol85, p.65])

2. Examples Revisited

Before proceeding with the abstract theory, we pause for a moment and examine the concrete semigroups from Section I.3 and the semigroup constructions established in Section I.1.b. In each case, we try to identify the corresponding

generator, its *spectrum* and *resolvent*,

so that our abstract definitions gain a more concrete meaning. However, the impatient reader might skip these examples and proceed with Section 3.

a. Standard Constructions

Let $(T(t))_{t \geq 0}$ be a strongly continuous semigroup with generator $(A, D(A))$ on a Banach space X. For each of the semigroups constructed in Section I.1.b, we now characterize its generator and its resolvent.

2.1 Similar Semigroups. If V is an isomorphism from a Banach space Y onto X and $(S(t))_{t \geq 0}$ is the strongly continuous semigroup on Y given by $S(t) := V^{-1}T(t)V$, then its generator is

$$B = V^{-1}AV \quad \text{with domain} \quad D(B) = \{y \in Y : Vy \in D(A)\}.$$

Equality of the spectra

$$\sigma(A) = \sigma(B)$$

is clear, and the resolvent of B is $R(\lambda, B) = V^{-1}R(\lambda, A)V$ for $\lambda \in \rho(A)$.

A particularly important example of this situation is given by the Spectral Theorem I.3.9, which states that every normal or self-adjoint operator on a Hilbert space is similar to a multiplication operator on an L^2-space.

2.2 Rescaled Semigroups. The rescaled semigroup $\left(e^{\mu t}T(\alpha t)\right)_{t\geq0}$ for some fixed $\mu \in \mathbb{C}$ and $\alpha > 0$ has generator

$$B = \alpha A + \mu I \quad \text{with domain} \quad D(A) = D(B).$$

Moreover, $\sigma(B) = \alpha\sigma(A) + \mu$ and $R(\lambda, B) = \frac{1}{\alpha}R\left(\frac{\lambda - \mu}{\alpha}, A\right)$ for $\lambda \in \rho(B)$.

This shows that we can switch quite easily between the original and the rescaled objects.

2.3 Subspace Semigroups. Although in Paragraph I.1.11 we considered the subspace semigroup $\left(T(t)_{|_Y}\right)_{t\geq0}$ only for *closed* subspaces Y in X, we begin here with a more general situation.

Let Y be a Banach space that is continuously embedded in X (in symbols: $Y \hookrightarrow X$). Assume also that the restrictions $T(t)_|$ leave Y invariant and form a strongly continuous semigroup $\left(T(t)_|\right)_{t\geq0}$ on Y. In order to be able to identify the generator of $\left(T(t)_{|_Y}\right)_{t\geq0}$, we introduce the following concept.

Definition. *The* part of A in Y *is the operator $A_|$ defined by*

$$A_|y := Ay$$

with domain
$$D(A_|) := \left\{y \in D(A) \cap Y : Ay \in Y\right\}.$$

In other words, $A_|$ is the "maximal" operator induced by A on Y and, as shown next, coincides with the generator of the semigroup $\left(T(t)_|\right)_{t\geq0}$ on the subspace Y.

Proposition. *Let $\left(A, D(A)\right)$ be the generator of a strongly continuous semigroup $\left(T(t)\right)_{t\geq0}$ on X. If the restricted semigroup $\left(T(t)_|\right)_{t\geq0}$ is strongly continuous on some $\left(T(t)\right)_{t\geq0}$-invariant Banach space $Y \hookrightarrow X$, then the generator of $\left(T(t)_|\right)_{t\geq0}$ is the part $\left(A_|, D(A_|)\right)$ of A in Y.*

PROOF. Let $\left(C, D(C)\right)$ denote the generator of $\left(T(t)_|\right)_{t\geq0}$. Because Y is continuously embedded in X, we immediately have that C is a restriction of $A_|$. For the converse inclusion, choose $\lambda \in \mathbb{R}$ large enough such that both $R(\lambda, C)$ and $R(\lambda, A)$ are given by the integral representation from Theorem 1.10.(i). Then

$$R(\lambda, C)y = \int_0^\infty e^{-\lambda s}T(s)y\,ds = R(\lambda, A)y \qquad \text{for all} \qquad y \in Y.$$

For $x \in D(A_|)$, we obtain that

$$x = R(\lambda, A)(\lambda - A)x = R(\lambda, C)(\lambda - A)x \in D(C),$$

and hence $D(A_|) = D(C)$. $\qquad\qquad\square$

If Y is a $(T(t))_{t \geq 0}$-invariant closed subspace of X, then the strong continuity of $(T(t)_|)_{t \geq 0}$ is automatic. Moreover, the existence of

$$z := \lim_{t \downarrow 0} \frac{1}{t}(T(t)y - y) \in X$$

for some $y \in Y$ implies that $z \in Y$. Therefore, the part $A_|$ simply becomes the "restriction" of A.

Corollary. *If Y is a $(T(t))_{t \geq 0}$-invariant closed subspace of X, then the generator of $(T(t)_|)_{t \geq 0}$ is*

$$A_| y = Ay,$$

with domain

$$D(A_|) = D(A) \cap Y.$$

Example. A typical example for the situation considered here occurs when we take $X := L^1(\Gamma, m)$ and $Y := C(\Gamma)$. The rotation group from I.3.18 is strongly continuous on both spaces; hence its generator on $C(\Gamma)$ is the part of its generator on $L^1(\Gamma, m)$. The generator on $L^1(\Gamma, m)$ can now be obtained as the first derivative by modifying the arguments from the proposition in Paragraph 2.9.(ii) below.

2.4 Quotient Semigroup. Let Y be a $(T(t))_{t \geq 0}$-invariant closed subspace of X. Then the generator $(A_/, D(A_/))$ of the quotient semigroup $(T(t)_{/Y})_{t \geq 0}$ on the quotient space $X_/ := {}^X/_Y$ is given (with the notation from Paragraph I.1.12) by

$$A_/ q(x) = q(Ax) \qquad \text{with domain } D(A_/) = q(D(A)).$$

This follows from the fact that each element $\widehat{x} := q(x) \in D(A_/)$ can be written as

$$\widehat{x} = \int_0^\infty e^{-\lambda s} T(s)_/ \widehat{y}\, ds$$

for some $\widehat{y} = q(y) \in {}^X/_Y$ and some $\lambda > \omega_0$ (use 1.10.(i)). Therefore,

$$\widehat{x} = \int_0^\infty e^{-\lambda s} q(T(s)y)\, ds = q\left(\int_0^\infty e^{-\lambda s} T(s)y\, ds\right) = q(z)$$

with $z \in D(A)$. This means that for every $\widehat{x} \in D(A_/)$ there exists a representative $z \in X$ belonging to $D(A)$.

For a concrete example, we refer to Paragraph 2.10.

2.5 Adjoint Semigroups. Even though the adjoint semigroup $(T(t)')_{t\geq 0}$ is not necessarily strongly continuous on X', it is still possible to associate a "generator" with it. In fact, defining

$$A^\sigma x' := \sigma(X',X)\text{-}\lim_{h\downarrow 0}\frac{1}{h}\big(T(h)'x' - x'\big),$$

$$D(A^\sigma) := \left\{x' \in X' : \sigma(X',X)\text{-}\lim_{h\downarrow 0}\frac{1}{h}\big(T(h)'x' - x'\big) \text{ exists}\right\},$$

one obtains a linear operator called the *weak* generator* of $(T(t)')_{t\geq 0}$. It is a $\sigma(X',X)$-closed and $\sigma(X',X)$-densely defined operator and coincides with the *adjoint* A' of A (see Definition A.12); i.e.,

$$D(A^\sigma) = \left\{x' \in X' : \begin{array}{l}\text{there exists } y' \in X' \text{ such that}\\ \langle x,y'\rangle = \langle Ax,x'\rangle \text{ for all } x \in D(A)\end{array}\right\}$$

and

$$A^\sigma x' = A'x'.$$

(See Exercise 2.7.) By Corollary A.16 it then follows that $\sigma(A^\sigma) = \sigma(A) = \sigma(A')$ and $R(\lambda, A^\sigma) = R(\lambda, A') = R(\lambda, A)'$ for $\lambda \in \rho(A)$.

2.6 Product Semigroups. Let $(B, D(B))$ be the generator of a second strongly continuous semigroup $(S(t))_{t\geq 0}$ commuting with $(T(t))_{t\geq 0}$. It is easy to deduce some information on the generator $(C, D(C))$ of the product semigroup $(U(t))_{t\geq 0}$, defined by $U(t) := S(t)T(t)$ for $t \geq 0$; see Paragraph I.1.14.

We first show that $D(A) \cap D(B)$ satisfies the conditions of Proposition 1.7 and so is a core for C.

Because $(T(t))_{t\geq 0}$ and $(S(t))_{t\geq 0}$ commute, each domain $D(A)$ and $D(B)$ is invariant under both semigroups. Hence $D(A) \cap D(B)$ is $(U(t))_{t\geq 0}$-invariant. Take λ large enough such that $R(\lambda, A) = \int_0^\infty e^{-\lambda s}T(s)\,ds$ and $R(\lambda, B) = \int_0^\infty e^{-\lambda s}S(s)\,ds$. From these representations we deduce that the resolvent operators commute; i.e., $R(\lambda, A)R(\lambda, B) = R(\lambda, B)R(\lambda, A)$. Therefore, $R(\lambda, B)$ maps $D(A)$ into $D(A)$, and so $R(\lambda, B)R(\lambda, A)X$ is contained in $D(A) \cap D(B)$. Because both $R(\lambda, A)$ and $R(\lambda, B)$ are continuous and have dense range, we conclude that $D(A) \cap D(B)$ is dense in X, i.e., is a core for C.

Now, by Lemma A.19, the map $\mathbb{R}_+ \ni t \mapsto U(t)x$ is differentiable for all elements $x \in D(A) \cap D(B)$. Moreover, its derivative at $t = 0$ is

$$\left[\tfrac{d}{dt}U(t)x\right](0) = Cx = Ax + Bx,$$

which determines the generator C of $(U(t))_{t\geq 0}$ on the core $D(A) \cap D(B)$.

2.7 Exercise. Show that the operator A^σ defined in Paragraph 2.5 is $\sigma(X',X)$-closed, $\sigma(X',X)$-densely defined, and that it coincides with the adjoint A' of A.

b. Standard Examples

In this subsection we return to the examples of strongly continuous semi-groups introduced in Chapter I, Section 3, and identify the corresponding generators and resolvent operators. We start with multiplication semi-groups for which all operators involved can be computed explicitly.

2.8 Multiplication Semigroups. We saw in Proposition I.3.6 (or Proposition I.3.12) that strongly continuous multiplication semigroups on spaces $C_0(\Omega)$ (or $L^p(\Omega, \mu)$) are multiplications by e^{tq}, $t \geq 0$, for some continuous (or measurable) function $q : \Omega \to \mathbb{C}$ with real part (essentially) bounded above. It should be no surprise that this function also yields the generator of the semigroup.

Lemma. *The generator $\big(A, D(A)\big)$ of a strongly continuous multiplication semigroup $\big(T(t)\big)_{t \geq 0}$ on $X := C_0(\Omega)$ or $X := L^p(\Omega, \mu)$ defined by*

$$T_q(t)f := e^{tq} \cdot f, \quad f \in X \text{ and } t \geq 0,$$

is given by the multiplication operator

$$Af = M_q f := q \cdot f$$

with domain $D(A) = D(M_q) := \{f \in X : qf \in X\}$.

PROOF. Let $X := C_0(\Omega)$ and take $f \in D(A)$. Then

$$\lim_{t \downarrow 0} \frac{e^{tq} f - f}{t}(s) = \lim_{t \downarrow 0} \frac{e^{tq(s)} f(s) - f(s)}{t} = q(s)f(s)$$

exists for all $s \in \Omega$, and we obtain $qf \in C_0(\Omega)$. This shows that $D(A) \subseteq D(M_q)$ and that $Af = M_q f$. Because by Theorem 1.10.(ii) and Proposition I.3.2.(iv), respectively, $A - \lambda$ and $M_q - \lambda$ are both invertible for λ sufficiently large, this implies $A = M_q$. The proof for $X := L^p(\Omega, \mu)$ is left as Exercise 2.13.(2).   □

This lemma, in combination with Propositions I.3.5 and I.3.6 (or Propositions I.3.11 and I.3.12), completely characterizes the generators of strongly continuous multiplication semigroups. We restate this in the following result by identifying the closed (or the essential) range of q with the spectrum of M_q; see Proposition I.3.2.(iv) (or Proposition I.3.10.(iv)).

Proposition. *For an operator $\big(A, D(A)\big)$ on the Banach space $C_0(\Omega)$ or $L^p(\Omega, \mu)$, $1 \leq p < \infty$, the following assertions are equivalent.*

(a) *$\big(A, D(A)\big)$ is the generator of a strongly continuous multiplication semigroup.*

(b) *$\big(A, D(A)\big)$ is a multiplication operator such that*

$$\{\lambda \in \mathbb{C} : \operatorname{Re}\lambda > w\} \subseteq \rho(A) \qquad \text{for some } w \in \mathbb{R}.$$

The remarkable feature of this proposition is the fact that condition (b), which corresponds to the spectral condition (ii) from Theorem 1.10, already guarantees the existence of a corresponding semigroup. This is in sharp contrast to the situation for general semigroups (see Generation Theorems 3.5 and 3.8 below).

2.9 Translation Semigroups. As seen in Paragraph I.3.15, the (left) translation operators

$$T_l(t)f(s) := f(s+t), \quad s, t \in \mathbb{R},$$

define a strongly continuous (semi) group on the spaces $C_{ub}(\mathbb{R})$ and $L^p(\mathbb{R})$, $1 \leq p < \infty$. In each case, the generator $(A, D(A))$ is given by differentiation, but we have to adapt its domain to the underlying space.

Proposition 1. *The generator of the (left) translation semigroup $(T_l(t))_{t \geq 0}$ on the space X is given by*

$$Af := f'$$

with domain:
 (i)

$$D(A) = \{f \in C_{ub}(\mathbb{R}) : f \text{ differentiable and } f' \in C_{ub}(\mathbb{R})\},$$

if $X := C_{ub}(\mathbb{R})$, and
 (ii)

$$D(A) = \{f \in L^p(\mathbb{R}) : f \text{ absolutely continuous and } f' \in L^p(\mathbb{R})\},$$

if $X := L^p(\mathbb{R})$, $1 \leq p < \infty$.

PROOF. It suffices to show that the generator $(B, D(B))$ of $(T_l(t))_{t \geq 0}$ is a restriction of the operator $(A, D(A))$ defined above. In fact, because $(T_l(t))_{t \geq 0}$ is a contraction semigroup on X, Theorem 1.10.(ii) implies $1 \in \rho(B)$. On the other hand, by Proposition 2 below, we know that $1 \in \rho(A)$, and therefore the inclusion $B \subseteq A$ will imply $A = B$.

(i) Fix $f \in D(B)$. Because δ_0 is a continuous linear form on $C_{ub}(\mathbb{R})$, the function

$$\mathbb{R}_+ \ni t \mapsto \delta_0(T_l(t)f) = f(t)$$

is differentiable by Lemma 1.1 and Definition 1.2, and

$$Bf = \left[\tfrac{d}{dt}T_l(t)f\right]_{t=0} = \left[\tfrac{d}{dt}f(t + \cdot)\right]_{t=0} = f'.$$

This proves $D(B) \subseteq D(A)$ and $A_{|D(B)} = B$. Hence, $A = B$ as mentioned above.

(ii) Take $f \in D(B)$ and set $g := Bf \in L^p(\mathbb{R})$. Because integration over compact intervals is continuous in $L^p(\mathbb{R})$, we obtain for every $a, b \in \mathbb{R}$ that

$$\frac{1}{h} \int_b^{b+h} f(s) \, ds - \frac{1}{h} \int_a^{a+h} f(s) \, ds = \int_a^b \frac{f(s+h) - f(s)}{h} \, ds$$

converges to $\int_a^b g(s) \, ds$ as $h \downarrow 0$. However, the left-hand side converges to $f(b) - f(a)$ for almost all a, b; see [Tay85, Thm. 9-8 VI]. By redefining f on a null set we obtain

$$f(b) = \int_a^b g(s) \, ds + f(a), \qquad b \in \mathbb{R},$$

which is an absolutely continuous function with derivative (almost everywhere) equal to g. Again this shows that $D(B) \subseteq D(A)$ and $A_{|D(B)} = B$. It follows that $A = B$ as above. $\square$

In order to finish this proof, we give an explicit formula for the resolvent of the differentiation operator A with "maximal" domain $D(A)$ as specified in the previous result. The simple proof is left as Exercise 2.13.(1).

Proposition 2. *The resolvent $R(\lambda, A)$ for $\operatorname{Re} \lambda > 0$ of the differentiation operator A with maximal domain $D(A)$ (i.e., of the generator of the left translation semigroup) on any of the above spaces X is given by*

$$(2.1) \qquad (R(\lambda, A)f)(s) = \int_s^\infty e^{-\lambda(\tau - s)} f(\tau) \, d\tau \quad \text{for } f \in X, \ s \in \mathbb{R}.$$

Clearly, there are many other function spaces on which the translations define a strongly continuous semigroup. As soon as they are contained in $L^p(\mathbb{R})$ or $C_{ub}(\mathbb{R})$, for example, Proposition 2.3 allows us to identify the generator as the part of the differentiation operator. This, and the quotient construction from Paragraph 2.4, yield the generators of the translation semigroups on $\mathbb{R}_+$ and on finite intervals (see Paragraphs I.3.16 and I.3.17).

We present an example for this argument.

2.10 Translation Semigroups (Continued). Consider the (left) translation (semi) group from Paragraph 2.9 on the space $X := L^1(\mathbb{R})$. Then the closed subspace

$$Y := \{ f \in L^1(\mathbb{R}) : f(s) = 0 \text{ for } s \geq 1 \},$$

which is isomorphic to $L^1(-\infty, 1)$, is $(T(t))_{t \geq 0}$-invariant. The generator of the subspace semigroup $(T(t)_|)_{t \geq 0}$ is

$$A_| f = f'$$

with domain

$$D(A_|) = \left\{ f \in L^1(\mathbb{R}) : \begin{array}{l} f \text{ is absolutely continuous,} \\ f' \in L^1(\mathbb{R}) \text{ and } f(s) = 0 \text{ for } s \geq 1 \end{array} \right\}.$$

In Y and for the subspace semigroup $(T(t)_|)_{t \geq 0}$, the space

$$Z := \{f \in Y : f(s) = 0 \text{ for } 0 \leq s \leq 1\}$$

is again closed and invariant. The quotient space $Y/_Z$ is isomorphic to $L^1[0,1]$, and the quotient semigroup is isomorphic to the nilpotent (left) translation semigroup from Paragraph I.3.17. By Paragraph 2.4, we obtain for its generator $A_{|/}$ that

$$A_{|/} f = f'$$

with domain

$$D(A_{|/}) = \left\{ f \in L^1[0,1] : \begin{array}{l} f \text{ is absolutely continuous,} \\ f' \in L^1[0,1] \text{ and } f(1) = 0 \end{array} \right\}.$$

As above, its resolvent can be determined explicitly using (1.13). We obtain for every $\lambda \in \mathbb{C}$ that

$$(2.2) \quad \big(R(\lambda, A_{|/})f\big)(s) = \int_s^1 e^{-\lambda(\tau - s)} f(\tau)\, d\tau \quad \text{for } f \in L^1[0,1], \ s \in [0,1].$$

In the previous examples we always started with an explicit semigroup and then identified its generator. In the final two examples we look at (second-order) differential operators and show by direct computation that they generate strongly continuous semigroups.

2.11 Diffusion Semigroups (One-Dimensional). Consider the Banach space $X := C[0,1]$ and the differential operator

$$Af := f''$$

with domain

$$D(A) := \{f \in C^2[0,1] : f'(0) = f'(1) = 0\}.$$

This domain is a dense subspace of X that is complete for the graph norm; hence $\big(A, D(A)\big)$ is a closed, densely defined operator. Each function

$$s \mapsto e_n(s) := \begin{cases} 1 & \text{if } n = 0, \\ \sqrt{2}\cos(\pi n s) & \text{if } n \geq 1, \end{cases}$$

belongs to $D(A)$ and satisfies

$$(2.3) \qquad\qquad A e_n = -\pi^2 n^2 e_n.$$

By the Stone–Weierstrass theorem and elementary trigonometric identities we conclude that

$$(2.4) \qquad\qquad Y := \lin\{e_n : n \geq 0\}$$

is a dense subalgebra of X. Consider the rank-one operators

$$e_n \otimes e_n : f \mapsto \langle f, e_n \rangle \, e_n := \left(\int_0^1 f(s) e_n(s) \, ds \right) e_n,$$

that satisfy

$$\| e_n \otimes e_n \| \leq 2$$

and

(2.5) $(e_n \otimes e_n) \, e_m = \delta_{nm} e_m$

for all $n, m \geq 0$. They can be used to define, for $t > 0$, the operators

(2.6) $$T(t) := \sum_{n=0}^{\infty} e^{-\pi^2 n^2 t} \cdot e_n \otimes e_n.$$

For $f \in C[0,1]$ and $s \in [0,1]$, this means that

(2.7) $$(T(t)f)(s) = \int_0^1 k_t(s, r) f(r) \, dr,$$

where

$$k_t(s, r) := 1 + 2 \sum_{n \in \mathbb{N}} e^{-\pi^2 n^2 t} \cos(\pi n s) \cdot \cos(\pi n r).$$

The Jacobi identity

$$w_t(s) := \frac{1}{\sqrt{4\pi t}} \sum_{n \in \mathbb{Z}} e^{-(s+2n)^2/4t} = \frac{1}{2} + \sum_{n \in \mathbb{N}} e^{-\pi^2 n^2 t} \cos(\pi n s)$$

(see [SD80, Kap. I, Satz 10.4]) and various trigonometric relations imply that for each $t > 0$, the kernel $k_t(\cdot, \cdot)$ satisfies

$$k_t(s, r) = w_t(s + r) + w_t(s - r).$$

Hence, $k_t(\cdot, \cdot)$ is a positive continuous function on $[0, 1]^2$, and we obtain

$$\| T(t) \| = \| T(t) \mathbb{1} \| = \sup_{s \in [0,1]} \int_0^1 k_t(s, r) \, dr = 1.$$

Using the identity (2.5), one easily verifies that on the one-dimensional subspaces generated by e_n, $n \geq 0$, the operators $T(t)$ satisfy the semigroup law (FE), which by continuity then holds on all of X. Similarly, the strong continuity holds on Y and hence, by Proposition I.1.3, on X.

These considerations already prove most of the following result.

Proposition. *The above operators $T(t)$, $t \geq 0$, with $T(0) = I$ form a strongly continuous semigroup on $X := C[0, 1]$ whose generator is given by*

$$Af = f'',$$

with domain

$$D(A) = \{ f \in C^2[0, 1] : f'(0) = f'(1) = 0 \}.$$

PROOF. It remains only to show that the generator B of $(T(t))_{t\geq 0}$ coincides with A. To this end, we first observe that the subspace Y defined by (2.4) is dense in X, contained in $D(B)$, and $(T(t))_{t\geq 0}$-invariant. Hence, by Proposition 1.7, it is a core for B. Next, using the definition of $T(t)$ and Formula (2.5), it follows that A and B coincide on Y. Therefore, we obtain that $B = \overline{A_{|Y}}$ and, in particular, that B is a restriction of A. From the theory of linear ordinary differential equations it follows that $1 \in \rho(A)$. Moreover, by Theorem 1.10.(ii), we know that $1 \in \rho(B)$, and therefore $A = B$. □

2.12 Diffusion Semigroups (n-Dimensional). The following classical example was one of the main sources for the development of semigroup theory. It describes heat flow, diffusion processes, or Brownian motion and bears names such as *heat semigroup*, *Gaussian semigroup*, or *diffusion semigroup*. We consider it on $X := L^p(\mathbb{R}^n)$, $1 \leq p < \infty$, where it is defined explicitly by

$$(2.8) \qquad T(t)f(s) := (4\pi t)^{-n/2} \int_{\mathbb{R}^n} e^{-|s-r|^2/4t} f(r)\, dr$$

for $t > 0$, $s \in \mathbb{R}^n$, and $f \in X$. By putting

$$\mu_t(s) := (4\pi t)^{-n/2} e^{-|s|^2/4t},$$

this can be written as

$$T(t)f(s) = \mu_t * f(s).$$

Proposition. *The above operators $T(t)$, for $t > 0$ and with $T(0) = I$, form a strongly continuous semigroup on $L^p(\mathbb{R}^n)$, $1 \leq p < \infty$, and its generator A coincides with the closure of the Laplace operator*

$$\Delta f(s) := \sum_{i=1}^{n} \frac{\partial^2}{\partial s_i^2} f(s_1, \ldots, s_n)$$

defined for every f in the Schwartz space

$$\mathscr{S}(\mathbb{R}^n) := \left\{ f \in C^\infty(\mathbb{R}^n) : \lim_{|x|\to\infty} |x|^k D^\alpha f(x) = 0 \text{ for all } k \in \mathbb{N} \text{ and } \alpha \in \mathbb{N}^n \right\}$$

(see [EN00, Def. VI.5.1]).

PROOF. The integral defining $T(t)f(s)$ exists for every $f \in L^p(\mathbb{R}^n)$, because $\mu_t \in \mathscr{S}(\mathbb{R}^n)$. Moreover,

$$\|T(t)f\|_p \leq \|\mu_t\|_1 \cdot \|f\|_p \leq \|f\|_p$$

by Young's inequality (see [RS75, p. 28]). Hence, each $T(t)$ is a contraction on L^p. Because $\mathscr{S}(\mathbb{R}^n)$ is dense in L^p and invariant under $T(t)$, it suffices to study $T(t)_{|\mathscr{S}(\mathbb{R}^n)}$. This is done using the Fourier transformation $\mathcal{F}$, which leaves $\mathscr{S}(\mathbb{R}^n)$ invariant. By the usual properties of $\mathcal{F}$ (see [Rud73, Thm. 7.2]) one obtains

$$\mathcal{F}(\mu_t * f) = \mathcal{F}(\mu_t) \cdot \mathcal{F}(f)$$

for each $f \in \mathscr{S}(\mathbb{R}^n)$. Because

$$\mathcal{F}(\mu_t)(\xi) = e^{-|\xi|^2 t}$$

for $\xi \in \mathbb{R}^n$, where $|\xi| := \left(\sum_{i=1}^{n} \xi_i^2\right)^{1/2}$, we see that $\mathcal{F}$ transforms the semigroup $\left(T(t)_{|\mathscr{S}(\mathbb{R}^n)}\right)_{t \geq 0}$ into a multiplication semigroup on $\mathscr{S}(\mathbb{R}^n)$, which is pointwise continuous for the usual topology on $\mathscr{S}(\mathbb{R}^n)$. Moreover, direct computations as in Lemma 2.8 show that the right derivative at $t = 0$ is the multiplication operator

$$Bg(\xi) := -|\xi|^2 g(\xi)$$

for $\xi \in \mathbb{R}^n$, $g \in \mathscr{S}(\mathbb{R}^n)$. Pulling this information back via the inverse Fourier transformation shows that $\left(T(t)\right)_{t \geq 0}$ satisfies the semigroup law. Because the topology of $\mathscr{S}(\mathbb{R}^n)$ is finer than the one induced from $L^p(\mathbb{R}^n)$, we also obtain strong continuity on $\mathscr{S}(\mathbb{R}^n)$, hence on $L^p(\mathbb{R}^n)$. Finally, we observe that the inverse Fourier transformation of the multiplication operator B is the Laplace operator. Because $\mathscr{S}(\mathbb{R}^n)$ is dense and $\left(T(t)\right)_{t \geq 0}$-invariant, by Proposition 1.7 we have therefore determined the generator A of $\left(T(t)\right)_{t \geq 0}$ on a core of its domain. $\square$

For generalizations of this example we refer to [ABHN01, Expl. 3.7.6 and Chap. 8], [EN00, Sect. VI.5], and [Lun95, Chap. 3]. In particular, we mention that (2.8) also defines a strongly continuous semigroup on $C_{ub}(\mathbb{R}^n)$. Its generator is given by the closure of the Laplacian Δ with domain $C_b^\infty(\mathbb{R}^n)$.

2.13 Exercises. (1) Compute the resolvent operators of the generators of the various translation semigroups on $\mathbb{R}$, $\mathbb{R}_+$, or on finite intervals. In particular, deduce the resolvent representation (2.1). (Hint: Use the integral representation (1.14).) Determine from this the generator and its domain as already found in Paragraphs 2.9 and 2.10.

(2) Prove the lemma in Paragraph 2.8 for $X := L^p(\Omega, \mu)$.

(3) Let $X := L^\infty(\mathbb{R})$. Show that

(i) A multiplication semigroup on X is strongly continuous if and only if it is uniformly continuous, and

(ii) The translation (semi) group is not strongly continuous.

Remark that Lotz in [Lot85] showed that a strongly continuous semigroup on a class of Banach spaces containing all L^∞-spaces is necessarily uniformly continuous. See also [Nag86, A-II.3].

(4*) Consider the translation (semi) group $(T(t))_{t\in\mathbb{R}}$ on $X := L^\infty(\mathbb{R})$ and the closed, $(T(t))_{t\in\mathbb{R}}$-invariant subspace $Y := C_{ub}(\mathbb{R})$. The quotient operators $T(t)_/$ define a contraction (semi) group on X/Y whose orbits $t \mapsto T(t)_/$ are continuous (differentiable) only if $\widehat{f} = 0$. Note that in this way we obtained a noncontinuous, but not pathological, solution of Problem I.2.13. (Hint: See [NP94].)

c. Sobolev Towers

In the spirit of Section 2.a, we continue to associate new semigroups on new spaces with a given strongly continuous semigroup. The constructions here are inspired by the classical Sobolev and distribution spaces and yield an important tool for abstract theory as well as for concrete applications (see [Haa06, Chap. 6], [HHK06], [Sin05]).

We start by considering a strongly continuous semigroup $(T(t))_{t\geq 0}$ with generator $(A, D(A))$ on a Banach space X. After applying the rescaling procedure, and hence without loss of generality (see Paragraph 2.2 and Exercise 2.22.(1)), we can assume that its growth bound ω_0 is negative. Therefore, its generator A is invertible with $A^{-1} \in \mathcal{L}(X)$. On the domains $D(A^n)$ of its powers A^n, we now introduce new norms $\|\cdot\|_n$.

2.14 Definition. *For each $n \in \mathbb{N}$ and $x \in D(A^n)$, we define the n-norm*

$$\|x\|_n := \|A^n x\|$$

and call

$$X_n := (D(A^n), \|\cdot\|_n)$$

the Sobolev space of order n associated with $(T(t))_{t\geq 0}$. The operators $T(t)$ restricted to X_n are denoted by

$$T_n(t) := T(t)_{|X_n}.$$

It turns out that the restrictions $T_n(t)$ behave surprisingly well on X_n.

2.15 Proposition. *With the above definitions, the following hold.*

(i) *Each X_n is a Banach space.*

(ii) *The operators $T_n(t)$ form a strongly continuous semigroup $\big(T_n(t)\big)_{t\geq 0}$ on X_n.*

(iii) *The generator A_n of $\big(T_n(t)\big)_{t\geq 0}$ is given by the part of A in X_n; i.e.,*

$$A_n x = Ax \qquad \text{for } x \in D(A_n) \text{ with}$$
$$D(A_n) := \{x \in X_n : Ax \in X_n\} = D(A^{n+1}) = X_{n+1}.$$

PROOF. The assertion follows by induction if we prove the case $n = 1$. Assertion (i) follows, because A is a closed operator and $\|\cdot\|_1$ is equivalent to the graph norm, as can be seen from the estimate

$$\|x\|_A = \left\|A^{-1}Ax\right\| + \|Ax\| \leq \left(\left\|A^{-1}\right\| + 1\right)\cdot\|x\|_1 \leq \left(\left\|A^{-1}\right\| + 1\right)\cdot\|x\|_A$$

for $x \in X_1$. From Lemma 1.3.(ii), we know that $T(t)$ maps X_1 into X_1. Each $T_1(t)$ is bounded, because

$$\|T_1(t)x\|_1 = \|T(t)Ax\| \leq \|T(t)\|\cdot\|x\|_1 \qquad \text{for } x \in X_1,$$

so $\big(T_1(t)\big)_{t\geq 0}$ is a semigroup on X_1. The strong continuity follows from

$$\|T_1(t)x - x\|_1 = \|T(t)Ax - Ax\| \to 0 \qquad \text{for } t \downarrow 0 \qquad \text{and } x \in X_1.$$

Finally, (iii) follows from the proposition in Paragraph 2.3 on subspace semigroups. $\qquad\square$

We visualize the above spaces and semigroups by a diagram. Before doing so, we point out that by definition, A_n is an isometry (with inverse A_n^{-1}) from X_{n+1} onto X_n. Moreover, we write $X_0 := X$, $T_0(t) := T(t)$ and $A_0 := A$.

$$
\begin{array}{ccc}
X_0 & \xrightarrow{\ T_0(t)\ } & X_0 \\
\Big\uparrow{\scriptstyle A_0} & & \Big\downarrow{\scriptstyle A_0^{-1}} \\
X_1 & \xrightarrow{\ T_1(t)\ } & X_1 = \big(D(A_0),\|\cdot\|_1\big) \\
\Big\uparrow{\scriptstyle A_1} & & \Big\downarrow{\scriptstyle A_1^{-1}} \\
X_2 & \xrightarrow{\ T_2(t)\ } & X_2 = \big(D(A_1),\|\cdot\|_2\big) = \big(D(A_0^2),\|\cdot\|_2\big) \\
\Big\uparrow{\scriptstyle A_2} & & \Big\downarrow{\scriptstyle A_2^{-1}} \\
\vdots & & \vdots
\end{array}
$$

Observe that each X_{n+1} is densely embedded in X_n but also, via A_n, isometrically isomorphic to X_n. In addition, the semigroup $(T_{n+1}(t))$ is the restriction of $(T_n(t))_{t\geq0}$, but also similar to $(T_n(t))_{t\geq0}$. We state this important property explicitly.

2.16 Corollary. *All the strongly continuous semigroups $(T_n(t))_{t\geq0}$ on the spaces X_n are similar. More precisely, one has*

$$T_{n+1}(t) = A_n^{-1}T_n(t)A_n = T_n(t)_{|X_{n+1}} \quad \text{for all } n \geq 0.$$

This similarity implies that spectrum, spectral bound, growth bound, etc. coincide for all the semigroups $(T_n(t))_{t\geq0}$.

In our construction, we obtained the $(n+1)$st Sobolev space from the nth Sobolev space. However, because X_{n+1} is a dense subspace of X_n (by Theorem 1.4), it is possible to invert this procedure and obtain X_n from X_{n+1} as the completion of X_{n+1} for the norm

$$\|x\|_n := \left\|A_{n+1}^{-1}x\right\|_{n+1}.$$

This observation permits us to extend the above diagram to the negative integers and to define *extrapolation spaces* or *Sobolev spaces of negative order*.

2.17 Definition. *For each $n \in \mathbb{N}$ and $x \in X_{-n+1}$, we define (recursively) the norm*

$$\|x\|_{-n} := \left\|A_{-n+1}^{-1}x\right\|_{-n+1}$$

and call the completion

$$X_{-n} := \left(X_{-n+1}, \|\cdot\|_{-n}\right)^{\sim}$$

the Sobolev space of order $-n$ associated with $(T_0(t))_{t\geq0}$. Moreover, we denote the continuous extensions of the operators $T_{-n+1}(t)$ to the extrapolated space X_{-n} by $T_{-n}(t)$.

Note that these extended operators $T_{-n}(t)$ have properties analogous to the ones stated in Proposition 2.15; hence our previous results hold for all $n \in \mathbb{Z}$.

2.18 Theorem. *With the above definitions, the following hold for all $m \geq n \in \mathbb{Z}$.*

(i) *Each X_n is a Banach space containing X_m as a dense subspace.*

(ii) *The operators $T_n(t)$ form a strongly continuous semigroup $(T_n(t))_{t\geq0}$ on X_n.*

(iii) *The generator A_n of $(T_n(t))_{t\geq0}$ has domain $D(A_n) = X_{n+1}$ and is the unique continuous extension of $A_m : X_{m+1} \to X_m$ to an isometry from X_{n+1} onto X_n.*

PROOF. It suffices to prove the assertions for $m = 0$ and $n = -1$ only. In this case, (i) holds true by definition. From

$$\|T_0(t)x\|_{-1} = \left\|T_0(t)A_0^{-1}x\right\|_0 \leq \|T_0(t)\| \cdot \|x\|_{-1},$$

we see that $T_0(t)$ extends continuously to X_{-1}. The semigroup property holds for $(T_0(t))_{t\geq 0}$ on X_0, hence for $(T_{-1}(t))_{t\geq 0}$ on X_{-1}. Similarly, the strong continuity follows, because it holds on the dense subset X_0 (even for the stronger norm $\|\cdot\|_0$).

To prove (iii), we observe first that A_{-1} extends A_0, because $T_{-1}(t)$ extends $T_0(t)$. The closedness of A_{-1} then implies $X_0 \subseteq D(A_{-1})$. Because X_0 is dense in X_{-1} and $(T_{-1}(t))_{t\geq 0}$-invariant, it is a core for A_{-1} by Proposition 1.7. Now, on X_0 the graph norm $\|\cdot\|_{A_{-1}}$ is equivalent to $\|\cdot\|$; hence X_0 is a Banach space for $\|\cdot\|_{A_{-1}}$, and therefore $X_0 = D(A_{-1})$.

The remaining assertions follow from the fact that $A_0 : D(A_0) \subset X_0 \to X_{-1}$, by definition of the norms, is an isometry. $\qquad\square$

So, we have constructed a two-sided infinite sequence of Banach spaces and strongly continuous semigroups thereon. Again we visualize this *Sobolev tower* associated with the semigroup $(T_0(t))_{t\geq 0}$ by a diagram. Note that Corollary 2.16 now holds for all $n \in \mathbb{Z}$. In addition, if we start this construction from any level, i.e., from the semigroup $(T_k(t))_{t\geq 0}$ on the space X_k for some $k \in \mathbb{Z}$, we will obtain the same scale of spaces and semigroups.

2.19 Diagram.

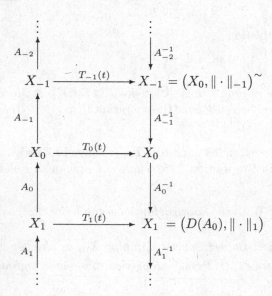

We point out again that each space X_n is the completion (unique up to isomorphism) of any of its subspaces X_m whenever $m \geq n \in \mathbb{Z}$.

For multiplication semigroups it is easy to identify all Sobolev spaces with concrete function spaces.

2.20 Example. We take $X_0 := C_0(\Omega)$ and $q : \Omega \to \mathbb{C}$ continuous assuming, for simplicity, that $\sup_{s \in \Omega} \operatorname{Re} q(s) < 0$. As in Section I.3.a, we define $M_q f := q \cdot f$ and the corresponding multiplication semigroup by

$$T_q(t)f := e^{tq} \cdot f$$

for $t \geq 0$, $f \in X$. The Sobolev spaces X_n are then given by

$$(2.9) \qquad X_n = \left\{ q^{-n} \cdot f : f \in X \right\} = \left\{ g \in C(\Omega) : q^n \cdot g \in X_0 \right\}$$

for all $n \in \mathbb{Z}$.

Note that the analogous statement holds if we start from

$$X_0 := L^p(\Omega, \mu) \qquad \text{for } 1 \leq p < \infty,$$

a measurable function $q : \Omega \to \mathbb{C}$ satisfying $\operatorname{ess\,sup}_{s \in \Omega} \operatorname{Re} q(s) < 0$, and the corresponding multiplication semigroup $(T_q(t))_{t \geq 0}$ (cf. Section I.3.b). In particular, (2.9) becomes

$$(2.10) \qquad X_n = L^p\left(\Omega, |q|^{np} \cdot \mu\right)$$

for all $n \in \mathbb{Z}$.

These abstract Sobolev spaces look quite familiar if we consider the translation semigroups and their generators from Paragraph 2.9.

2.21 Examples. (i) First, we look at the (left) translation group $(T_l(t))_{t \in \mathbb{R}}$ on $X := L^2(\mathbb{R})$ as discussed in Paragraph 2.9. If by $\mathcal{F}$ we denote the Fourier transform, then by Plancherel's equation $(2\pi)^{-1/2} \mathcal{F}$ maps $L^2(\mathbb{R})$ isometrically onto $L^2(\mathbb{R})$ and transforms $(T_l(t))_{t \in \mathbb{R}}$ into the multiplication group $(\widehat{T}(t))_{t \in \mathbb{R}}$ given by

$$\widehat{T}(t)f(\xi) = e^{it\xi} \cdot f(\xi) \qquad \text{for } f \in L^2(\mathbb{R}), \ \xi \in \mathbb{R}.$$

(Note that this is a concrete version of the Spectral Theorem I.3.9.) The generator of $(\widehat{T}(t))_{t \in \mathbb{R}}$ is the multiplication operator given by the function $\widehat{q} : \xi \mapsto i\xi$; hence the associated Sobolev spaces have been determined in Example 2.20 as

$$\widehat{X}_n = \left\{ \xi \mapsto (1 - i\xi)^{-n} \cdot f(\xi) : f \in L^2(\mathbb{R}) \right\}$$

for all $n \in \mathbb{Z}$. If we now apply the inverse Fourier transform (and its extension to the space of distributions), we obtain the Sobolev spaces associated with the translation group as

$$X_n = \left\{ (1 - D)^{-n} f : f \in L^2(\mathbb{R}) \right\},$$

where D denotes the distributional derivative. Hence, X_n coincides with the usual Sobolev space $W^{2,n}(\mathbb{R})$ for all $n \in \mathbb{Z}$.

(ii) In the case of the translation group $\big(T_l(t)\big)_{t\in\mathbb{R}}$ on $X := C_0(\mathbb{R})$, we can avoid the use of the Fourier transform and work in the space of test functions $\mathscr{D}(\mathbb{R})$ and its dual $\mathscr{D}(\mathbb{R})'$ (see [Rud73, Chap. 6]) to obtain an analogous characterization of X_n. For $n \geq 1$, the spaces X_n are easy to identify as

$$X_n = \left\{ f \in C_0(\mathbb{R}) : \begin{array}{l} f \text{ is } n\text{-times differentiable and} \\ f^{(k)} \in C_0(\mathbb{R}) \text{ for } k = 1, \ldots, n \end{array} \right\}.$$

To find the Sobolev spaces of negative order, we only consider the case $n = -1$ and recall that X_{-1} is the set of (equivalence classes of) Cauchy sequences in X for the norm $\|f\|_{-1} := \|R(1, A)f\|$ for $f \in X$ and A the generator of $\big(T_l(t)\big)_{t\in\mathbb{R}}$. Then each such $\|\cdot\|_{-1}$-Cauchy sequence $(f_n)_{n\in\mathbb{N}}$ defines a distribution $F \in \mathscr{D}(\mathbb{R})'$ by

$$\langle F, \varphi \rangle := \left\langle \lim_{n\to\infty} R(1, A)f_n, \varphi + \varphi' \right\rangle$$

for $\varphi \in \mathscr{D}(\mathbb{R})$. This shows that X_{-1} is continuously embedded in the space $\big(\mathscr{D}'(\mathbb{R}), \sigma(\mathscr{D}', \mathscr{D})\big)$. Because A_{-1} is the continuous extension of the classical derivative defined on X_1, it coincides with the distributional derivative D, and hence

$$X_{-1} = \big\{ F \in \mathscr{D}' : F = f - Df \text{ for some } f \in C_0(\mathbb{R}) \big\}.$$

2.22 Exercises. (1) Let $\big(A, D(A)\big)$ be a closed densely defined operator on X such that $\rho(A) \neq \emptyset$. Prove the following.

(i) For each fixed $n \in \mathbb{N}$, all the norms

$$\|x\|_{n,\lambda} := \big\|(\lambda - A)^n x\big\|, \quad x \in D(A^n),$$

are equivalent for $\lambda \in \rho(A)$.

(ii) For each fixed $n \in \mathbb{N}$, all the norms

$$\|x\|_{-n,\lambda} := \big\|R(\lambda, A)^n x\big\|, \quad x \in X,$$

are equivalent for $\lambda \in \rho(A)$.

(iii) Now take $\lambda = 0 \in \rho(A)$ and define the *Sobolev spaces* X_n, $n \in \mathbb{Z}$, as in Definition 2.14 and Definition 2.17. Then the operator A can be restricted/extended to an isometry from X_{n+1} onto X_n for each $n \in \mathbb{Z}$.

(2) Identify the abstract Sobolev spaces X_n in Example 2.20 assuming only that $\sup_{s\in\mathbb{R}} \operatorname{Re} q(s) < \infty$.

(3) Show that an operator $\big(A, D(A)\big)$ on X with $\rho(A) \neq \emptyset$ is bounded if and only if $X_n = X$ for all $n \in \mathbb{Z}$.

(4) Take an operator $(A, D(A))$ with $\rho(A) \neq \emptyset$ on the Banach space X. Show that the dual of the extrapolated Sobolev space X_{-1} is canonically isomorphic to the domain $D(A')$ of the adjoint A' in X' endowed with the graph norm.

(5) Show that for two densely defined operators $(A, D(A))$ with $\rho(A) \neq \emptyset$ and $(B, D(B))$ on the Banach space X the following assertions are equivalent.

   (i) $D(A') \subseteq D(B')$.

  (ii) $\overline{R(\lambda, A)B} \in \mathcal{L}(X)$ for one (hence, all) $\lambda \in \rho(A)$.

 (iii) $B : D(B) \subseteq X \to X_{-1}^A$ is bounded; i.e., B can be extended to a bounded operator from X to X_{-1}^A.

3. Generation Theorems

We now turn to the fundamental problem of semigroup theory, which is to find arrows in Diagram 1.14 leading from the generator (or its resolvent) to the semigroup. This means that we discuss the following problem.

3.1 Problem. *Characterize those linear operators that are generators of some strongly continuous semigroup, and describe how the semigroup is generated.*

a. Hille–Yosida Theorems

In Theorems 1.4 and 1.10, we already saw that generators
- Are necessarily closed operators,
- Have dense domain, and
- Have their spectrum contained in some proper left half-plane.

These conditions, however, are not sufficient.

3.2 Example. On the space

$$X := \{f \in C_0(\mathbb{R}_+) : f \text{ continuously differentiable on } [0, 1]\}$$

endowed with the norm

$$\|f\| := \sup_{s \in \mathbb{R}_+} |f(s)| + \sup_{s \in [0,1]} |f'(s)|,$$

we consider the operator $(A, D(A))$ defined by

$$Af := f' \quad \text{for } f \in D(A) := \{f \in C_0^1(\mathbb{R}_+) : f' \in X\}.$$

Then A is closed and densely defined, its resolvent exists for $\operatorname{Re}\lambda > 0$, and can be expressed by

$$\big(R(\lambda, A)f\big)(s) = \int_s^\infty e^{-\lambda(\tau-s)} f(\tau)\, d\tau \qquad \text{for } f \in X,\ s \geq 0$$

(compare (2.1)). Assume now that A generates a strongly continuous semigroup $\big(T(t)\big)_{t\geq 0}$ on X. For $f \in D(A)$ and $0 \leq s, t$ we define

$$\xi(\tau) := \big(T(t-\tau)f\big)(s+\tau), \qquad 0 \leq \tau \leq t,$$

which is a differentiable function. Its derivative satisfies

$$\dot{\xi}(\tau) := -\big(T(t-\tau)Af\big)(s+\tau) + \big(T(t-\tau)f'\big)(s+\tau) = 0,$$

and hence

$$\big(T(t)f\big)(s) = \xi(0) = \xi(t) = f(s+t).$$

This proves that $\big(T(t)\big)_{t\geq 0}$ must be the (left) translation semigroup. The translation operators, however, do not map X into itself.

This indicates that we need more assumptions on A, and the norm estimate

- $\|R(\lambda, A)\| \leq \frac{M}{\operatorname{Re}\lambda - w}, \quad \operatorname{Re}\lambda > w,$

proved in Theorem 1.10.(iii) may serve for this purpose.

To tackle the above problem, it is helpful to recall the results from the introduction and to think of the semigroup generated by an operator A as an "exponential function"

$$t \mapsto e^{tA}.$$

3.3 Exponential Formulas. We pursue this idea by recalling the various ways by which we can define "exponential functions." Each of these formulas and each method is then checked for a possible generalization to infinite-dimensional Banach spaces and, in particular, to unbounded operators. Here are some more or less promising formulas for "e^{tA}."

Formula (i) As in the matrix case (see Section I.2.a) we might use the power series and define

$$(3.1) \qquad\qquad e^{tA} := \sum_{n=0}^{\infty} \frac{t^n}{n!} A^n.$$

Comment. For unbounded A, it is unrealistic to expect convergence of this series. In fact, there exist strongly continuous semigroups such that for its generator A the series

$$\sum_{n=0}^{\infty} \frac{t^n}{n!} A^n x$$

converges only for $t = 0$ or $x = 0$. See Exercise 3.12.(2).

Formula (ii) We might use the Cauchy integral formula and define

$$(3.2) \qquad \mathrm{e}^{tA} := \frac{1}{2\pi\mathrm{i}} \int_{+\partial U} \mathrm{e}^{\lambda t} R(\lambda, A)\, d\lambda.$$

Comment. As already noted, the generator A, hence also its spectrum $\sigma(A)$, may be unbounded. Therefore, the path $+\partial U$ surrounding $\sigma(A)$ will be unbounded, and so we need extra conditions to make the integral converge. See Section 4 for a class of semigroups for which this approach does work.

Formula (iii) At least in the one-dimensional case, the formulas

$$\mathrm{e}^{tA} = \lim_{n\to\infty} \left(1 + \frac{t}{n}A\right)^n = \lim_{n\to\infty} \left(1 - \frac{t}{n}A\right)^{-n}$$

are well known (indeed Euler already used them; see [EN00, Sect. VII.3]).

Comment. Whereas the first formula again involves powers of the unbounded operator A and therefore will rarely converge, we can rewrite the second (using the resolvent operators $R(\lambda, A) := (\lambda - A)^{-1}$) as

$$(3.3) \qquad \mathrm{e}^{tA} = \lim_{n\to\infty} \left[\frac{n}{t}R\left(\frac{n}{t}, A\right)\right]^n.$$

This yields a formula involving only powers of bounded operators. It was Hille's idea (in 1948) to use this formula and to prove that under appropriate conditions, the limit exists and defines a strongly continuous semigroup. We return to this idea later (see Corollary IV.2.5 below).

Formula (iv) Because it is well understood how to define the exponential function for bounded operators (see Section I.2.b), one can try to approximate A by a sequence $(A_n)_{n\in\mathbb{N}}$ of bounded operators and hope that

$$(3.4) \qquad \mathrm{e}^{tA} := \lim_{n\to\infty} \mathrm{e}^{tA_n}$$

exists and is a strongly continuous semigroup.

Comment. This was Yosida's idea (also in 1948) and is now examined in detail in order to obtain strongly continuous semigroups.

We start with an important convergence property for the resolvent under the assumption that $\|\lambda R(\lambda, A)\|$ remains bounded as $\lambda \to \infty$.

3.4 Lemma. *Let $(A, D(A))$ be a closed, densely defined operator. Suppose there exist $w \in \mathbb{R}$ and $M > 0$ such that $[w, \infty) \subset \rho(A)$ and $\|\lambda R(\lambda, A)\| \leq M$ for all $\lambda \geq w$. Then the following convergence statements hold for $\lambda \to \infty$.*

(i) $\lambda R(\lambda, A)x \to x$ for all $x \in X$.

(ii) $\lambda A R(\lambda, A)x = \lambda R(\lambda, A)Ax \to Ax$ for all $x \in D(A)$.

PROOF. If $y \in D(A)$, then $\lambda R(\lambda, A)y = R(\lambda, A)Ay + y$ by (1.1) in Chapter V. This expression converges to y as $\lambda \to \infty$, because $\|R(\lambda, A)Ay\| \leq {}^{M}/_{\lambda} \|Ay\|$. Because $\|\lambda R(\lambda, A)\|$ is uniformly bounded for all $\lambda \geq w$, statement (i) follows by Proposition A.3. The second statement is then an immediate consequence of the first one. $\square$

This lemma suggests immediately which bounded operators A_n should be chosen to approximate the unbounded operator A. Because for contraction semigroups the technical details of the subsequent proof become much easier (and because the general case can then be deduced from this one), we first give the characterization theorem for generators in this special case.

3.5 Generation Theorem. (*Contraction Case,* HILLE, YOSIDA, 1948). *For a linear operator $(A, D(A))$ on a Banach space X, the following properties are all equivalent.*

(a) $(A, D(A))$ *generates a strongly continuous contraction semigroup.*

(b) $(A, D(A))$ *is closed, densely defined, and for every $\lambda > 0$ one has $\lambda \in \rho(A)$ and*

$$(3.5) \qquad \|\lambda R(\lambda, A)\| \leq 1.$$

(c) $(A, D(A))$ *is closed, densely defined, and for every $\lambda \in \mathbb{C}$ with $\operatorname{Re} \lambda > 0$ one has $\lambda \in \rho(A)$ and*

$$(3.6) \qquad \|R(\lambda, A)\| \leq \frac{1}{\operatorname{Re} \lambda}.$$

PROOF. In view of Theorem 1.4 and Theorem 1.10, it suffices to show (b) $\Rightarrow$ (a). To that purpose, we define the so-called *Yosida approximants*

$$(3.7) \qquad A_n := nAR(n, A) = n^2 R(n, A) - nI, \quad n \in \mathbb{N},$$

which are bounded, mutually commuting operators for each $n \in \mathbb{N}$. Consider then the uniformly continuous semigroups given by

$$(3.8) \qquad T_n(t) := e^{tA_n}, \quad t \geq 0.$$

Because A_n converges to A pointwise on $D(A)$ (by Lemma 3.4.(ii)), we anticipate that the following properties hold.

(i) $T(t)x := \lim_{n \to \infty} T_n(t)x$ exists for each $x \in X$.

(ii) $(T(t))_{t \geq 0}$ is a strongly continuous contraction semigroup on X.

(iii) This semigroup has generator $(A, D(A))$.

By establishing these statements we complete the proof.

(i) Each $\big(T_n(t)\big)_{t\geq 0}$ is a contraction semigroup, because

$$\|T_n(t)\| \leq e^{-nt}e^{\|n^2 R(n,A)\|t} \leq e^{-nt}e^{nt} = 1 \qquad \text{for } t \geq 0.$$

So, again by Proposition A.3, it suffices to prove convergence just on $D(A)$. By (the vector-valued version of) the fundamental theorem of calculus, applied to the functions

$$s \mapsto T_m(t-s)T_n(s)x$$

for $0 \leq s \leq t$, $x \in D(A)$, and $m, n \in \mathbb{N}$, and using the mutual commutativity of the semigroups $\big(T_n(t)\big)_{t\geq 0}$ for all $n \in \mathbb{N}$, one has

$$
\begin{aligned}
T_n(t)x - T_m(t)x &= \int_0^t \tfrac{d}{ds}\big(T_m(t-s)T_n(s)x\big)\,ds \\
&= \int_0^t T_m(t-s)T_n(s)(A_n x - A_m x)\,ds.
\end{aligned}
$$

Accordingly,

(3.9) $$\|T_n(t)x - T_m(t)x\| \leq t\,\|A_n x - A_m x\|.$$

By Lemma 3.4.(ii), $(A_n x)_{n\in\mathbb{N}}$ is a Cauchy sequence for each $x \in D(A)$. Therefore, $\big(T_n(t)x\big)_{n\in\mathbb{N}}$ converges for each $x \in D(A)$, hence for each $x \in X$ and even uniformly on each interval $[0, t_0]$.

(ii) The pointwise convergence of $\big(T_n(t)x\big)_{n\in\mathbb{N}}$ implies that the limit family $\big(T(t)\big)_{t\geq 0}$ satisfies the functional equation (FE), hence is a semigroup, and consists of contractions. Moreover, for each $x \in D(A)$, the corresponding orbit map

$$\xi : t \mapsto T(t)x, \quad 0 \leq t \leq t_0,$$

is the uniform limit of continuous functions (use (3.9)) and so is continuous itself. This suffices to obtain strong continuity via Proposition I.1.3.

(iii) Denote by $\big(B, D(B)\big)$ the generator of $\big(T(t)\big)_{t\geq 0}$ and fix $x \in D(A)$. On each compact interval $[0, t_0]$, the functions

$$\xi_n : t \mapsto T_n(t)x$$

converge uniformly to $\xi(\cdot)$ by (3.9), and the differentiated functions

$$\dot{\xi}_n : t \mapsto T_n(t)A_n x$$

converge uniformly to

$$\eta : t \mapsto T(t)Ax.$$

This implies differentiability of ξ with $\dot{\xi}(0) = \eta(0)$; i.e., $D(A) \subset D(B)$ and $Ax = Bx$ for $x \in D(A)$.

Now choose $\lambda > 0$. Then $\lambda - A$ is a bijection from $D(A)$ onto X, because $\lambda \in \rho(A)$ by assumption. On the other hand, B generates a contraction semigroup, and so $\lambda \in \rho(B)$ by Theorem 1.10. Hence, $\lambda - B$ is also a bijection from $D(B)$ onto X. But we have seen that $\lambda - B$ coincides with $\lambda - A$ on $D(A)$. This is possible only if $D(A) = D(B)$ and $A = B$. $\square$

If a strongly continuous semigroup $(T(t))_{t \geq 0}$ with generator A satisfies, for some $w \in \mathbb{R}$, an estimate

$$\|T(t)\| \leq e^{wt} \qquad \text{for } t \geq 0,$$

then we can apply the above characterization to the rescaled contraction semigroup given by

$$S(t) := e^{-wt} T(t) \qquad \text{for } t \geq 0.$$

Because the generator of $(S(t))_{t \geq 0}$ is $B = A - w$ (see Paragraph 2.2), Generation Theorem 3.5 takes the following form.

3.6 Corollary. *Let $w \in \mathbb{R}$. For a linear operator $(A, D(A))$ on a Banach space X the following conditions are equivalent.*

(a) *$(A, D(A))$ generates a strongly continuous semigroup $(T(t))_{t \geq 0}$ satisfying*

$$(3.10) \qquad\qquad \|T(t)\| \leq e^{wt} \qquad \text{for } t \geq 0.$$

(b) *$(A, D(A))$ is closed, densely defined, and for each $\lambda > w$ one has $\lambda \in \rho(A)$ and*

$$(3.11) \qquad\qquad \|(\lambda - w) R(\lambda, A)\| \leq 1.$$

(c) *$(A, D(A))$ is closed, densely defined, and for each $\lambda \in \mathbb{C}$ with $\operatorname{Re} \lambda > w$ one has $\lambda \in \rho(A)$ and*

$$(3.12) \qquad\qquad \|R(\lambda, A)\| \leq \frac{1}{\operatorname{Re} \lambda - w}.$$

Semigroups satisfying (3.10) are also called *quasi-contractive*.

Note, by Paragraph 3.11 below, that an operator A generates a strongly continuous *group* if and only if both A and $-A$ are generators. Therefore, we can combine the conditions of the Generation Theorem 3.5 for A and $-A$ simultaneously and obtain a characterization of generators of contraction groups, i.e., of groups of isometries.

3.7 Corollary. *For a linear operator* $(A, D(A))$ *on a Banach space X the following properties are equivalent.*

(a) $(A, D(A))$ *generates a strongly continuous group of isometries.*

(b) $(A, D(A))$ *is closed, densely defined, and for every $\lambda \in \mathbb{R} \setminus \{0\}$ one has $\lambda \in \rho(A)$ and*

$$(3.13) \qquad \|\lambda R(\lambda, A)\| \leq 1.$$

(c) $(A, D(A))$ *is closed, densely defined, and for every $\lambda \in \mathbb{C} \setminus i\mathbb{R}$ one has $\lambda \in \rho(A)$ and*

$$(3.14) \qquad \|R(\lambda, A)\| \leq \frac{1}{|\operatorname{Re}\lambda|}.$$

It is now a pleasant surprise that the characterization of generators of arbitrary strongly continuous semigroups can be deduced from the above result for contraction semigroups. However, norm estimates for *all* powers of the resolvent are needed.

3.8 Generation Theorem. (*General Case*, FELLER, MIYADERA, PHILLIPS, 1952). *Let $(A, D(A))$ be a linear operator on a Banach space X and let $w \in \mathbb{R}$, $M \geq 1$ be constants. Then the following properties are equivalent.*

(a) $(A, D(A))$ *generates a strongly continuous semigroup $(T(t))_{t \geq 0}$ satisfying*

$$(3.15) \qquad \|T(t)\| \leq Me^{wt} \qquad \text{for } t \geq 0.$$

(b) $(A, D(A))$ *is closed, densely defined, and for every $\lambda > w$ one has $\lambda \in \rho(A)$ and*

$$(3.16) \qquad \left\| \left[(\lambda - w)R(\lambda, A) \right]^n \right\| \leq M \qquad \text{for all } n \in \mathbb{N}.$$

(c) $(A, D(A))$ *is closed, densely defined, and for every $\lambda \in \mathbb{C}$ with $\operatorname{Re}\lambda > w$ one has $\lambda \in \rho(A)$ and*

$$(3.17) \qquad \|R(\lambda, A)^n\| \leq \frac{M}{(\operatorname{Re}\lambda - w)^n} \qquad \text{for all } n \in \mathbb{N}.$$

PROOF. The implication (a) $\Rightarrow$ (c) has been proved in Corollary 1.11, and (c) $\Rightarrow$ (b) is trivial. To prove (b) $\Rightarrow$ (a) we use, as for Corollary 3.6, the rescaling technique from Paragraph 2.2. So, without loss of generality, we assume that $w = 0$; i.e.,

$$\|\lambda^n R(\lambda, A)^n\| \leq M \qquad \text{for all } \lambda > 0, \; n \in \mathbb{N}.$$

For every $\mu > 0$, define a new norm on X by

$$\|x\|_\mu := \sup_{n \geq 0} \|\mu^n R(\mu, A)^n x\|.$$

These norms have the following properties.

(i) $\|x\| \leq \|x\|_\mu \leq M \|x\|$; i.e., they are all equivalent to $\|\cdot\|$.

(ii) $\|\mu R(\mu, A)\|_\mu \leq 1$.

(iii) $\|\lambda R(\lambda, A)\|_\mu \leq 1$ for all $0 < \lambda \leq \mu$.

(iv) $\|\lambda^n R(\lambda, A)^n x\| \leq \|\lambda^n R(\lambda, A)^n x\|_\mu \leq \|x\|_\mu$ for all $0 < \lambda \leq \mu$ and $n \in \mathbb{N}$.

(v) $\|x\|_\lambda \leq \|x\|_\mu$ for $0 < \lambda \leq \mu$.

We only give the proof of (iii). Due to the Resolvent Equation in Paragraph V.1.2, we have that

$$y := R(\lambda, A)x = R(\mu, A)x + (\mu - \lambda)R(\mu, A)R(\lambda, A)x = R(\mu, A)(x + (\mu - \lambda)y).$$

This implies, by using (ii), that

$$\|y\|_\mu \leq \frac{1}{\mu}\|x\|_\mu + \frac{\mu - \lambda}{\mu}\|y\|_\mu \,, \qquad \text{whence} \qquad \lambda\|y\|_\mu \leq \|x\|_\mu.$$

On the basis of these properties one can define still another norm by

$$(3.18) \qquad\qquad \|x\| := \sup_{\mu > 0} \|x\|_\mu \,,$$

which evidently satisfies

(vi) $\|x\| \leq \|x\| \leq M\|x\|$ and

(vii) $\|\lambda R(\lambda, A)\| \leq 1$ for all $\lambda > 0$.

Thus, the operator $(A, D(A))$ satisfies Condition (3.5) for the equivalent norm $\|\cdot\|$ and so, by the Generation Theorem 3.5, generates a $\|\cdot\|$-contraction semigroup $(T(t))_{t \geq 0}$. Using (vi) again, we obtain $\|T(t)\| \leq M$. $\qquad\square$

3.9 Comment. As a general rule, we point out that for an operator $(A, D(A))$ to be a generator one needs

- Conditions on the location of the spectrum $\sigma(A)$ in some left half-plane and
- Growth estimates of the form

$$\|R(\lambda, A)^n\| \leq \frac{M}{(\operatorname{Re}\lambda - w)^n}$$

for *all* powers of the resolvent $R(\lambda, A)$ in some right half-plane (or on some semiaxis (w, ∞)). See Exercise 3.12.(3) for an example that the estimate with $n = 1$ does not suffice.

This last condition is rather complicated and can be checked for non-trivial examples only in the (quasi) contraction case, i.e., only if $n = 1$ is sufficient as in Generation Theorem 3.5 and Corollary 3.6.

On the other hand, every strongly continuous semigroup can be rescaled (see Paragraph I.1.10) to become bounded. For a bounded semigroup, we can find an equivalent norm making it a contraction semigroup. This does not help much in concrete examples, because only in rare cases will it be possible to compute this new norm. However, this fact is extremely helpful in abstract considerations and is stated explicitly.

3.10 Lemma. *Let $(T(t))_{t \geq 0}$ be a bounded, strongly continuous semigroup on a Banach space X. Then the norm*

$$\|x\| := \sup_{t \geq 0} \|T(t)x\|, \qquad x \in X,$$

is equivalent to the original norm on X, and $(T(t))_{t \geq 0}$ becomes a contraction semigroup on $(X, \|\cdot\|)$.

The proof is left as Exercise 3.12.(1).

3.11 Generators of Groups. This paragraph is devoted to the question of which operators are generators of strongly continuous *groups* (see the explanation following Definition I.1.1). In order to make this more precise we first adapt Definition 1.2 to this situation.

Definition. *The generator $A : D(A) \subseteq X \to X$ of a strongly continuous group $(T(t))_{t \in \mathbb{R}}$ on a Banach space X is the operator*

$$Ax := \lim_{h \to 0} \tfrac{1}{h}\big(T(h)x - x\big)$$

defined for every x in its domain

$$D(A) := \Big\{ x \in X : \lim_{h \to 0} \tfrac{1}{h}\big(T(h)x - x\big) \text{ exists} \Big\}.$$

Given a strongly continuous group $(T(t))_{t\in\mathbb{R}}$ with generator $(A, D(A))$, we can define $T_+(t) := T(t)$ and $T_-(t) := T(-t)$ for $t \geq 0$. Then, from the previous definition, it is clear that $(T_+(t))_{t\geq0}$ and $(T_-(t))_{t\geq0}$ are strongly continuous semigroups with generators A and $-A$, respectively. Therefore, if A is the generator of a group, then both A and $-A$ generate strongly continuous semigroups. The next result shows that the converse of this statement is also true.

Generation Theorem for Groups. *Let $w \in \mathbb{R}$ and $M \geq 1$ be constants. For a linear operator $(A, D(A))$ on a Banach space X the following properties are equivalent.*

(a) *$(A, D(A))$ generates a strongly continuous group $(T(t))_{t\in\mathbb{R}}$ satisfying the growth estimate*

$$\|T(t)\| \leq Me^{w|t|} \qquad \text{for } t \in \mathbb{R}.$$

(b) *$(A, D(A))$ and $(-A, D(A))$ are the generators of strongly continuous semigroups $(T_+(t))_{t\geq0}$ and $(T_-(t))_{t\geq0}$, respectively, which satisfy*

$$\|T_+(t)\|, \ \|T_-(t)\| \leq Me^{wt} \qquad \text{for all } t \geq 0.$$

(c) *$(A, D(A))$ is closed, densely defined, and for every $\lambda \in \mathbb{R}$ with $|\lambda| > w$ one has $\lambda \in \rho(A)$ and*

$$\left\|\left[(|\lambda| - w)R(\lambda, A)\right]^n\right\| \leq M \qquad \text{for all } n \in \mathbb{N}.$$

(d) *$(A, D(A))$ is closed, densely defined, and for every $\lambda \in \mathbb{C}$ with $|\operatorname{Re}\lambda| > w$ one has $\lambda \in \rho(A)$ and*

$$(3.19) \qquad \|R(\lambda, A)^n\| \leq \frac{M}{(|\operatorname{Re}\lambda| - w)^n} \qquad \text{for all } n \in \mathbb{N}.$$

PROOF. (a) implies (b) as already mentioned above.

(b) $\Rightarrow$ (d). We first recall, by Theorem 1.4, that the generator $(A, D(A))$ is closed and densely defined. Moreover, using the assumptions on A, we obtain from Generation Theorem 3.8 the estimate (3.19) for the case $\operatorname{Re}\lambda > w$. In order to verify (3.19) for $\operatorname{Re}\lambda < -w$, observe that $R(-\lambda, A) = -R(\lambda, -A)$ for all $\lambda \in -\rho(A) = \rho(-A)$. Then, using the conditions on $-A$, the required estimate follows as above.

Because the implication (d) $\Rightarrow$ (c) is trivial, it suffices to prove that (c) $\Rightarrow$ (a). To this end we first note, by Generation Theorem 3.8, that both A and $-A$ are generators of strongly continuous semigroups $(T_+(t))_{t\geq0}$ and $(T_-(t))_{t\geq0}$, respectively, which satisfy $\|T_\pm(t)\| \leq Me^{wt}$ for $t \geq 0$. Moreover, the Yosida approximants (cf. (3.7)) $A_{+,n}$ and $A_{-,n}$ of A and

$-A$, respectively, commute. Because as in the contractive case (cf. (i)–(iii) in the proof of Generation Theorem 3.5, p. 66), we have

$$T_+(t)x = \lim_{n\to\infty} \exp(tA_{+,n})x \quad \text{and} \quad T_-(t)x = \lim_{n\to\infty} \exp(tA_{-,n})x$$

for all $x \in X$, we see that $(T_+(t))_{t\geq 0}$ and $(T_-(t))_{t\geq 0}$ commute. Hence, by what was shown in Paragraph 2.6, the products

$$U(t) := T_+(t)T_-(t), \quad t \geq 0,$$

define a strongly continuous semigroup with generator C that satisfies

$$Cx = Ax - Ax = 0$$

for all $x \in D(A) \cap D(-A) = D(A) \subset D(C)$. From (1.6) in Lemma 1.3 we then obtain $U(t)x = x$ for all $x \in X$; i.e., $T_-(t) = T_+(t)^{-1}$. Finally, the operators

$$T(t) := \begin{cases} T_+(t) & \text{if } t \geq 0, \\ T_-(-t) & \text{if } t < 0, \end{cases}$$

form a one-parameter group $(T(t))_{t\in\mathbb{R}}$ and satisfy the estimate $\|T(t)\| \leq Me^{w|t|}$. Because the map $\mathbb{R} \ni t \mapsto T(t)$ is strongly continuous if and only if it is strongly continuous at some arbitrary point $t_0 \in \mathbb{R}$, the group $(T(t))_{t\in\mathbb{R}}$ is strongly continuous. This completes the proof. $\qquad\square$

The following result is quite useful in order to check whether a given semigroup can be embedded in a group.

Proposition. *Let $(T(t))_{t\geq 0}$ be a strongly continuous semigroup on a Banach space X. If there exists some $t_0 > 0$ such that $T(t_0)$ is invertible, then $(T(t))_{t\geq 0}$ can be embedded in a group $(T(t))_{t\in\mathbb{R}}$ on X.*

PROOF. First, we show that $T(t)$ is invertible for all $t \geq 0$. This follows for $t \in [0, t_0]$ from

$$T(t_0) = T(t_0 - t)T(t) = T(t)T(t_0 - t),$$

because by assumption, $T(t_0)$ is bijective. If $t \geq t_0$, we write $t = nt_0 + s$ for $n \in \mathbb{N}$, $s \in [0, t_0)$ and conclude from

$$T(t) = T(t_0)^n T(s)$$

that $T(t)$ is invertible. Hence, we can extend $(T(t))_{t\geq 0}$ to all of $\mathbb{R}$ by

$$T(t) := T(-t)^{-1} \quad \text{for } t \leq 0,$$

thereby obtaining a group $(T(t))_{t\in\mathbb{R}}$. Because the map $\mathbb{R} \ni t \mapsto T(t)$ is strongly continuous if and only if it is strongly continuous at some arbitrary point, the proof is complete. $\qquad\square$

3.12 Exercises. (1) Prove Lemma 3.10 dealing with the renorming of bounded semigroups.

(2) For a strongly continuous semigroup $(T(t))_{t\geq 0}$ with generator A on a Banach space X, we call a vector $x \in D(A^\infty)$ *entire* if the power series

(3.20)
$$\sum_{n=0}^{\infty} \frac{t^n}{n!} A^n x$$

converges for every $t \in \mathbb{R}$. Show the following properties.

(i) If x is an entire vector of $(T(t))_{t\geq 0}$, then $T(t)x$ is given by (3.20) for every $t \geq 0$.

(ii) If $(T(t))_{t\geq 0}$ is nilpotent, then the set of entire vectors consists of $x = 0$ only.

(iii) If $(T(t))_{t\in\mathbb{R}}$ is a strongly continuous group, then the set of entire vectors is dense in X. Moreover, if x is an entire vector of $(T(t))_{t\in\mathbb{R}}$, then $T(t)x$ is given by (i) for every $t \in \mathbb{R}$. (Hint: For given $x \in X$ consider the sequence $x_n := (n/2\pi)^{1/2} \int_{-\infty}^{\infty} e^{-ns^2/2} T(s)x\, ds$. See also [Gel39].)

(3) Let M_q be a multiplication operator on $X := C_0(\mathbb{R}_+)$ and define the operator $\mathcal{A} := \left(\begin{smallmatrix} M_q & M_q \\ 0 & M_q \end{smallmatrix}\right)$ with domain $D(\mathcal{A}) := D(M_q) \times D(M_q)$ on $\mathcal{X} := X \times X$.

(i) If $q(s) := \mathrm{i}s$, $s \geq 0$, then $\mathcal{A}$ satisfies $\|R(\lambda,\mathcal{A})\| \leq 2/\lambda$ for $\lambda > 0$, but is *not* the generator of a strongly continuous semigroup on $\mathcal{X}$.

(ii) Find an unbounded function q on $\mathbb{R}_+$ such that $\mathcal{A}$ becomes a generator.

(iii) Find necessary and sufficient conditions on q such that $\mathcal{A}$ becomes a generator on $\mathcal{X}$. (Hint: Compare Exercise 4.14.(7).)

(4) Let $(T(t))_{t\geq 0}$ be a strongly continuous semigroup on a Banach space X. Show that $(T(t))_{t\geq 0}$ can be embedded in a group $(T(t))_{t\in\mathbb{R}}$ if there exists $t_0 > 0$ such that $I - T(t_0)$ is compact. (Hint: By the proposition in Paragraph 3.11 and the compactness assumption, it suffices to show that 0 is not an eigenvalue of $T(t_0)$.)

b. The Lumer–Phillips Theorem

Due to their importance, we now return to the study of contraction semigroups and look for a characterization of their generator that does not require explicit knowledge of the resolvent. The following is a key notion towards this goal.

3.13 Definition. *A linear operator* $(A, D(A))$ *on a Banach space* X *is called* dissipative *if*

(3.21) $$\|(\lambda - A)x\| \geq \lambda \|x\|$$

for all $\lambda > 0$ *and* $x \in D(A)$.

To familiarize ourselves with these operators we state some of their basic properties.

3.14 Proposition. *For a dissipative operator* $(A, D(A))$ *the following properties hold.*

(i) $\lambda - A$ *is injective for all* $\lambda > 0$ *and*

$$\|(\lambda - A)^{-1}z\| \leq \frac{1}{\lambda} \|z\|$$

   for all z *in the range* $\mathrm{rg}(\lambda - A) := (\lambda - A)D(A)$.

(ii) $\lambda - A$ *is surjective for some* $\lambda > 0$ *if and only if it is surjective for each* $\lambda > 0$. *In that case, one has* $(0, \infty) \subset \rho(A)$.

(iii) A *is closed if and only if the range* $\mathrm{rg}(\lambda - A)$ *is closed for some (hence all)* $\lambda > 0$.

(iv) *If* $\mathrm{rg}(A) \subseteq \overline{D(A)}$, *e.g., if* A *is densely defined, then* A *is closable. Its closure* $\overline{A}$ *is again dissipative and satisfies* $\mathrm{rg}(\lambda - \overline{A}) = \overline{\mathrm{rg}(\lambda - A)}$ *for all* $\lambda > 0$.

PROOF. (i) is just a reformulation of estimate (3.21).

To show (ii) we assume that $(\lambda_0 - A)$ is surjective for some $\lambda_0 > 0$. In combination with (i), this yields $\lambda_0 \in \rho(A)$ and $\|R(\lambda_0, A)\| \leq 1/\lambda_0$. The series expansion for the resolvent (see Proposition V.1.3.(i)) yields $(0, 2\lambda_0) \subset \rho(A)$, and the dissipativity of A implies that

$$\|R(\lambda, A)\| \leq \frac{1}{\lambda}$$

for $0 < \lambda < 2\lambda_0$. Proceeding in this way, we see that $\lambda - A$ is surjective for all $\lambda > 0$, and therefore $(0, \infty) \subset \rho(A)$.

(iii) The operator A is closed if and only if $\lambda - A$ is closed for some (hence all) $\lambda > 0$. This is again equivalent to

$$(\lambda - A)^{-1} : \mathrm{rg}(\lambda - A) \to D(A)$$

being closed. By (i), this operator is bounded. Hence, by Theorem A.10, it is closed if and only if its domain, i.e., $\mathrm{rg}(\lambda - A)$, is closed.

(iv) Take a sequence $(x_n)_{n\in\mathbb{N}} \subset D(A)$ satisfying $x_n \to 0$ and $Ax_n \to y$. By Proposition A.8, we have to show that $y = 0$. The inequality (3.21) implies that

$$\|\lambda(\lambda - A)x_n + (\lambda - A)w\| \geq \lambda \|\lambda x_n + w\|$$

for every $w \in D(A)$ and all $\lambda > 0$. Passing to the limit as $n \to \infty$ yields

$$\|-\lambda y + (\lambda - A)w\| \geq \lambda \|w\|, \quad \text{and hence} \quad \left\|-y + w - \frac{1}{\lambda}Aw\right\| \geq \|w\|.$$

For $\lambda \to \infty$ we obtain that

$$\|-y + w\| \geq \|w\|,$$

and by choosing w from the domain $D(A)$ arbitrarily close to $y \in \overline{\mathrm{rg}(A)}$, we see that

$$0 \geq \|y\|.$$

Hence $y = 0$.

In order to verify that $\overline{A}$ is dissipative, take $x \in D\left(\overline{A}\right)$. By definition of the closure of a linear operator, there exists a sequence $(x_n)_{n\in\mathbb{N}} \subset D(A)$ satisfying $x_n \to x$ and $Ax_n \to \overline{A}x$ when $n \to \infty$. Because A is dissipative and the norm is continuous, this implies that $\|(\lambda - \overline{A})x\| \geq \lambda \|x\|$ for all $\lambda > 0$. Hence $\overline{A}$ is dissipative. Finally, observe that the range $\mathrm{rg}(\lambda - A)$ is dense in $\mathrm{rg}\left(\lambda - \overline{A}\right)$. Because by assertion (iii) $\mathrm{rg}\left(\lambda - \overline{A}\right)$ is closed in X, we obtain the final assertion in (iv). $\square$

From the resolvent estimate (3.5) in Generation Theorem 3.5, it is evident that the generator of a contraction semigroup satisfies the estimate (3.21), and hence is dissipative. On the other hand, many operators can be shown directly to be dissipative and densely defined. We therefore reformulate Generation Theorem 3.5 in such a way as to single out the property that ensures that a densely defined, dissipative operator is a generator.

3.15 Theorem. (LUMER, PHILLIPS, 1961). *For a densely defined, dissipative operator* $\left(A, D(A)\right)$ *on a Banach space* X *the following statements are equivalent.*

(a) *The closure* $\overline{A}$ *of* A *generates a contraction semigroup.*

(b) $\mathrm{rg}(\lambda - A)$ *is dense in* X *for some (hence all)* $\lambda > 0$.

PROOF. (a) $\Rightarrow$ (b). Generation Theorem 3.5 implies that $\mathrm{rg}(\lambda - \overline{A}) = X$ for all $\lambda > 0$. Because $\mathrm{rg}(\lambda - \overline{A}) = \overline{\mathrm{rg}(\lambda - A)}$, by Proposition 3.14.(iv), we obtain (b).

(b) $\Rightarrow$ (a). By the same argument, the density of the range $\text{rg}(\lambda - A)$ implies that $(\lambda - \overline{A})$ is surjective. Proposition 3.14.(ii) shows that $(0, \infty) \subset \rho(\overline{A})$, and dissipativity of A implies the estimate

$$\|R(\lambda, \overline{A})\| \le \frac{1}{\lambda} \quad \text{for } \lambda > 0.$$

This was required in Generation Theorem 3.5 to assure that $\overline{A}$ generated a contraction semigroup. $\qquad\Box$

The above theorem gains its significance when viewed in the context of the abstract Cauchy problem associated with an operator A (see Section 6).

3.16 Remark. Assume that the operator A is known to be closed, densely defined, and dissipative. Then Theorem 3.15 in combination with Proposition 6.2 below yields the following fact.

In order to solve the (time-dependent) initial value problem

(ACP) $\dot{x}(t) = Ax(t), \ x(0) = x$

for all $x \in D(A)$, it is sufficient to solve the (stationary) resolvent equation

(RE) $x - Ax = y$

for all y in some dense subset in the Banach space X.

In many examples (RE) can be solved explicitly whereas (ACP) cannot, cf. Paragraph 3.29 or [EN00, Sect. VI.6].

The following result, in combination with the characterization of dissipativity in Proposition 3.23 below, gives an even simpler condition for an operator to generate a contraction semigroup.

3.17 Corollary. *Let $(A, D(A))$ be a densely defined operator on a Banach space X. If both A and its adjoint A' are dissipative, then the closure $\overline{A}$ of A generates a contraction semigroup on X.*

PROOF. By the Lumer–Phillips Theorem 3.15, it suffices to show that the range $\text{rg}(I - A)$ is dense in X. By way of contradiction, assume that $\overline{\text{rg}(I - A)} \ne X$. By the Hahn–Banach theorem there exists $0 \ne x' \in X'$ such that
$$\langle (I - A)x, x' \rangle = 0 \qquad \text{for all } x \in D(A).$$
It follows that $x' \in D(A')$ and
$$\langle x, (I - A')x' \rangle = 0 \qquad \text{for all } x \in D(A).$$

Because $D(A)$ is dense in X, we conclude that $(I - A')x' = 0$, thereby contradicting Proposition 3.14.(i). $\qquad\Box$

At this point we insert various considerations concerning the density of the domain, which up to now was a more or less standard assumption in our results. In the next two corollaries we show how dissipativity can be used to get around this hypothesis. However, based on the properties stated in Proposition 3.14, we assume that the dissipative operator A is such that $\lambda - A$ is surjective for some $\lambda > 0$. Hence $(0, \infty) \subset \rho(A)$.

3.18 Corollary. *Let $\big(A, D(A)\big)$ be a dissipative operator on the Banach space X such that $\lambda - A$ is surjective for some $\lambda > 0$. Then the part $A_|$ of A in the subspace $X_0 := \overline{D(A)}$ is densely defined and generates a contraction semigroup in X_0.*

PROOF. We recall from Definition 2.3 that

$$A_| x := Ax$$

for $x \in D(A_|) := \{x \in D(A) : Ax \in X_0\} = R(\lambda, A)X_0$. Because $R(\lambda, A)$ exists for $\lambda > 0$, this implies that $R(\lambda, A)_| = R(\lambda, A_|)$. Hence $(0, \infty) \subset \rho(A_|)$. Due to the Generation Theorem 3.5, it remains to show that $D(A_|)$ is dense in X_0. Take $x \in D(A)$ and set $x_n := nR(n, A)x$. Then $x_n \in D(A)$ and $\lim_{n \to \infty} x_n = \lim_{n \to \infty} R(n, A)Ax + x = x$, because $\|R(n, A)\| \leq 1/n$ (see Proposition 3.14.(i) and Lemma 3.4). Therefore, the operators $nR(n, A)$ converge pointwise on $D(A)$ to the identity. Because $\|nR(n, A)\| \leq 1$ for all $n \in \mathbb{N}$, we obtain convergence of

$$y_n := nR(n, A)y \to y$$

for all $y \in X_0$. Because each y_n is in $D(A_|)$, the density of $D(A_|)$ in X_0 is proved. □

We now give two rather typical examples for dissipative operators with nondense domains, one concrete and one abstract.

3.19 Examples. (i) Let $X := C[0, 1]$ and consider the operator

$$Af := -f'$$

with domain
$$D(A) := \{f \in C^1[0, 1] : f(0) = 0\}.$$

It is a closed operator whose domain is not dense. However, it is dissipative, because its resolvent can be computed explicitly as

$$R(\lambda, A)f(t) := \int_0^t e^{-\lambda(t-s)} f(s)\, ds$$

for $t \in [0, 1]$, $f \in C[0, 1]$. Moreover,

$$\|R(\lambda, A)\| \leq \frac{1}{\lambda}$$

for all $\lambda > 0$. Therefore, $\big(A, D(A)\big)$ is dissipative.

Let $X_0 := \overline{D(A)} = \{f \in C[0,1] : f(0) = 0\}$, and consider the part $A_|$ of A in X_0; i.e.,

$$A_|f = -f',$$
$$D(A_|) = \{f \in C^1[0,1] : f(0) = f'(0) = 0\}.$$

By the above corollary, this operator generates a semigroup on X_0. In fact, this semigroup $(T_0(t))_{t\geq 0}$ can be identified as the nilpotent right translation semigroup (cf. Paragraph I.3.17) given by

$$T_0(t)f(s) := \begin{cases} f(s-t) & \text{for } t \leq s, \\ 0 & \text{for } t > s. \end{cases}$$

Observe that the same definition applied to an arbitrary function $f \in C[0,1]$ does not necessarily yield a continuous function again. Therefore, the semigroup $(T_0(t))_{t\geq 0}$ does not extend to the space $C[0,1]$.

(ii) Consider a strongly continuous contraction semigroup $(T(t))_{t\geq 0}$ on a Banach space X. Its generator A is dissipative with $(0,\infty) \subset \rho(A)$. The same holds for its adjoint A', because $R(\lambda, A') = R(\lambda, A)'$ and $\|R(\lambda, A')\| = \|R(\lambda, A)\|$ for all $\lambda > 0$. The domain $D(A')$ of the adjoint is not dense in X' in general (see the example in [EN00, Sect. II.2.6]). However, taking the part of A' in $X^\odot := \overline{D(A')} \subset X'$, we obtain the generator of a contraction semigroup (given by the restrictions of $T(t)'$ to $X^\odot$; see [EN00, Sect. II.2.6] on so-called sun dual semigroups).

In the next corollary we show that the phenomenon discussed in Corollary 3.18 and Example 3.19 cannot occur in reflexive Banach spaces.

3.20 Corollary. *Let $(A, D(A))$ be a dissipative operator on a reflexive Banach space such that $\lambda - A$ is surjective for some $\lambda > 0$. Then A is densely defined and generates a contraction semigroup.*

PROOF. We only have to show the density of $D(A)$. Take $x \in X$ and define $x_n := nR(n, A)x \in D(A)$. The element $y := R(1, A)x$ also belongs to $D(A)$. Moreover, by the proof of Corollary 3.18 the operators $nR(n, A)$ converge towards the identity pointwise on $X_0 := \overline{D(A)}$. It follows that

$$y_n := R(1, A)x_n = nR(n, A)R(1, A)x \to y \quad \text{for } n \to \infty.$$

Because X is reflexive and $\{x_n : n \in \mathbb{N}\}$ is bounded, there exists a subsequence, still denoted by $(x_n)_{n\in\mathbb{N}}$, that converges weakly to some $z \in X$. Because $x_n \in D(A)$, Proposition A.1.(i) implies that $z \in \overline{D(A)}$. On the other hand, the elements $x_n = (1 - A)y_n$ converge weakly to z, so the weak closedness of A (see Definition A.5) implies that $y \in D(A)$ and $x = (1 - A)y = z \in \overline{D(A)}$. $\qquad\square$

In Corollary 3.18 and Corollary 3.20, we considered not necessarily densely defined operators and showed that dissipativity and the range condition $\mathrm{rg}(\lambda - A) = X$ for some $\lambda > 0$ imply certain generation properties. It is now a direct consequence of the renorming trick used in the proof of Generation Theorem 3.8 that these results also hold for all operators satisfying the Hille–Yosida resolvent estimates (3.16). We state this extension of Generation Theorem 3.8.

3.21 Corollary. *Let $w \in \mathbb{R}$ and $\big(A, D(A)\big)$ be an operator on a Banach space X. Suppose that $(w, \infty) \subset \rho(A)$ and*

$$(3.22) \qquad \|R(\lambda, A)^n\| \leq \frac{M}{(\lambda - w)^n}$$

for all $n \in \mathbb{N}$, $\lambda > w$ and some $M \geq 1$. Then the part $A_|$ of A in $X_0 := \overline{D(A)}$ generates a strongly continuous semigroup $\big(T_0(t)\big)_{t \geq 0}$ satisfying $\|T_0(t)\| \leq M e^{wt}$ for all $t \geq 0$. If in addition the Banach space X is reflexive, then $X_0 = X$.

PROOF. As in many previous cases we may assume that $w = 0$. Then the renorming procedure (3.18) from the proof of the implication (b) $\Rightarrow$ (a) in Generation Theorem 3.8 yields an equivalent norm for which A is a dissipative operator. The assertions then follow from Corollary 3.18 (after returning to the original norm) and, in the reflexive case, from Corollary 3.20. □

It is sometimes convenient to use the following terminology.

3.22 Definition. *Operators satisfying the assumptions of Corollary 3.21 and, in particular, the resolvent estimate (3.22) are called Hille–Yosida operators.*

Observe that Corollary 3.21 states that Hille–Yosida operators satisfy *all* assumptions of the Hille–Yosida Generation Theorem 3.8 on the closure of their domains.

We now return to dissipative operators, which represent, up to renorming, the most general case. When introducing them we had aimed for an easy (or at least more direct) way to characterizing generators. However, up to now, the only way to arrive at the norm inequality (3.21) was explicit computation of the resolvent and then deducing the norm estimate

$$\|R(\lambda, A)\| \leq \frac{1}{\lambda} \qquad \text{for } \lambda > 0.$$

This was done in Example 3.19.(i). Fortunately, there is a simpler method that works particularly well in concrete function spaces such as $C_0(\Omega)$ or $L^p(\mu)$.

To introduce this method we start with a Banach space X and its dual space X'. By the Hahn–Banach theorem, for every $x \in X$ there exists $x' \in X'$ such that

$$\langle x, x' \rangle = \|x\|^2 = \|x'\|^2.$$

Hence, for every $x \in X$ the following set, called its *duality set*,

$$(3.23) \qquad \mathcal{J}(x) := \left\{ x' \in X' : \langle x, x' \rangle = \|x\|^2 = \|x'\|^2 \right\},$$

is nonempty. Such sets allow a new characterization of dissipativity.

3.23 Proposition. *An operator $(A, D(A))$ is dissipative if and only if for every $x \in D(A)$ there exists $j(x) \in \mathcal{J}(x)$ such that*

$$(3.24) \qquad\qquad \mathrm{Re}\, \langle Ax, j(x) \rangle \leq 0.$$

If A is the generator of a strongly continuous contraction semigroup, then (3.24) holds for all $x \in D(A)$ and arbitrary $x' \in \mathcal{J}(x)$.

PROOF. Assume (3.24) is satisfied for $x \in D(A)$, $\|x\| = 1$, and some $j(x) \in \mathcal{J}(x)$. Then $\langle x, j(x) \rangle = \|j(x)\|^2 = 1$ and

$$\|\lambda x - Ax\| \geq |\langle \lambda x - Ax, j(x) \rangle|$$
$$\geq \mathrm{Re}\, \langle \lambda x - Ax, j(x) \rangle \geq \lambda$$

for all $\lambda > 0$. This proves one implication.

To show the converse, we take $x \in D(A)$, $\|x\| = 1$, and assume that $\|\lambda x - Ax\| \geq \lambda$ for all $\lambda > 0$. Choose $y'_\lambda \in \mathcal{J}(\lambda x - Ax)$ and consider the normalized elements

$$z'_\lambda := \frac{y'_\lambda}{\|y'_\lambda\|}.$$

Then the inequalities

$$\lambda \leq \|\lambda x - Ax\| = \langle \lambda x - Ax, z'_\lambda \rangle$$
$$= \lambda \, \mathrm{Re}\, \langle x, z'_\lambda \rangle - \mathrm{Re}\, \langle Ax, z'_\lambda \rangle$$
$$\leq \min\{\lambda - \mathrm{Re}\, \langle Ax, z'_\lambda \rangle,\ \lambda \, \mathrm{Re}\, \langle x, z'_\lambda \rangle + \|Ax\|\}$$

are valid for each $\lambda > 0$. This yields

$$\mathrm{Re}\, \langle Ax, z'_\lambda \rangle \leq 0 \qquad \text{and} \qquad 1 - \frac{1}{\lambda} \|Ax\| \leq \mathrm{Re}\, \langle x, z'_\lambda \rangle.$$

Let z' be a weak* accumulation point of z'_λ as $\lambda \to \infty$. Then

$$\|z'\| \leq 1, \qquad \mathrm{Re}\, \langle Ax, z' \rangle \leq 0, \qquad \text{and} \qquad \mathrm{Re}\, \langle x, z' \rangle \geq 1.$$

Combining these facts, it follows that z' belongs to $\mathcal{J}(x)$ and satisfies (3.24).

Finally, assume that A generates a contraction semigroup $(T(t))_{t\geq 0}$ on X. Then, for every $x \in D(A)$ and arbitrary $x' \in \mathcal{J}(x)$, we have

$$\mathrm{Re}\,\langle Ax, x'\rangle = \lim_{h\downarrow 0}\left(\frac{\mathrm{Re}\,\langle T(h)x, x'\rangle}{h} - \frac{\mathrm{Re}\,\langle x, x'\rangle}{h}\right)$$

$$\leq \varlimsup_{h\downarrow 0}\left(\frac{\|T(h)x\|\cdot\|x'\|}{h} - \frac{\|x\|^2}{h}\right) \leq 0.$$

This completes the proof. $\qquad\qquad\square$

Using the previous results we easily arrive at the following characterization of unitary groups on Hilbert spaces. Its discovery by Stone was one of the major steps towards the construction of the exponential function in infinite dimensions, hence towards the solution of Problem I.2.13; cf. [EN00, Chap. VII].

3.24 Theorem. (STONE, 1932). *Let $(A, D(A))$ be a densely defined operator on a Hilbert space H. Then A generates a unitary group $(T(t))_{t\in\mathbb{R}}$ on H if and only if A is skew-adjoint; i.e., $A^* = -A$.*

PROOF. First, assume that A generates a unitary group $(T(t))_{t\in\mathbb{R}}$. By Paragraph 3.11, we have

$$T(t)^* = T(t)^{-1} = T(-t) \qquad \text{for all } t \in \mathbb{R}.$$

Moreover, by Paragraphs I.1.13 and 2.5 on adjoint semigroups, the generator of $(T(t)^*)_{t\in\mathbb{R}}$ is given by A^*. This implies that $A^* = -A$.

On the other hand, if $A^* = -A$, then we conclude from

$$(Ax\,|\,x) = (x\,|\,A^*x) = -(x\,|\,Ax) = -\overline{(Ax\,|\,x)} \qquad \text{for all } x \in D(A) = D(A^*)$$

that $(Ax\,|\,x) \in i\mathbb{R}$. Combining Proposition 3.23 with the identification of the duality set as $\mathcal{J}(x) = \{x\}$ (see Exercise 3.25.(i) below), this shows that both $\pm A$ are dissipative and closed. From Corollary 3.17 and the characterization of group generators in Paragraph 3.11, it follows that the operator A generates a contraction group $(T(t))_{t\in\mathbb{R}}$. Because $T(t)^{-1} = T(-t)$, we conclude that each $T(t)$ is a surjective isometry and therefore unitary (see [Ped89, Sect. 3.2.15]). $\qquad\square$

3.25 Exercise. Prove the following statements for a Hilbert space H.

(i) For every $x \in H$, one has $\mathcal{J}(x) = \{x\}$.

(ii) If A is a normal operator on H, then A is a generator of a strongly continuous semigroup if and only if

$$s(A) < \infty.$$

(iii) Prove Stone's theorem by arguing via multiplication semigroups.

(Hint: For (ii) and (iii) use the Spectral Theorem I.3.9 and the results of Paragraph 3.11.)

c. More Examples

We close this section with a discussion of all of these notions and results for concrete examples. We begin by identifying the sets $\mathcal{J}(x)$ for some standard function spaces.

3.26 Examples. (i) Consider $X := C_0(\Omega)$, Ω locally compact. For $0 \neq f \in X$, the set $\mathcal{J}(f) \subset X'$ contains (multiples of) all point measures supported by those points $s_0 \in \Omega$ where $|f|$ reaches its maximum. More precisely,

$$(3.25) \qquad \left\{ \overline{f(s_0)} \cdot \delta_{s_0} : s_0 \in \Omega \text{ and } |f(s_0)| = \|f\| \right\} \subset \mathcal{J}(f).$$

(ii) Let $X := L^p(\Omega, \mu)$ for $1 \leq p < \infty$, and $0 \neq f \in L^p(\Omega, \mu)$. Then

$$\varphi \in \mathcal{J}(f) \subset L^q(\Omega, \mu), \ \ 1/p + 1/q = 1,$$

where φ is defined by

$$(3.26) \qquad \varphi(s) := \begin{cases} \overline{f(s)} \cdot |f(s)|^{p-2} \cdot \|f\|^{2-p} & \text{if } f(s) \neq 0, \\ 0 & \text{otherwise.} \end{cases}$$

Note that for the reflexive L^p-spaces, as for every Banach space with a strictly convex dual, the sets $\mathcal{J}(f)$ are singletons (see [Bea82]). Hence, for $1 < p < \infty$, one has $\mathcal{J}(f) = \{\varphi\}$, whereas for $p = 1$ every function $\varphi \in L^\infty(\Omega, \mu)$ satisfying

$$(3.27) \qquad \|\varphi\|_\infty \leq \|f\|_1 \quad \text{and} \quad \varphi(s)\,|f(s)| = \overline{f(s)}\,\|f\|_1 \ \ \text{if } f(s) \neq 0$$

belongs to $\mathcal{J}(f)$.

(iii) It is easy, but important, to state the result for Hilbert spaces H. After the canonical identification of H with its dual H', the duality set of $x \in H$ is

$$(3.28) \qquad\qquad \mathcal{J}(x) = \{x\};$$

cf. Exercise 3.25.(i). Hence, a linear operator on H is dissipative if and only if

$$(3.29) \qquad\qquad \operatorname{Re}(Ax \mid x) \leq 0$$

for all $x \in D(A)$.

These examples suggest that dissipativity for concrete operators on such function spaces can be verified via the inequality (3.24). In the following examples we do this and establish the dissipativity and generation property for various operators. We start with a concrete version of Theorem 3.24.

3.27 Example. (*Self-Adjoint Operators*). Consider on the Hilbert space $H := L^2(\Omega, \mu)$ the multiplication operator $A := M_q$ for some (measurable) function $q : \Omega \to \mathbb{C}$. Because its adjoint is $A^* = M_{\overline{q}}$, this operator is self-adjoint if and only if q is real-valued. In this case, it follows by Theorem 3.24 that the group $(T_{iq}(t))_{t \in \mathbb{R}}$ generated by M_{iq} is unitary.

However, this can be seen more directly by inspection of the corresponding multiplication group $\big(T_{iq}(t)\big)_{t\in\mathbb{R}}$, for which we have

$$T_{iq}(t)^* = T_{\overline{iq}}(t) = T_{-iq}(t) = T_{iq}(-t) \qquad \text{for all } t \in \mathbb{R}.$$

It is this argument for multiplication operators and semigroups that can be used to give a simple proof of Stone's Theorem 3.24. In fact, an application of the Spectral Theorem I.3.9 transforms the unitary group $\big(T(t)\big)_{t\in\mathbb{R}}$ and its (skew-adjoint) generator A on an arbitrary Hilbert space into multiplication operators on some L^2-space. See Exercise 3.25.(iii).

The same argument, i.e., passing from a self-adjoint operator to a (real-valued) multiplication operator, yields the following characterization of self-adjoint semigroups.

Proposition. *A self-adjoint operator $\big(A, D(A)\big)$ on a Hilbert space H generates a strongly continuous semigroup (of self-adjoint operators) if and only if it is bounded above; i.e., there exists $w \in \mathbb{R}$ such that*

$$(Ax \mid x) \leq w \, \|x\|^2 \qquad \text{for all } x \in D(A).$$

PROOF. It suffices to consider the multiplication operator M_q that is isomorphic, via the Spectral Theorem I.3.9, to A. Then the boundedness condition $(Ax \mid x) \leq w \, \|x\|^2$ for all $x \in D(A)$ means that the real-valued function q satisfies

$$\operatorname*{ess\,sup}_{s\in\Omega} \operatorname{Re} q(s) \leq w.$$

This, however, is exactly what is needed for M_q to generate a semigroup (see Propositions I.3.11 and I.3.12). $\square$

3.28 First-Order Differential Operators and Flows. We begin by considering a continuously differentiable vector field $F : \mathbb{R}^n \to \mathbb{R}^n$ satisfying the estimate $\sup_{s\in\mathbb{R}^n} \|DF(s)\| < \infty$ for the derivative $DF(s)$ of F at $s \in \mathbb{R}$. With this vector field we associate the following operator on the space $X := C_0(\mathbb{R}^n)$.

Definition 1. *The first-order differential operator on $C_0(\mathbb{R}^n)$ corresponding to the vector field $F : \mathbb{R}^n \to \mathbb{R}^n$ is*

$$Af(s) := \langle \operatorname{grad} f(s), F(s) \rangle$$

$$= \sum_{i=1}^{n} F_i(s) \frac{\partial f}{\partial s_i}(s)$$

for $f \in C_c^1(\mathbb{R}^n) := \{ f \in C^1(\mathbb{R}^n) : f \text{ has compact support} \}$ and $s \in \mathbb{R}^n$.

Using Example 3.26.(i) and the fact that $\partial f(s_0)/\partial s_i = 0$ if $|f(s_0)| = \|f\|$, it is immediate that A is dissipative. However, in order to show that the closure of A is a generator, there is a natural and explicit choice for what the semigroup generated by A should be. By writing it down, one simply checks that its generator is the closure of A.

Because F is globally Lipschitz, it follows from standard results on ordinary differential equations that there exists a *continuous flow* $\Phi : \mathbb{R} \times \mathbb{R}^n \to \mathbb{R}^n$; i.e., Φ is continuous with $\Phi(t+r, s) = \Phi(t, \Phi(r, s))$ and $\Phi(0, s) = s$ for every $r, t \in \mathbb{R}$ and $s \in \mathbb{R}^n$, which solves the differential equation

$$\frac{\partial}{\partial t} \Phi(t, s) = F(\Phi(t, s))$$

for all $t \in \mathbb{R}$, $s \in \mathbb{R}^n$ (see [Ama90, Thm. 10.3]). With such a flow we associate a one-parameter group of linear operators on $C_0(\mathbb{R}^n)$ as follows.

Definition 2. *The group defined by the operators*

$$T(t)f(s) := f(\Phi(t, s))$$

for $f \in C_0(\mathbb{R}^n)$, $s \in \mathbb{R}^n$, and $t \in \mathbb{R}$, is called the group induced by the flow Φ on the Banach space $C_0(\mathbb{R}^n)$.

The group property and the strong continuity follow immediately from the corresponding properties of the flow; we refer to Exercise 3.31.(2) for a closer look at the relations between (nonlinear) semiflows and (linear) semigroups. We now determine the generator of $(T(t))_{t \in \mathbb{R}}$.

Proposition. *The generator of the group $(T(t))_{t \in \mathbb{R}}$ on $C_0(\mathbb{R}^n)$ is the closure of the first-order differential operator*

$$Af(s) := \langle \operatorname{grad} f(s), F(s) \rangle$$

with domain

$$D(A) := C_c^1(\mathbb{R}^n).$$

PROOF. Let $(B, D(B))$ denote the generator of $(T(t))_{t \in \mathbb{R}}$. For $f \in C_c^1(\mathbb{R}^n)$ consider $g := f - Af \in C_c(\mathbb{R}^n)$ and compute the resolvent using the integral representation (1.13) in Chapter II. This yields

$$[R(1, B)g](s) = \int_0^\infty e^{-t} f(\Phi(t, s)) \, dt$$
$$- \int_0^\infty e^{-t} \langle \operatorname{grad} f(\Phi(t, s)), F(\Phi(t, s)) \rangle \, dt$$
$$= f(s)$$

after an integration by parts. Accordingly, $C_c^1(\mathbb{R}^n) \subset D(B)$ and $A \subset B$. On the other hand, $C_c^1(\mathbb{R}^n)$ is dense in $C_0(\mathbb{R}^n)$ and invariant under the group $(T(t))_{t \in \mathbb{R}}$ induced by the flow. So, $C_c^1(\mathbb{R}^n)$ is a core by Proposition 1.7, and the assertion is proved. $\square$

Analogous results on first-order differential operators on bounded domains $\Omega \subset \mathbb{R}^n$ need so-called boundary conditions and have been obtained, e.g., in [Ulm92]. In the next paragraph we discuss an example of such a boundary condition in a very simple situation.

3.29 Delay Differential Operators. On the space $X := C[-1, 0]$, consider the operator

$$Af := f'$$

with domain

$$D(A) := \{f \in C^1[-1, 0] : f'(0) = Lf\},$$

where L is a continuous linear form on $C[-1, 0]$. This can be rewritten as

$$D(A) = \ker \varphi,$$

where φ is the linear form on $C^1[-1, 0]$ defined by

$$C^1[-1, 0] \ni f \mapsto f'(0) - Lf \in \mathbb{C}.$$

Because this functional is bounded on the Banach space $C^1[-1, 0]$ but unbounded for the sup-norm, we deduce that $D(A)$ is dense in $C[-1, 0]$ and closed in $C^1[-1, 0]$; cf. Proposition A.9.

Next, we show that the rescaled operator $A - \|L\| \cdot I$ is dissipative. To this end, take $f \in D(A)$. As seen in Example 3.26.(i), the linear form $\overline{f(s_0)}\, \delta_{s_0}$ belongs to $\mathcal{J}(f)$ if $|f(s_0)| = \|f\|$ for some $s_0 \in [-1, 0]$. This means that $A - \|L\| I$ is dissipative, provided that

$$(3.30) \quad \operatorname{Re} \left\langle f' - \|L\| f, \overline{f(s_0)}\, \delta_{s_0} \right\rangle \leq 0 \quad \text{or} \quad \operatorname{Re} \overline{f(s_0)} f'(s_0) \leq \|L\| \cdot \|f\|^2.$$

In the case $-1 < s_0 < 0$ we have $f'(s_0) = 0$, so that (3.30) certainly holds. The same is true if $s_0 = -1$, because then $2 \operatorname{Re} \overline{f(-1)} f'(-1) = (f \cdot \overline{f})'(-1) \leq 0$. It remains to consider the case where $s_0 = 0$. Here, we use $f'(0) = Lf$ for $f \in D(A)$ to obtain

$$\operatorname{Re} \overline{f(0)} f'(0) = \operatorname{Re} \overline{f(0)} Lf \leq \|f\| \cdot \|L\| \cdot \|f\|.$$

So, we are now well prepared to apply Theorem 3.15 to conclude that A is a generator.

Proposition. *Let* $L \in C[-1, 0]'$. *The delay differential operator*

$$Af := f' \quad \text{with} \quad D(A) := \{f \in C^1[-1, 0] : f'(0) = Lf\}$$

on the Banach space $C[-1, 0]$ *generates a strongly continuous semigroup* $(T(t))_{t \geq 0}$ *satisfying*

$$\|T(t)\| \leq e^{\|L\| t} \quad \text{for } t \geq 0.$$

PROOF. By the rescaling technique, the assertion follows from Theorem 3.15 and the above consideration, provided that $\lambda - A$ is surjective for some $\lambda > \|L\|$. This means we have to show that for every $g \in C[-1,0]$ there exists $f \in C^1[-1,0]$ satisfying both

$$\lambda f - f' = g$$

and

$$f'(0) = Lf; \quad \text{i.e.,} \quad f \in D(A).$$

The first equation has

$$f(s) := c e^{\lambda s} - \int_0^s e^{\lambda(s-\tau)} g(\tau)\, d\tau$$
$$=: c\varepsilon_\lambda(s) - h(s), \quad s \in [-1,0],$$

as a solution for every constant $c \in \mathbb{C}$. If $\lambda > \|L\|$, then we can choose this constant as

$$c := \frac{g(0) - Lh}{\lambda - L\varepsilon_\lambda}$$

in order to obtain $f \in D(A)$. $\qquad\square$

The importance (and name) of this operator stems from the fact that the semigroup it generates solves a *delay differential equation* of the form

$$\begin{cases} \dot{u}(t) = Lu_t & \text{for } t \geq 0, \\ u(s) = f(s) & \text{for } -1 \leq s \leq 0, \end{cases}$$

where f is an initial function from $C[-1,0]$. Here, $u_t \in C[-1,0]$ is defined by $u_t(s) := u(t+s)$ for $s \in [-1,0]$. In [EN00, Sect. VI.6] and more systematically in [BP05] it is shown how these and more general equations can be solved via semigroups.

3.30 Second-Order Differential Operators. (i) We first reconsider the operator from Paragraph 2.11; i.e., we take on $X := C[0,1]$ the operator

$$Af := f'', \quad D(A) := \{f \in C^2[0,1] : f'(0) = f'(1) = 0\}.$$

This time, instead of constructing the generated semigroup, we verify the conditions of Theorem 3.15. It is simple to show that $(A, D(A))$ is densely defined and closed. To show dissipativity, we take $f \in D(A)$ and $s_0 \in [0,1]$ such that $|f(s_0)| = \|f\|$. By Example 3.26.(i) we have

$$\overline{f(s_0)}\, \delta_{s_0} \in \mathfrak{I}(f).$$

Because $t \mapsto \operatorname{Re} \overline{f(s_0)} \cdot f(t)$ takes its maximum at s_0, it follows that

$$\operatorname{Re} \left\langle f'', \overline{f(s_0)}\, \delta_{s_0} \right\rangle = \left(\operatorname{Re} \overline{f(s_0)} f \right)'' (s_0) \leq 0,$$

where we need to use the boundary condition

$$f'(0) = f'(1) = 0$$

if $s_0 = 0$ or $s_0 = 1$. We finally show that $\lambda^2 - A$ is surjective for $\lambda > 0$. Take $g \in \mathrm{C}[0,1]$ and define

$$k(s) := \frac{1}{2\lambda} \left[e^{\lambda s} \int_s^1 e^{-\lambda \tau} g(\tau) \, d\tau - e^{-\lambda s} \int_s^1 e^{\lambda \tau} g(\tau) \, d\tau \right] \qquad \text{for } s \in [0,1].$$

Then k is in $\mathrm{C}^2[0,1]$ and satisfies

$$\lambda^2 k - k'' = g.$$

On the other hand, for each $a, b \in \mathbb{C}$, the function

$$h_{a,b}(s) := a \, e^{\lambda s} + b \, e^{-\lambda s}, \quad s \in [0,1],$$

satisfies

$$\lambda^2 h_{a,b} - h_{a,b}'' = 0.$$

It is now an exercise in linear algebra to determine $\widetilde{a}, \widetilde{b} \in \mathbb{C}$ such that the function

$$f := k + h_{\widetilde{a},\widetilde{b}}$$

satisfies $f'(0) = f'(1) = 0$. Then $f \in D(A)$ and $\lambda^2 f - f'' = g$; i.e., $\lambda^2 - A$ is surjective. It follows from Theorem 3.15 that $\big(A, D(A)\big)$ generates a contraction semigroup on $\mathrm{C}[0,1]$.

(ii) The above method is now applied to the same differential operator on a different space and with different boundary conditions. Let $X := \mathrm{L}^2[0,1]$ and

$$Af := f'', \quad D(A) := \{ f \in \mathrm{C}^2[0,1] : f(0) = f(1) = 0 \}.$$

Then $D(A)$ is dense in X, and for $f \in D(A)$ one has

$$(3.31) \qquad (Af \mid f) = \int_0^1 f'' \overline{f} \, ds = f' \overline{f} \Big|_0^1 - \int_0^1 f' \overline{f'} \, ds \leq 0.$$

By Example 3.26.(iii), this means that A is dissipative on the Hilbert space $\mathrm{L}^2[0,1]$. As in the previous case, for every $g \in \mathrm{C}^2[0,1]$ and $\lambda > 0$ there exists a function $f \in \mathrm{C}^2[0,1]$ satisfying $f(0) = f(1) = 0$ and

$$\lambda^2 f - f'' = g;$$

i.e., $\mathrm{rg}(\lambda^2 - A)$ is dense. Again by Theorem 3.15 we conclude that $\big(\overline{A}, D(\overline{A})\big)$ generates a contraction semigroup on $\mathrm{L}^2[0,1]$. Here the domain of the closure $\overline{A}$ is given by $D(\overline{A}) = \mathrm{H}_0^2[0,1]$; see Exercise 3.31.(1).

(iii) As a somewhat less canonical second-order differential operator on $X := \mathrm{C}[0,1]$, consider $\big(A, D(A)\big)$ defined by

$$Af(s) := s(1-s) f''(s), \qquad s \in [0,1],$$

for $f \in D(A) := \big\{ f \in \mathrm{C}[0,1] \cap \mathrm{C}^2(0,1) : \lim_{s \to 0,1} s(1-s) f''(s) = 0 \big\}$. We show that it generates a strongly continuous contraction semigroup by verifying the conditions of Theorem 3.15.

As above, it is easy to show that $(A, D(A))$ is closed, densely defined, and dissipative. Therefore, it suffices to prove that $\lambda - A$ is surjective for some $\lambda > 0$. Observe first that the functions $h_0 : s \mapsto 1$ and $h_1 : s \mapsto s$ belong to $D(A)$ and satisfy

$$(3.32) \qquad (\lambda - A)h_i = \lambda h_i, \qquad i = 0, 1 \text{ and } \lambda > 0.$$

Hence, it suffices to consider the part A_0 of A in the closed subspace $X_0 := \{f \in X : f(0) = f(1) = 0\}$ with domain $D(A_0) := \{f \in X_0 \cap C^2(0,1) : \lim_{s \to 0,1} s(1 - s)f''(s) = 0\}$. Then $(A_0, D(A_0))$ is still dissipative, but is now surjective. Its inverse R can be computed as

$$Rf(s) = \int_0^1 \sigma(s,t) \frac{f(t)}{t(1-t)}\, dt,$$

where

$$\sigma(s,t) := \begin{cases} s(t-1) & \text{for } 0 \le s \le t \le 1, \\ t(s-1) & \text{for } 0 \le t \le s \le 1, \end{cases}$$

and $f \in X_0$. This shows that $0 \in \rho(A_0)$ and hence $[0, \infty) \subset \rho(A_0)$. From (3.32) we conclude that $(0, \infty) \subset \rho(A)$. Accordingly, A is a generator.

3.31 Exercises. (1) Show that the domain of $\overline{A}$ in Paragraph 3.30.(ii) is given by $D(\overline{A}) = H_0^2[0,1] := \{f \in W^{2,2}[0,1] : f(0) = f(1) = 0\}$. (Hint: Show first that the second derivative D_2 on $L^2[0,1]$ with domain $H_0^2[0,1]$ is invertible. The assertion then follows from the fact that $A \subset D_2$.)

(2) Let Ω be a compact space and take $X := C(\Omega)$. A *semiflow* $\Phi : \mathbb{R}_+ \times \Omega \to \Omega$ is defined by the properties

$$(3.33) \qquad \begin{aligned} \Phi(t+r, s) &= \Phi\big(t, \Phi(r,s)\big), \\ \Phi(0, s) &= s \end{aligned}$$

for every $s \in \Omega$ and $r, t \in \mathbb{R}_+$. Establish the following facts.

(i) The semiflow Φ is continuous if and only if it induces a strongly continuous semigroup $(T(t))_{t \ge 0}$ on X by the formula

$$(3.34) \qquad \big(T(t)f\big)(s) := f\big(\Phi(t, s)\big) \qquad \text{for } s \in \Omega, \ t \ge 0, \ f \in X.$$

(ii) The generator A of $(T(t))_{t \ge 0}$ is a derivation (cf. Exercise 1.15.(5)).

(iii*) Every strongly continuous semigroup $(T(t))_{t \ge 0}$ on X that consists of algebra homomorphisms originates, via (3.34), from a continuous semiflow on Ω. (Hint: See [Nag86, B-II, Thm. 3.4].)

(3) Show that the semigroup $(T(t))_{t \ge 0}$ on $X := C[-1, 0]$ generated by the delay differential operator from Paragraph 3.29 satisfies the *translation property*; i.e.,

$$(\text{TP}) \qquad \big(T(t)f\big)(s) = \begin{cases} f(t+s) & \text{if } t + s \le 0, \\ [T(t+s)f](0) & \text{if } t + s > 0, \end{cases}$$

for all $f \in X$ (cf. also [EN00, Thm. VI.6.2]).

4. Analytic Semigroups

Up to now, we have classified semigroups only as being *strongly continuous* in the general case or being *uniformly continuous* as a somewhat uninteresting case. Between these two extreme cases there is room for a wide range of continuity properties; see [EN00, Sect. II.4]. Here we introduce just one more class of semigroups enjoying a rather strong *regularity* property. Other natural regularity properties for semigroups are discussed in Section 5 below.

We start our discussion by reconsidering the exponential Formula (3.2), but now impose conditions on the operator A (and its resolvent $R(\lambda, A)$) that make the contour integrals converge even if A and $\sigma(A)$ are unbounded.

4.1 Definition. *A closed linear operator* $(A, D(A))$ *in a Banach space* X *is called sectorial (of angle δ) if there exists $0 < \delta \leq \pi/2$ such that the sector*

$$\Sigma_{\pi/2+\delta} := \left\{ \lambda \in \mathbb{C} : |\arg \lambda| < \frac{\pi}{2} + \delta \right\} \setminus \{0\}$$

is contained in the resolvent set $\rho(A)$, and if for each $\varepsilon \in (0, \delta)$ there exists $M_\varepsilon \geq 1$ such that

$$(4.1) \qquad \|R(\lambda, A)\| \leq \frac{M_\varepsilon}{|\lambda|} \quad \text{for all } 0 \neq \lambda \in \overline{\Sigma}_{\pi/2+\delta-\varepsilon}.$$

For densely defined sectorial operators and appropriate paths γ, the exponential function "e^{tA}" can now be defined via the Cauchy integral formula as used in the Dunford functional calculus for bounded operators (see, e.g., [DS58, Sect. VII.3], [TL80, Sect. V.8]).

4.2 Definition. *Let* $(A, D(A))$ *be a densely defined sectorial operator of angle δ. Define $T(0) := I$ and operators $T(z)$, for $z \in \Sigma_\delta$, by*

$$(4.2) \qquad T(z) := \frac{1}{2\pi i} \int_\gamma e^{\mu z} R(\mu, A) \, d\mu,$$

where γ is any piecewise smooth curve in $\Sigma_{\pi/2+\delta}$ going from $\infty \, e^{-i(\pi/2+\delta')}$ to $\infty \, e^{i(\pi/2+\delta')}$ for some $\delta' \in (|\arg z|, \delta)$.[2]

As a first step, we need to justify this definition. In particular, we show that the essential properties of the analytic functional calculus for bounded operators (cf. Definition I.2.10) prevail in this situation.

---

[2] See Figure 1.

4.3 Proposition. Let $(A, D(A))$ be a densely defined sectorial operator of angle δ. Then, for all $z \in \Sigma_\delta$, the maps $T(z)$ are bounded linear operators on X satisfying the following properties.

(i) $\|T(z)\|$ is uniformly bounded for $z \in \Sigma_{\delta'}$ if $0 < \delta' < \delta$.

(ii) The map $z \mapsto T(z)$ is analytic in Σ_δ.

(iii) $T(z_1 + z_2) = T(z_1)T(z_2)$ for all $z_1, z_2 \in \Sigma_\delta$.

(iv) The map $z \mapsto T(z)$ is strongly continuous in $\Sigma_{\delta'} \cup \{0\}$ if $0 < \delta' < \delta$.

PROOF. We first verify that for $z \in \Sigma_{\delta'}$, with $\delta' \in (0, \delta)$ fixed, the integral in (4.2) defining $T(z)$ converges uniformly in $\mathcal{L}(X)$ with respect to the operator norm. Because the integrand is analytic in $\mu \in \Sigma_{\pi/2+\delta}$, this integral, if it exists, is by Cauchy's integral theorem independent of the particular choice of γ. Hence, we may choose $\gamma = \gamma_r$ as in Figure 1; i.e., γ consists of the three parts

$$
\begin{aligned}
&\gamma_{r,1} : \left\{ -\rho e^{-i(\pi/2+\delta-\varepsilon)} : -\infty \le \rho \le -r \right\}, \\
(4.3) \quad &\gamma_{r,2} : \left\{ re^{i\alpha} : -(\pi/2 + \delta - \varepsilon) \le \alpha \le (\pi/2 + \delta - \varepsilon) \right\}, \\
&\gamma_{r,3} : \left\{ \rho e^{i(\pi/2+\delta-\varepsilon)} : r \le \rho \le \infty \right\},
\end{aligned}
$$

where $\varepsilon := (\delta - \delta')/2 > 0$ and $r := 1/|z|$.

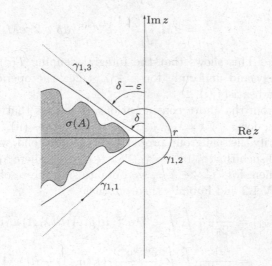

Figure 1

Then, for $\mu \in \gamma_{r,3}$, $z \in \Sigma_{\delta'}$, we can write

$$
\mu z = |\mu z|\, e^{i(\arg \mu + \arg z)},
$$

where $\pi/2 + \varepsilon \leq \arg \mu + \arg z \leq \frac{3\pi}{2} - \varepsilon$. Hence, we have

$$\frac{1}{|\mu z|} \operatorname{Re}(\mu z) = \cos(\arg \mu + \arg z) \leq \cos(\pi/2 + \varepsilon) = -\sin \varepsilon,$$

and therefore

(4.4) $$|e^{\mu z}| \leq e^{-|\mu z| \sin \varepsilon}$$

for all $z \in \Sigma_{\delta'}$ and $\mu \in \gamma_{r,3}$. Similarly, one shows that (4.4) is true for $z \in \Sigma_{\delta'}$ and $\mu \in \gamma_{r,1}$, from which we conclude

(4.5) $$\|e^{\mu z} R(\mu, A)\| \leq e^{-|\mu z| \sin \varepsilon} \frac{M_\varepsilon}{|\mu|}$$

for all $z \in \Sigma_{\delta'}$ and $\mu \in \gamma_{r,1} \cup \gamma_{r,3}$. On the other hand, the estimate

(4.6) $$\|e^{\mu z} R(\mu, A)\| \leq e \frac{M_\varepsilon}{|\mu|} = e M_\varepsilon |z|$$

holds for all $z \in \Sigma_{\delta'}$ and $\mu \in \gamma_{r,2}$. Using the estimates (4.5) and (4.6), we then conclude

$$\left\| \int_{\gamma_r} e^{\mu z} R(\mu, A) \, d\mu \right\| \leq \sum_{k=1}^{3} \left\| \int_{\gamma_{r,k}} e^{\mu z} R(\mu, A) \, d\mu \right\|$$

$$\leq 2 M_\varepsilon \int_{1/|z|}^{\infty} \frac{1}{\rho} e^{-\rho |z| \sin \varepsilon} d\rho + e M_\varepsilon |z| \cdot \frac{2\pi}{|z|}$$

$$= 2 M_\varepsilon \int_{1}^{\infty} \frac{1}{\rho} e^{-\rho \sin \varepsilon} d\rho + 2\pi e M_\varepsilon$$

for all $z \in \Sigma_{\delta'}$. This shows that the integral defining $T(z)$ converges in $\mathcal{L}(X)$ absolutely and uniformly for $z \in \Sigma_{\delta'}$; i.e., the operators $T(z)$ are well-defined and satisfy (i).

Moreover, from the above considerations, it follows that the map $z \mapsto T(z)$ is analytic for $z \in \Sigma_\delta = \cup_{0 < \delta' < \delta} \Sigma_{\delta'}$, which proves (ii).

Next, we verify the semigroup property (iii). To this end, we choose some constant $c > 0$ such that $\gamma \cap \gamma' := \gamma_1 \cap (\gamma_1 + c) = \emptyset$, where γ_1 is as in (4.3) with $r = 1$. Then, for $z_1, z_2 \in \Sigma_{\delta'}$, we obtain using the resolvent equation in Paragraph V.1.2 and Fubini's theorem that

$$T(z_1) T(z_2) = \frac{1}{(2\pi i)^2} \int_\gamma \int_{\gamma'} e^{\mu z_1} e^{\lambda z_2} R(\mu, A) R(\lambda, A) \, d\lambda \, d\mu$$

$$= \frac{1}{(2\pi i)^2} \int_\gamma \int_{\gamma'} \frac{e^{\mu z_1} e^{\lambda z_2}}{\lambda - \mu} \left(R(\mu, A) - R(\lambda, A) \right) d\lambda \, d\mu$$

$$= \frac{1}{2\pi i} \int_\gamma e^{\mu z_1} R(\mu, A) \left(\frac{1}{2\pi i} \int_{\gamma'} \frac{e^{\lambda z_2}}{\lambda - \mu} \, d\lambda \right) d\mu$$

$$- \frac{1}{2\pi i} \int_{\gamma'} e^{\lambda z_2} R(\lambda, A) \left(\frac{1}{2\pi i} \int_\gamma \frac{e^{\mu z_1}}{\lambda - \mu} \, d\mu \right) d\lambda.$$

By closing the curves γ and γ' by circles with increasing diameter on the left and using the fact that γ lies to the left of γ', Cauchy's integral theorem implies

$$\frac{1}{2\pi i} \int_\gamma \frac{e^{\mu z_1}}{\lambda - \mu} \, d\mu = 0 \quad \text{and} \quad \frac{1}{2\pi i} \int_{\gamma'} \frac{e^{\lambda z_2}}{\lambda - \mu} \, d\lambda = e^{\mu z_2}.$$

Thus, we conclude

$$\begin{aligned} T(z_1)T(z_2) &= \frac{1}{2\pi i} \int_\gamma e^{\mu z_1} e^{\mu z_2} R(\mu, A) \, d\mu \\ &= T(z_1 + z_2) \end{aligned}$$

for all $z_1, z_2 \in \Sigma_{\delta'}$, which proves (iii).

It remains only to show (iv), i.e., that the map $z \mapsto T(z)$ is strongly continuous in $\Sigma_{\delta'} \cup \{0\}$ for every $0 < \delta' < \delta$. By (i) and (ii), it suffices, as usual, to verify that

$$(4.7) \qquad \lim_{\Sigma_{\delta'} \ni z \to 0} T(z)x - x = 0 \quad \text{for all } x \in D(A).$$

We start from estimate (4.4) and Cauchy's integral formula and obtain for $\gamma = \gamma_1$ that

$$\frac{1}{2\pi i} \int_\gamma \frac{e^{\mu z}}{\mu} \, d\mu = 1$$

for all $z \in \Sigma_{\delta'}$. Hence, the identity $R(\mu, A)Ax = \mu R(\mu, A)x - x$ for $x \in D(A)$ yields

$$\begin{aligned} T(z)x - x &= \frac{1}{2\pi i} \int_\gamma e^{\mu z} \left(R(\mu, A) - \frac{1}{\mu} \right) x \, d\mu \\ &= \frac{1}{2\pi i} \int_\gamma \frac{e^{\mu z}}{\mu} R(\mu, A) Ax \, d\mu \end{aligned}$$

for all $z \in \Sigma_{\delta'}$. Now, by (4.1) and (4.5), we have

$$\left\| \frac{e^{\mu z}}{\mu} R(\mu, A) Ax \right\| \leq \frac{M_\varepsilon}{|\mu|^2} \left(1 + e^{|z|} \right) \|Ax\|$$

for all $\mu \in \gamma$ and $z \in \Sigma_{\delta'}$. Using this estimate and because $\lim_{z \to 0} e^{\mu z} = 1$, Lebesgue's dominated convergence theorem implies

$$\lim_{\Sigma_{\delta'} \ni z \to 0} T(z)x - x = \frac{1}{2\pi i} \int_\gamma \frac{1}{\mu} R(\mu, A) Ax \, d\mu = 0,$$

where the second equality follows from Cauchy's integral theorem by closing the path γ by circles with increasing diameter on the right. This proves (4.7), and the proof is complete. $\qquad \square$

If in Definition 4.2 we only consider values $z \in \mathbb{R}_+$, we obtain, by the previous proposition, a strongly continuous semigroup $(T(t))_{t \geq 0}$ on X. It turns out that its generator is the operator from which we started.

4.4 Proposition. *The generator of the strongly continuous semigroup defined by* (4.2) *is the sectorial operator* $(A, D(A))$.

PROOF. Denoting by $(B, D(B))$ the generator of $(T(t))_{t \geq 0}$, it suffices to show that

(4.8) $$R(\lambda, A) = R(\lambda, B)$$

for $\lambda = |\omega_0| + 2$, where ω_0 denotes the growth bound of $(T(t))_{t \geq 0}$, cf. Definition I.1.5. However, from Theorem 1.10 we know that the resolvent of B in λ is given as the integral

$$R(\lambda, B)x = \int_0^\infty e^{-\lambda t} T(t) x \, dt \qquad \text{for all } x \in X.$$

Take now $t_0 > 0$ and choose $\gamma = \gamma_1$ as in (4.3). Then, by Fubini's theorem, we obtain

$$\int_0^{t_0} e^{-\lambda t} T(t) x \, dt = \frac{1}{2\pi i} \int_\gamma \frac{e^{t_0(\mu - \lambda)} - 1}{\mu - \lambda} R(\mu, A) x \, d\mu$$

$$= R(\lambda, A)x + \frac{1}{2\pi i} \int_\gamma \frac{e^{t_0(\mu - \lambda)}}{\mu - \lambda} R(\mu, A) x \, d\mu.$$

Here, we used the formula $\int_\gamma \frac{R(\mu, A)}{\mu - \lambda} x \, d\mu = -2\pi i R(\lambda, A) x$, which can be verified using Cauchy's integral formula and by closing γ on the right by circles of diameter converging to ∞. Because $\operatorname{Re}(\mu - \lambda) \leq -1$, for $\varepsilon = (\delta - \delta')/2$ we can estimate

$$\left\| \int_\gamma \frac{e^{t_0(\mu - \lambda)}}{\mu - \lambda} R(\mu, A) x \, d\mu \right\| \leq e^{-t_0} \cdot \|x\| \int_\gamma \frac{M_\varepsilon}{|\mu - \lambda| \cdot |\mu|} |d\mu|$$

and obtain (4.8) by taking the limit as $t_0 \to \infty$. □

  Combining the two previous results, we see that a densely defined sectorial operator is always the generator of a strongly continuous semigroup that can be extended analytically to some sector Σ_δ containing $\mathbb{R}_+$. At this point, we remark that sectorial operators are characterized by the single resolvent estimate (4.1), whereas the Hille–Yosida Generation Theorem 3.8 requires estimates on all powers of the resolvent.

  Semigroups that can be extended analytically enjoy many nice properties; see, e.g., Theorem III.2.10, [EN00, Cors. IV.3.12, VI.3.6 and VI.7.17] and [Lun95]. Therefore, we give various characterizations of these *analytic* semigroups. First, we introduce the appropriate terminology.

4.5 Definition. *A family of operators* $(T(z))_{z \in \Sigma_\delta \cup \{0\}} \subset \mathcal{L}(X)$ *is called an analytic semigroup (of angle* $\delta \in (0, \pi/2]$*) if*

(i) $T(0) = I$ *and* $T(z_1 + z_2) = T(z_1)T(z_2)$ *for all* $z_1, z_2 \in \Sigma_\delta$.

(ii) *The map* $z \mapsto T(z)$ *is analytic in* Σ_δ.

(iii) $\lim_{\Sigma_{\delta'} \ni z \to 0} T(z)x = x$ *for all* $x \in X$ *and* $0 < \delta' < \delta$.

If, in addition,

(iv) $\|T(z)\|$ *is bounded in* $\Sigma_{\delta'}$ *for every* $0 < \delta' < \delta$,

we call $(T(z))_{z \in \Sigma_\delta \cup \{0\}}$ *a bounded analytic semigroup.*

In our next result, we give various equivalences characterizing generators of bounded analytic semigroups.

4.6 Theorem. *For an operator* $(A, D(A))$ *on a Banach space* X, *the following statements are equivalent.*

(a) *A generates a bounded analytic semigroup* $(T(z))_{z \in \Sigma_\delta \cup \{0\}}$ *on* X.

(b) *There exists* $\vartheta \in (0, \pi/2)$ *such that the operators* $e^{\pm i\vartheta}A$ *generate bounded strongly continuous semigroups on* X.

(c) *A generates a bounded strongly continuous semigroup* $(T(t))_{t \geq 0}$ *on X such that* $\mathrm{rg}(T(t)) \subset D(A)$ *for all* $t > 0$, *and*

$$(4.9) \qquad M := \sup_{t > 0} \|tAT(t)\| < \infty.$$

(d) *A generates a bounded strongly continuous semigroup* $(T(t))_{t \geq 0}$ *on X, and there exists a constant* $C > 0$ *such that*

$$(4.10) \qquad \|R(r + is, A)\| \leq \frac{C}{|s|}$$

for all $r > 0$ *and* $0 \neq s \in \mathbb{R}$.

(e) *A is densely defined and sectorial.*

PROOF. We show that (a) $\Rightarrow$ (b) $\Rightarrow$ (d) $\Rightarrow$ (e) $\Rightarrow$ (c) $\Rightarrow$ (a).

(a) $\Rightarrow$ (b). For $\vartheta \in (0, \delta)$, we define $T_\vartheta(t) := T(e^{i\vartheta}t)$. Then, by Definition 4.5, the operator family $(T_\vartheta(t))_{t \geq 0} \subset \mathcal{L}(X)$ is a bounded strongly continuous semigroup on X. In order to determine its generator, we define $\gamma : [0, \infty) \to \mathbb{C}$ by $\gamma(r) := e^{i\vartheta}r$. Then, by analyticity and Cauchy's integral theorem, we obtain

$$R(1, A)x = \int_0^\infty e^{-t}T(t)x\, dt = \int_\gamma e^{-r}T(r)x\, dr$$

$$= e^{i\vartheta} \int_0^\infty e^{-e^{i\vartheta}r}T_\vartheta(r)x\, dr = e^{i\vartheta}R(e^{i\vartheta}, A_\vartheta)x$$

for all $x \in X$, hence $A_\vartheta = e^{i\vartheta}A$. Similarly, it follows that $(T(e^{-i\vartheta}t))_{t \geq 0}$ is a bounded strongly continuous semigroup with generator $e^{-i\vartheta}A$; i.e., (b) is proved.

(b) $\Rightarrow$ (d). Let $e^{-i\vartheta} = a - ib$ for $a, b > 0$. Then, applying the Hille–Yosida Generation Theorem 3.8 to the generator $e^{-i\vartheta}A$, we obtain a constant $\widetilde{C} \geq 1$ such that

$$\|R(r + is, A)\| = \left\|e^{-i\vartheta}R\bigl(e^{-i\vartheta}(r + is), e^{-i\vartheta}A\bigr)\right\|$$

$$\leq \frac{\widetilde{C}}{ar + bs} \leq \frac{C}{s}$$

for all $r, s > 0$ and $C := \widetilde{C}/b$. For $s < 0$, we obtain a similar estimate using the fact that $e^{i\vartheta}A$ is a generator on X.

(d) $\Rightarrow$ (e). By assumption, A generates a bounded strongly continuous semigroup, and hence is densely defined by Proposition 1.7. Moreover, by Theorem 1.10 we have $\Sigma_{\pi/2} \subset \rho(A)$. From Corollary V.1.14, we know that

$$\|R(\lambda, A)\| \geq \frac{1}{\text{dist}(\lambda, \sigma(A))} \qquad \text{for all } \lambda \in \rho(A).$$

Therefore, the estimate (4.10) implies $i\mathbb{R} \setminus \{0\} \subset \rho(A)$ and, by continuity of the resolvent map,

$$(4.11) \qquad \|R(\mu, A)\| \leq \frac{C}{|\mu|} \qquad \text{for all } 0 \neq \mu \in i\mathbb{R}.$$

We now develop the resolvent of A in $0 \neq \mu \in i\mathbb{R}$ in its Taylor series (see Proposition V.1.3),

$$(4.12) \qquad R(\lambda, A) = \sum_{n=0}^{\infty}(\mu - \lambda)^n R(\mu, A)^{n+1}.$$

This series converges uniformly in $\mathcal{L}(X)$, provided that $|\mu - \lambda| \cdot \|R(\mu, A)\| \leq q < 1$ for some fixed $q \in (0, 1)$. In particular, for $\mu = i\,\text{Im}\,\lambda$, we see from (4.11) that this is the case if $|\text{Re}\,\lambda| \leq q/C\,|\text{Im}\,\lambda|$. Because this is true for arbitrary $0 < q < 1$, we conclude that

$$\left\{\lambda \in \mathbb{C} : \text{Re}\,\lambda \leq 0 \text{ and } \left|\frac{\text{Re}\,\lambda}{\text{Im}\,\lambda}\right| < \frac{1}{C}\right\} \subset \rho(A),$$

and hence $\Sigma_{\pi/2 + \delta} \subseteq \rho(A)$ for $\delta := \arctan 1/C$.

It remains to estimate $\|R(\lambda, A)\|$ for $\lambda \in \Sigma_{\pi/2 + \delta - \varepsilon}$ and $\varepsilon \in (0, \delta)$. We assume first that $\text{Re}\,\lambda > 0$. Then, by the Hille–Yosida Generation Theorem 3.8 for the bounded semigroup $(T(t))_{t \geq 0}$, there exists a constant $\widetilde{M} \geq 1$ such that $\|R(\lambda, A)\| \leq \tilde{M}/\text{Re}\,\lambda$. Moreover, by (4.10), we have $\|R(\lambda, A)\| \leq C/|\text{Im}\,\lambda|$; hence there exists $M \geq 1$ such that

$$\|R(\lambda, A)\| \leq \frac{M}{|\lambda|} \qquad \text{if } \text{Re}\,\lambda > 0.$$

In the case $\operatorname{Re}\lambda \leq 0$, we choose $q \in (0,1)$ such that $\delta - \varepsilon = \arctan(q/C)$. Then $|\operatorname{Re}\lambda/\operatorname{Im}\lambda| \leq q/C$, and from estimate (4.11) combined with the Taylor expansion (4.12) for $\mu = i\operatorname{Im}\lambda$ we obtain

$$\|R(\lambda, A)\| \leq \sum_{n=0}^{\infty} |\operatorname{Re}\lambda|^n \frac{C^{n+1}}{|\operatorname{Im}\lambda|^{n+1}}$$

$$\leq \frac{1}{1-q} \cdot \frac{C}{|\operatorname{Im}\lambda|} \leq \frac{\sqrt{C^2+1}}{1-q} \cdot \frac{1}{|\lambda|}.$$

(e) $\Rightarrow$ (c). By Propositions 4.3 and 4.4, A generates a bounded strongly continuous semigroup $(T(t))_{t\geq 0}$, and the map

$$(0,\infty) \ni t \mapsto T(t)x \in X$$

is differentiable for all $x \in X$. In particular, the limit

$$\lim_{h\downarrow 0} \frac{T(t+h) - T(t)}{h} x = \lim_{h\downarrow 0} \frac{T(h) - I}{h} T(t)x$$

exists for all $x \in X$ and $t > 0$; hence $\operatorname{rg}(T(t)) \subset D(A)$ for $t > 0$.

Because for $t > 0$ the operator $AT(t)$ is closed with domain $D(AT(t)) = X$, it is bounded by the closed graph theorem.

To estimate its norm, we use the integral representation (4.2) of $T(t)$ and obtain, using the closedness of A, the resolvent equation, and Cauchy's integral theorem that

$$AT(t) = A \frac{1}{2\pi i} \int_\gamma e^{\mu t} R(\mu, A) \, d\mu$$

$$= \frac{1}{2\pi i} \int_\gamma e^{\mu t} \big(\mu R(\mu, A) - I\big) \, d\mu$$

$$= \frac{1}{2\pi i} \int_\gamma \mu e^{\mu t} R(\mu, A) \, d\mu.$$

Because by analyticity we may choose $\gamma = \gamma_r$ for $r := 1/t$ as in the proof of Proposition 4.3, we conclude, using (4.5) and (4.6), that

$$\left\| \int_\gamma \mu e^{\mu t} R(\mu, A) \, d\mu \right\| \leq 2M_\varepsilon \int_{1/t}^{\infty} e^{-\rho t \sin \varepsilon} d\rho + \frac{2\pi e M_\varepsilon}{t}$$

$$\leq 2M_\varepsilon \left(\frac{1}{\sin \varepsilon} + \pi e \right) \cdot \frac{1}{t},$$

where $\varepsilon := (\delta - \delta')/2$ for some $\delta' \in (0, \delta)$. This proves (c).

(c) ⇒ (a). We claim first that the map $t \mapsto T(t)x \in X$ is infinitely many times differentiable for all $t > 0$ and $x \in X$. In fact, using the formula $AT(s)y = T(s)Ay$, valid for $s \geq 0$ and $y \in D(A)$ (see Lemma 1.3), one easily verifies by induction that $\mathrm{rg}(T(t)) \subset D(A^\infty) = \cap_{n \in \mathbb{N}} D(A^n)$ and

$$A^n T(t) = \left(AT(t/n)\right)^n$$

for all $t > 0$ and $n \in \mathbb{N}$. We now fix some $\varepsilon \in (0, t)$. Then, by Lemma 1.3,

$$\begin{aligned} A^n T(t)x &= AT(t - \varepsilon)A^{n-1}T(\varepsilon)x \\ &= \tfrac{d}{dt}T(t - \varepsilon)A^{n-1}T(\varepsilon)x \\ &\quad\vdots \\ &= \tfrac{d^n}{dt^n}T(t)x \end{aligned}$$

for all $x \in X$. This establishes our claim. Combining this with (4.9) and the inequality[3] $n!\,e^n \geq n^n$, we obtain, while writing $T^{(n)}(t) := \frac{d^n}{dt^n}T(t)$,

$$(4.13) \qquad \frac{1}{n!}\left\|T^{(n)}(t)\right\| \leq \left(\frac{eM}{t}\right)^n \qquad \text{for all } n \in \mathbb{N} \text{ and } t > 0.$$

Next, we develop $T(t)$ in its Taylor series. To this end, we choose $t > 0$ and $x \in X$ arbitrary. Then, by Taylor's theorem, we have for $|h| < t$ and all $n \in \mathbb{N}$

$$(4.14) \quad T(t+h)x = \sum_{k=0}^{n} \frac{h^k}{k!} T^{(k)}(t)x + \frac{1}{n!}\int_t^{t+h}(t+h-s)^n T^{(n+1)}(s)x\,ds.$$

Denoting the integral term on the right-hand side of (4.14) by $R_{n+1}(t+h)x$, we see from (4.13) that

$$\lim_{n \to \infty}\|R_{n+1}(t+h)\| = 0$$

uniformly for $|h| \leq q \cdot t/eM$ for every fixed $q \in (0, 1)$. On the other hand, the series

$$(4.15) \qquad T(z) := \sum_{k=0}^{\infty} \frac{(z-t)^k}{k!} T^{(k)}(t)$$

converges uniformly for all $z \in \mathbb{C}$ satisfying $|z - t| \leq q \cdot t/eM$; hence it extends the given semigroup $(T(t))_{t \geq 0}$ analytically to the sector Σ_δ for $\delta := \arctan(1/eM)$. This proves (ii) of Definition 4.5.

---

[3] Taking logarithms, this inequality can be restated as $1/n \sum_{k=1}^{n} \log k/n \geq -1$, which follows from $\int_0^1 \log x\,dx = -1$.

In order to verify the semigroup property for $(T(z))_{z \in \Sigma_\delta \cup \{0\}}$, we first take some $t > 0$. Then the map $\Sigma_\delta \ni z \mapsto T(t)T(z) \in \mathcal{L}(X)$ is analytic and satisfies $T(t)T(z) = T(t+z)$ for $z \geq 0$. Hence, by the identity theorem for analytic functions, we conclude that $T(t)T(z) = T(t+z)$ for all $z \in \Sigma_\delta$. Now fix some $z_1 \in \Sigma_\delta$ and consider the map $\Sigma_\delta \ni z \mapsto T(z_1)T(z) \in \mathcal{L}(X)$. This map is analytic as well and satisfies $T(z_1)T(z) = T(z_1 + z)$ for $z \geq 0$. Using the analyticity again, we obtain the functional equation $T(z_1)T(z_2) = T(z_1 + z_2)$ for all $z_1, z_2 \in \Sigma_\delta$.

To verify that $z \mapsto T(z)$ is uniformly bounded on the sector $\Sigma_{\delta'}$ for every $0 < \delta' < \delta$, we choose $q \in (0,1)$ such that $\delta' := \arctan(q/eM)$. Then, by equations (4.13) and (4.15),

$$\|T(z)\| = \left\| \sum_{k=0}^{\infty} \frac{(i \operatorname{Im} z)^k}{k!} T^{(k)}(\operatorname{Re} z) \right\|$$

(4.16)
$$\leq \sum_{k=0}^{\infty} |\operatorname{Im} z|^k \left(\frac{eM}{\operatorname{Re} z} \right)^k \leq \frac{1}{1-q}.$$

It remains only to prove that the map

$$\Sigma_{\delta'} \cup \{0\} \ni z \mapsto T(z) \in \mathcal{L}(X)$$

is strongly continuous in $z = 0$. To this end, we choose $x \in X$ and $\varepsilon > 0$. Because $(T(t))_{t \geq 0}$ is strongly continuous, there exists $h_0 > 0$ such that $\|T(h)x - x\| < \varepsilon(1-q)$ for all $0 < h < h_0$. Then, using (4.16), we obtain

$$\|T(z)x - x\| \leq \|T(z)(x - T(h)x)\| + \|T(z+h)x - T(h)x\| + \|T(h)x - x\|$$
$$< 2\varepsilon + \|T(z+h) - T(h)\| \cdot \|x\|$$

for all $h \in (0, h_0)$. Because the map $z \mapsto T(z+h) \in \mathcal{L}(X)$ is analytic in some neighborhood of $z = 0$, we have $\lim_{z \to 0} \|T(z+h) - T(h)\| = 0$, which completes the proof of the implication (c) $\Rightarrow$ (a). $\qquad \square$

4.7 Remarks. (i) We point out that from the previous proof it follows that for an analytic semigroup $(T(t))_{t \geq 0}$ and its generator A we always have

$$\operatorname{rg}(T(t)) \subset D(A^\infty)$$

and (by (4.13)) for every $n \in \mathbb{N}$

$$\overline{\lim_{t \downarrow 0}} \, t^n \|A^n T(t)\| < \infty.$$

(ii) We note that in concrete applications one usually verifies condition (d) in Theorem 4.6 in order to show that an operator generates an analytic semigroup. In the case where the semigroup $(T(t))_{t \geq 0}$ is already known one can also try to verify condition (c).

Next we give some abstract and concrete examples of analytic semigroups.

4.8 Corollary. *If A is a normal operator on a Hilbert space H satisfying*

$$(4.17) \qquad\qquad \sigma(A) \subseteq \{z \in \mathbb{C} : |\arg(-z)| < \delta\}$$

for some $\delta \in [0, \pi/2)$, then A generates a bounded analytic semigroup.

PROOF. Because A is normal, the same is true for $R(\lambda, A)$ for all $\lambda \in \rho(A)$. Hence, by [TL80, Thm. VI.3.5] or [Wei80, Thm. 5.44], we have

$$\|R(\lambda, A)\| = \mathrm{r}(R(\lambda, A)),$$

and the assertion follows from Theorem 4.6.(d) combined with the Spectral Mapping Theorem for the Resolvent in Paragraph V.1.13. $\qquad\qquad\square$

A different proof of the previous result is indicated in Exercise 4.14.(8).

In particular, Corollary 4.8 shows that the semigroup generated by a self-adjoint operator A that is bounded above, which means that there exists $w \in \mathbb{R}$ such that

$$(Ax \,|\, x) \leq w \,\|x\|^2 \qquad \text{for all } x \in D(A),$$

is analytic of angle $\pi/2$. Moreover, this semigroup is bounded if and only if $w \leq 0$.

4.9 Example. In Paragraph II.3.30.(ii) we showed that the closure $\overline{A}$ of the operator

$$Af := f'', \quad D(A) := \{f \in \mathrm{C}^2[0,1] : f(0) = f(1) = 0\}$$

generates a strongly continuous contraction semigroup $(T(t))_{t \geq 0}$ on the Hilbert space $H = \mathrm{L}^2[0,1]$. Because it is not difficult to show that

$$\overline{A}f := f'', \quad D(\overline{A}) := \{f \in \mathrm{H}^2[0,1] : f(0) = f(1) = 0\}$$

is self-adjoint, the semigroup $(T(t))_{t \geq 0}$ is analytic. See Exercise 4.14.(9) and, for more general operators, [EN00, Sect. VI.4].

It is, however, even simpler to verify the inequality in (3.29) with A replaced by $e^{\pm i\vartheta}\overline{A}$ for some $\vartheta \in (0, \pi/2)$ in order to conclude that $e^{\pm i\vartheta}\overline{A}$ are dissipative. Because $\rho(e^{\pm i\vartheta}\overline{A}) = e^{\pm i\vartheta}\rho(\overline{A})$, we then conclude by the Lumer–Phillips Theorem 3.15 that $e^{\pm i\vartheta}\overline{A}$ are generators of contraction semigroups. Hence, Theorem 4.6.(b) implies that the operator $\overline{A}$ generates a bounded analytic semigroup on H.

Another important class of generators of analytic semigroups is provided by squares of group generators.

4.10 Corollary. *Let A be the generator of a strongly continuous group $(T(t))_{t \in \mathbb{R}}$. Then A^2 generates an analytic semigroup $(S(t))_{t \geq 0}$ of angle $\pi/2$. Moreover, if $(T(t))_{t \in \mathbb{R}}$ is bounded this semigroup is given by*

$$S(t) = \frac{1}{\sqrt{4\pi t}} \int_{\mathbb{R}} e^{-s^2/4t}\, T(s)\, ds, \quad t > 0.$$

PROOF. We first show that A^2 generates an analytic semigroup where we assume that $(T(t))_{t \in \mathbb{R}}$ is bounded. For the general case we refer to [Nag86, A-II, Thm. 1.15].

Take some $0 < \delta' < \pi/2$ and $\lambda \in \Sigma_{\pi/2+\delta'}$. Then there exists a square root $re^{i\alpha}$ of λ with $0 < r$ and $|\alpha| < (\pi/2+\delta')/2 < \pi/2$, and we obtain

$$(\lambda - A^2) = (re^{i\alpha} - A)(re^{i\alpha} + A).$$

This implies $\lambda \in \rho(A^2)$ and $R(\lambda, A^2) = R(re^{i\alpha}, A)R(re^{i\alpha}, -A)$. Because A generates a bounded group, there exists a constant $\widetilde{M} \geq 1$ such that

$$\|R(\mu, \pm A)\| \leq \frac{\widetilde{M}}{\operatorname{Re}\mu} \qquad \text{for all } \mu \in \Sigma_{\pi/2}.$$

Consequently, one has

$$\|R(\lambda, A^2)\| \leq \frac{\widetilde{M}^2}{(r\cos\alpha)^2} \leq \frac{1}{r^2}\left(\frac{\widetilde{M}}{\cos\left(\frac{\pi/2+\delta'}{2}\right)}\right)^2$$

$$= \frac{M}{|\lambda|} \qquad \text{for all } \lambda \in \Sigma_{\pi/2+\delta'},$$

and the assertion follows from Propositions 4.3 and 4.4.

The explicit representation of $(S(t))_{t\geq 0}$ can be proved by verifying that the Laplace transform of $S(\cdot)$ is given by $R(\cdot, A^2)$. For the details see [ABHN01, Cor. 3.7.15]. $\qquad\square$

4.11 Example. It is immediately clear from the discussion of the translation groups in Paragraph 2.9 that starting from $Af := f'$ (and appropriate domain) on $\mathrm{C}_0(\mathbb{R})$ or $\mathrm{L}^p(\mathbb{R})$, $1 \leq p < \infty$, the operator

$$A^2 f = f''$$

generates a bounded analytic semigroup.

We now consider the slightly more involved case of several space dimensions; i.e., we consider the spaces $\mathrm{C}_0(\mathbb{R}^n)$ or $\mathrm{L}^p(\mathbb{R}^n)$, $1 \leq p < \infty$. Denote by $(U_i(t))_{t\in\mathbb{R}}$ the strongly continuous group given by

$$(U_i(t)f)(x) := f(x_1, \ldots, x_{i-1}, x_i + t, \ldots, x_n),$$

where $x \in \mathbb{R}^n$, $t \in \mathbb{R}$, and $1 \leq i \leq n$, and let A_i be its generator. Obviously, these semigroups commute as do the resolvents of A_i and hence of A_i^2. Denote by $(T_i(t))_{t\geq 0}$ the semigroup generated by A_i^2, which by Corollary 4.10 has an analytic extension $(T_i(z))_{z\in\Sigma_{\pi/2}}$. These extensions also commute, and therefore

$$T(z) := T_1(z)\cdots T_n(z), \quad z \in \Sigma_{\pi/2},$$

defines a bounded analytic semigroup of angle $\frac{\pi}{2}$. The domain $D(A)$ of its generator A contains $D(A_1^2) \cap \cdots \cap D(A_n^2)$ by Paragraph 2.6. In particular, it contains

$$D_0 := \{f \in X \cap \mathrm{C}^2(\mathbb{R}^n) : D^\alpha f \in X \text{ for every multi-index } \alpha \text{ with } |\alpha| \leq 2\},$$

and for every $f \in D_0$ the generator is given by

$$Af = (A_1^2 + \cdots + A_n^2)f = \sum_{i=1}^n \frac{\partial^2}{\partial x_i^2} f = \Delta f.$$

Finally, we note that $(T(t))_{t\geq 0}$ is given by (2.8) in Paragraph II.2.12.

We close this section by studying the analyticity of multiplication semi-groups and characterize it in terms of the function defining its generator.

4.12 Multiplication Semigroups. As in Definition I.3.3, we consider a multiplication operator

$$M_q : f \mapsto q \cdot f$$

on $X := C_0(\Omega)$ (or, if one prefers, on $L^p(\Omega, \mu)$) for some continuous function $q : \Omega \to \mathbb{C}$. If $\sup_{s \in \Omega} \operatorname{Re} q(s) < \infty$, then

$$T_q(t)f := e^{tq} \cdot f$$

defines a strongly continuous semigroup (see Proposition I.3.5) for which the following holds.

Theorem. *Let $(T_q(t))_{t \geq 0}$ be the strongly continuous multiplication semi-group on X generated by the multiplication operator M_q. Then $(T_q(t))_{t \geq 0}$ is bounded and analytic if and only if the spectrum $\sigma(M_q) = \overline{q(\Omega)}$ satisfies the conditions stated in Theorem 4.6. More precisely, $(T_q(t))_{t \geq 0}$ is bounded analytic of angle δ if and only if*

$$\Sigma_{\delta + \pi/2} \subset \mathbb{C} \setminus \overline{q(\Omega)} = \rho(M_q).$$

PROOF. The condition is necessary by Theorem 4.6. Conversely, if $\Sigma_{\delta + \pi/2}$ is contained in $\mathbb{C} \setminus \overline{q(\Omega)}$, it follows that the functions $q_\pm := e^{\pm i\delta} \cdot q$ still have nonpositive real part. By Proposition I.3.5, this implies that

$$e^{\pm i\delta} \cdot M_q$$

are both generators of bounded strongly continuous semigroups. By Theorem 4.6.(b), this proves that M_q generates a bounded analytic semigroup. □

4.13 Comment. We point out that in most of the above results the density of the domain of A is not needed. In fact, the integral (4.2) exists even for nondensely defined sectorial operators and yields an *analytic semi-group* without, however, the strong continuity in Proposition 4.3.(iv). This is treated in detail in [Lun95].

We close this subsection by adding an arrow to Diagram 1.14 in the case of analytic semigroups.

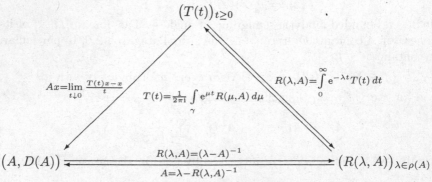

4.14 Exercises. (1) Let X be a Banach space and consider a function $F : \Omega \to \mathcal{L}(X)$ defined on an open set $\Omega \subseteq \mathbb{C}$. Show that the following assertions are equivalent.

(a) $F : \Omega \to \mathcal{L}(X)$ is analytic.

(b) $F(\cdot)x : \Omega \to X$ is analytic for all $x \in X$.

(c) $\langle F(\cdot)x, x' \rangle : \Omega \to \mathbb{C}$ is analytic for all $x \in X$ and $x' \in X'$.

(Hint: Use Cauchy's integral formula and the uniform boundedness principle.)

(2) Show that an analytic semigroup $(T(z))_{z \in \Sigma_\delta \cup \{0\}}$ for every $0 < \delta' < \delta$ is exponentially bounded on $\Sigma_{\delta'}$.

(3) Show that the generator A of an analytic semigroup $(T(z))_{z \in \Sigma_\delta \cup \{0\}}$ coincides with the "complex" generator; i.e.,

$$Ax = \lim_{\Sigma_{\delta'} \ni z \to 0} \frac{T(z)x - x}{z}, \quad D(A) = \left\{ x \in X : \lim_{\Sigma_{\delta'} \ni z \to 0} \frac{T(z)x - x}{z} \text{ exists} \right\}$$

for every $0 < \delta' < \delta$

(4) Show that for an analytic semigroup $(T(z))_{z \in \Sigma_\delta \cup \{0\}}$ on a Banach space X one always has $T(t)X \subset D(A^\infty)$ for all $t > 0$.

(5*) Give a proof of Corollary 4.10 in the case where the group $(T(t))_{t \in \mathbb{R}}$ is not necessarily bounded. (Hint: See [Nag86, A-II, Thm. 1.15].)

(6) Let $(A, D(A))$ be a closed, densely defined linear operator on a Banach space X. If there exist constants $\delta > 0$, $r > 0$, and $M \geq 1$ such that $\Sigma := \{\lambda \in \mathbb{C} : |\lambda| > r \text{ and } |\arg(\lambda)| < \pi/2 + \delta\} \subseteq \rho(A)$ and $\|R(\lambda, A)\| \leq M/|\lambda|$ for all $\lambda \in \Sigma$, then $A - w$ is sectorial for w sufficiently large. In particular, A generates an analytic semigroup.

(7) For an operator $(A, D(A))$ on a Banach space X define on $\mathfrak{X} := X \times X$ the operator matrix

$$\mathcal{A} := \begin{pmatrix} A & A \\ 0 & A \end{pmatrix} \quad \text{with domain} \quad D(\mathcal{A}) := D(A) \times D(A).$$

Show that the following assertions are equivalent.

(i) A generates an analytic semigroup on X.

(ii) $\mathcal{A}$ generates a strongly continuous semigroup on $\mathfrak{X}$.

(iii) $\mathcal{A}$ generates an analytic semigroup on $\mathfrak{X}$.

(Hint: If A generates the semigroup $(T(t))_{t \geq 0}$, then the candidate for the semigroup $(\mathfrak{T}(t))_{t \geq 0}$ generated by $\mathcal{A}$ is given by $\mathfrak{T}(t) = \begin{pmatrix} T(t) & tAT(t) \\ 0 & T(t) \end{pmatrix}$. Now use Theorem 4.6.(c).)

(8) Give an alternative proof of Corollary 4.8 based on the Spectral Theorem I.3.9 and the results on multiplication semigroups from Section I.3.b. (Hint: Observe the theorem in Paragraph 4.12.)

(9) Show that the closure of the operator A in Example 4.9 is self-adjoint.

(10*) Show that for every closed and densely defined operator T on a Hilbert space H the operator T^*T is self-adjoint and positive semidefinite. (Hint: See [Ped89, Thm. 5.1.9].)

(11) Consider the first derivative $D := {}^d/_{dx}$ on $L^2[a,b]$ with the domains $D(D_0) := H_0^1[a,b] := \{f \in H^1[a,b] : f(a) = 0 = f(b)\}$ and $D(D_m) := H^1[a,b]$.

   (i) Show that $(D_0)^* = -D_m$ and $(D_m)^* = -D_0$.

   (ii) Show that $\Delta_D := D_m D_0$ and $\Delta_N := D_0 D_m$ generate bounded analytic semigroups. Write down these operators explicitly. Compare this with Example 4.9. (Hint: Use Exercise (10).)

(12) Show that the operator $A := {}^{d^2}/_{dx^2}$ with domain $D(A) := \{f \in C^2[0,1] : f'(0) = 0 = f'(1)\}$ generates an analytic contraction semigroup $(T(t))_{t \geq 0}$ on $X := C[0,1]$. In addition, show that $T(t)f \geq 0$ for every $f \geq 0$; i.e., $(T(t))_{t \geq 0}$ is positive. (Hint: Observe Paragraphs II.2.11 and II.3.30.)

5. Further Regularity Properties of Semigroups

We have seen in the previous section that requiring the orbit maps $t \mapsto T(t)x$ to be analytic and not just continuous yields a new and important class of semigroups. In this section we discuss some regularity (or smoothness) properties lying between strong continuity and analyticity.

For the first of these concepts we weaken analyticity (on a sector) to differentiability (on an interval).

5.1 Definition. *A strongly continuous semigroup $(T(t))_{t \geq 0}$ on a Banach space X is called* eventually differentiable *if there exists $t_0 \geq 0$ such that the orbit maps $\xi_x : t \mapsto T(t)x$ are differentiable on (t_0, ∞) for every $x \in X$. The semigroup is called* immediately differentiable *if t_0 can be chosen as $t_0 = 0$.*

In analogy to the Hille–Yosida Theorem it is possible to characterize eventual/immediate differentiability by certain estimates on the resolvent of the generator (see [EN00, Sect. II.4.b] for precise statements). More important for both theoretical and practical purposes is the following class of semigroups where strong continuity becomes uniform after some time.

5.2 Definition. *A strongly continuous semigroup $(T(t))_{t \geq 0}$ is called* eventually norm-continuous *if there exists $t_0 \geq 0$ such that the function*

$$t \mapsto T(t)$$

is norm-continuous from (t_0, ∞) into $\mathcal{L}(X)$. The semigroup is called immediately norm-continuous *if t_0 can be chosen to be $t_0 = 0$.*

A Hille–Yosida type characterization of these semigroups is still open. However, immediately norm-continuous semigroups on Hilbert spaces can be characterized via a growth condition on the resolvent of the generator; see [EN00, Thm. II.4.20]. As a necessary condition in Banach spaces we show that the spectrum of their generators is bounded along imaginary lines in the complex plane.

5.3 Theorem. *Let* $(A, D(A))$ *be the generator of an eventually norm-continuous semigroup* $(T(t))_{t\geq 0}$. *Then, for every* $b \in \mathbb{R}$, *the set*

$$\{\lambda \in \sigma(A) : \operatorname{Re}\lambda \geq b\}$$

is bounded.

PROOF. Fix $a \in \mathbb{R}$ larger than the growth bound ω_0 of $(T(t))_{t\geq 0}$. If we show that for every $\varepsilon > 0$, there exist $n \in \mathbb{N}$ and $r_0 \geq 0$ such that

$$\|R(a + ir, A)^n\|^{1/n} < \varepsilon \qquad \text{for all } r \in \mathbb{R} \text{ with } |r| \geq r_0,$$

then the assertion follows from the inequality

$$\operatorname{dist}(a + ir, \sigma(A)) = \frac{1}{\mathrm{r}(R(a + ir, A))}$$

$$\geq \|R(a + ir, A)^n\|^{-1/n} > \frac{1}{\varepsilon}$$

(see Corollary V.1.14).

First, we obtain from the integral representation of the resolvent (see Corollary 1.11) that

$$R(\lambda, A)^{n+1}x = \frac{1}{n!}\int_0^\infty e^{-\lambda t}t^n T(t)x\, dt$$

for all $x \in X$, $n \in \mathbb{N}$, and $\operatorname{Re}\lambda > \omega_0$. Now choose $t_1 > 0$ such that $t \mapsto T(t)$ is norm-continuous on $[t_1, \infty)$ and choose $w \in (\omega_0, a)$, $M \geq 1$ such that $\|T(t)\| \leq Me^{wt}$ for $t \geq 0$. Finally, set $N := M \cdot \int_0^{t_1} e^{-at}e^{wt}\, dt$ and take $\varepsilon > 0$. Then there exist $n \in \mathbb{N}$ and $t_2 > t_1$ such that

$$\frac{N \cdot t_1^n}{n!} < \frac{\varepsilon^{n+1}}{3} \qquad \text{and} \qquad \frac{1}{n!}\int_{t_2}^\infty t^n e^{-at}\|T(t)\|\, dt < \frac{\varepsilon^{n+1}}{3}.$$

Now apply the Riemann–Lebesgue lemma (see Theorem A.20) to the norm-continuous function $t \mapsto t^n e^{-at}T(t)$ on $[t_1, t_2]$ to obtain $r_0 \geq 0$ such that

$$\left\|\frac{1}{n!}\int_{t_1}^{t_2} t^n e^{-irt}e^{-at}T(t)\, dt\right\| < \frac{\varepsilon^{n+1}}{3}$$

whenever $|r| \geq r_0$. The combination of these three estimates yields

$$\left\| R(a+ir,A)^{n+1}x \right\| = \frac{1}{n!} \left\| \int_0^\infty e^{-(a+ir)t} t^n T(t)x \, dt \right\|$$

$$< \frac{1}{n!} \int_0^{t_1} e^{-at} t^n \left\| T(t)x \right\| dt + \frac{1}{n!} \left\| \int_{t_1}^{t_2} t^n e^{-irt} e^{-at} T(t)x \, dt \right\|$$

$$+ \frac{1}{n!} \int_{t_2}^\infty e^{-at} t^n \left\| T(t)x \right\| dt$$

$$< \left(\frac{1}{n!} t_1^n \int_0^{t_1} e^{-at} M e^{wt} \, dt + \frac{2}{3} \varepsilon^{n+1} \right) \cdot \|x\|$$

$$= \left(\frac{1}{n!} t_1^n N + \frac{2}{3} \varepsilon^{n+1} \right) \cdot \|x\| < \varepsilon^{n+1} \cdot \|x\|$$

for all $x \in X$. $\qquad\qquad\qquad\qquad\qquad\qquad\qquad\qquad\qquad\qquad\qquad\qquad\square$

By analyzing the previous proof, one sees that in the case where $(T(t))_{t\geq 0}$ is immediately norm-continuous, one can choose $t_1 = 0$ and $n = 0$. This observation yields the following result.

5.4 Corollary. *If* $(A, D(A))$ *is the generator of an immediately norm-continuous semigroup* $(T(t))_{t\geq 0}$, *then*

$$(5.1) \qquad\qquad\qquad \lim_{r \to \pm\infty} \| R(a+ir,A) \| = 0$$

for all $a > \omega_0$.

Up to now we have classified the semigroups according to smoothness (or regularity) properties of the map $t \mapsto T(t)$. Next, we introduce a property of the semigroup based on the "regularity" of a single operator. We prepare for the definition with the following lemma.

5.5 Lemma. *Let* $(T(t))_{t\geq 0}$ *be a strongly continuous semigroup on a Banach space* X. *If* $T(t_0)$ *is compact for some* $t_0 > 0$, *then* $T(t)$ *is compact for all* $t \geq t_0$, *and the map* $t \mapsto T(t)$ *is norm-continuous on* $[t_0, \infty)$.

PROOF. The first assertion follows immediately from the semigroup law (FE). By Lemma I.1.2, we know that $\lim_{h\to 0} T(s+h)x = T(s)x$ for all $s \geq 0$ uniformly for x in any compact subset K of X. Let U be the unit ball in X. Because $T(t_0)$ is compact, we have that $K := \overline{T(t_0)U}$ is compact, and hence

$$\lim_{s\to t} \big(T(t)x - T(s)x \big) = \lim_{s\to t} \big(T(t-t_0) - T(s-t_0) \big) T(t_0)x = 0$$

for arbitrary $t \geq t_0$ and uniformly for $x \in U$. $\qquad\qquad\qquad\qquad\qquad\square$

5.6 Definition. *A strongly continuous semigroup $\big(T(t)\big)_{t \geq 0}$ is called immediately compact if $T(t)$ is compact for all $t > 0$ and eventually compact if there exists $t_0 > 0$ such that $T(t_0)$ is compact.*

From Lemma 5.5 we obtain that an immediately (eventually) compact semigroup is immediately (eventually) norm-continuous. In addition, one might expect some relation between the compactness of the semigroup and the compactness of the resolvent of its generator. Before introducing the appropriate terminology, we observe that due to the resolvent equation, a resolvent operator is compact for one $\lambda \in \rho(A)$ if and only if it is compact for all $\lambda \in \rho(A)$.

5.7 Definition. *A linear operator A with $\rho(A) \neq \emptyset$ has compact resolvent if its resolvent $R(\lambda, A)$ is compact for one (and hence all) $\lambda \in \rho(A)$.*

Operators with compact resolvent on infinite-dimensional Banach spaces are necessarily unbounded (see Exercise 5.13.(1)). For concrete operators, the following characterization is quite useful.

5.8 Proposition. *Let $\big(A, D(A)\big)$ be an operator on X with $\rho(A) \neq \emptyset$ and take $X_1 := \big(D(A), \|\cdot\|_A\big)$ (see Section 2.c and Exercise 2.22.(1)). Then the following assertions are equivalent.*

(a) *The operator A has compact resolvent.*

(b) *The canonical injection $i : X_1 \hookrightarrow X$ is compact.*

PROOF. Observe that for every $\lambda \in \rho(A)$, the graph norm $\|\cdot\|_A$ is equivalent to the norm
$$\|x\|_\lambda := \|(\lambda - A)x\|$$
(see the proof of Proposition 2.15.(i)). Therefore, the operator
$$R(\lambda, A) : X \to X_1$$
is an isomorphism with continuous inverse $\lambda - A$. The assertion then follows from the following factorization.

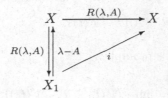

This proposition allows us to prove that differential operators on certain function spaces have compact resolvent. It suffices to apply appropriate *Sobolev embedding theorems*; see, e.g., [RR93, Sect. 6.4]. Here is a very simple example.

5.9 Example. Let Ω be a bounded domain in $\mathbb{R}^n$ and take $X = C_0(\Omega)$. Assume that $(A, D(A))$ is an operator on X such that $D(A)$ is a continuously embedded subspace of the Banach space

$$C_0^1(\Omega) := \{f \in C_0(\Omega) : f \text{ is differentiable and } f' \in C_0(\Omega)\}.$$

By the Arzelà–Ascoli theorem, the injection $i : C_0^1(\Omega) \hookrightarrow C_0(\Omega)$ is compact, whence A has compact resolvent whenever $\rho(A) \neq \emptyset$. See Exercise 5.13.(4) for the analogous L^p-result.

The relation between compactness of the semigroup and the resolvent is not simple. We show first what is not true.

5.10 Examples. (i) Consider the translation semigroup on the Banach space $L^1([0,1] \times [0,1])$ defined by

$$T(t)f(r,s) := \begin{cases} f(r+t,s) & \text{for } r+t \leq 1; \\ 0 & \text{for } r+t > 1. \end{cases}$$

This semigroup is nilpotent, hence eventually compact. However, its generator does not have compact resolvent. (See Exercise 5.13.(3).)

(ii) The generator of the periodic translation group (or rotation group, see Paragraph I.3.18) has compact resolvent. The group, however, does not have any of the smoothness properties defined above.

5.11 Lemma. *Let $(T(t))_{t \geq 0}$ be a strongly continuous semigroup with generator A. Moreover, assume that the map $t \mapsto T(t)$ is norm-continuous at some point $t_0 \geq 0$ and that $R(\lambda, A)T(t_0)$ is compact for some (and hence all) $\lambda \in \rho(A)$. Then the operators $T(t)$ are compact for all $t \geq t_0$.*

PROOF. As usual, we may assume that $0 \in \rho(A)$. For the operators $V(t)$ defined by $V(t)x := \int_0^t T(s)x \, ds$ for $x \in X$ and $t \geq 0$ one has

$$AV(t)x = T(t)x - x \qquad \text{for all } x \in X;$$

hence

$$V(t) = R(0, A)(I - T(t)).$$

The norm continuity for $t \geq t_0$ implies

$$T(t_0) = \lim_{h \downarrow 0} \frac{1}{h}\big(V(t_0 + h) - V(t_0)\big)$$

in operator norm. Because it follows from the assumptions that $V(t_0 + h) - V(t_0)$ is compact for all $h > 0$, this implies that $T(t_0)$ as the norm limit of compact operators is compact as well. $\qquad\square$

5.12 Theorem. *For a strongly continuous semigroup* $(T(t))_{t\geq 0}$ *the following properties are equivalent.*

(a) $(T(t))_{t\geq 0}$ *is immediately compact.*

(b) $(T(t))_{t\geq 0}$ *is immediately norm-continuous, and its generator has compact resolvent.*

PROOF. If $(T(t))_{t\geq 0}$ is immediately compact, it is immediately norm-continuous by Lemma 5.5. Therefore, the integral representation for the resolvent in Theorem 1.10.(i) exists in the norm topology; hence $R(\lambda, A)$ is compact. The converse implication follows from Lemma 5.11. $\qquad\square$

We close these considerations by visualizing the implications between the various classes of semigroups in the following diagram.

analytic $\implies$ immediately differentiable $\implies$ eventually differentiable

$\Downarrow$ $\Downarrow$

(5.2)   immediately norm-continuous $\implies$ eventually norm-continuous

$\Uparrow$ $\Uparrow$

immediately compact $\implies$ eventually compact

It can be shown (using multiplication semigroups and nilpotent semigroups) that all these classes are different; see [EN00, Sect. II.4.e].

5.13 Exercises. (1) Show that a bounded operator $A \in \mathcal{L}(X)$ has compact resolvent if and only if X is finite-dimensional.

(2) Let $(A, D(A))$ be an operator on a Banach space X having compact resolvent and let $B \in \mathcal{L}(X)$ be such that $\rho(A + B) \neq \emptyset$. Then $A + B$ has compact resolvent. (Hint: Use the formula $U^{-1} - V^{-1} = U^{-1}(V - U)V^{-1}$ valid for each pair of invertible operators having the same domain.)

(3) Show that the generator of the semigroup in Example 5.10.(i) does not have compact resolvent. (Hint: Compute the resolvent, using the integral representation (1.14), for functions of the form $f(r, s) := h(r)g(s)$ for $0 \leq r, s \leq 1$ and $h, g \in \mathrm{L}^1[0, 1]$.)

(4) Let $X := \mathrm{L}^p(\Omega)$ for $1 \leq p < \infty$ and a bounded domain $\Omega \subset \mathbb{R}^n$ with smooth boundary $\partial\Omega$. If $(A, D(A))$ is an operator on X satisfying $\rho(A) \neq \emptyset$ and $D(A) \subset \mathrm{W}^{1,p}(\Omega)$, then A has compact resolvent. (Hint: Use Corollary A.11 and Sobolev's embedding theorem.)

6. Well-Posedness for Evolution Equations

Only now we turn our attention to what could have been, in a certain perspective, our starting point: We want to solve a differential equation. More precisely, we look at abstract (i.e., Banach-space-valued) linear initial value problems of the form

$$\text{(ACP)} \qquad \begin{cases} \dot{u}(t) = Au(t) & \text{for } t \geq 0, \\ u(0) = x, \end{cases}$$

where the independent variable t represents time, $u(\cdot)$ is a function with values in a Banach space X, $A : D(A) \subset X \to X$ a linear operator, and $x \in X$ the initial value.

We start by introducing the necessary terminology.

6.1 Definition. (i) *The initial value problem* (ACP) *is called the abstract Cauchy problem associated with* $(A, D(A))$ *and the initial value* x.

(ii) *A function* $u : \mathbb{R}_+ \to X$ *is called a (classical) solution of* (ACP) *if* u *is continuously differentiable,* $u(t) \in D(A)$ *for all* $t \geq 0$, *and* (ACP) *holds.*

If the operator A is the generator of a strongly continuous semigroup, it follows from Lemma 1.3.(ii) that the semigroup yields solutions of the associated abstract Cauchy problem.

6.2 Proposition. *Let* $(A, D(A))$ *be the generator of the strongly continuous semigroup* $(T(t))_{t \geq 0}$. *Then, for every* $x \in D(A)$, *the function*

$$u : t \mapsto u(t) := T(t)x$$

is the unique classical solution of (ACP).

The important point is that (classical) solutions exist if (and, by the definition of $D(A)$, only if) the initial value x belongs to $D(A)$. However, one might substitute the differential equation by an integral equation, thereby obtaining a more general concept of "solution."

6.3 Definition. *A continuous function* $u : \mathbb{R}_+ \to X$ *is called a mild solution of* (ACP) *if* $\int_0^t u(s)\, ds \in D(A)$ *for all* $t \geq 0$ *and*

$$u(t) = A \int_0^t u(s)\, ds + x.$$

It follows from our previous (and elementary) results (use Lemma 1.3.(iv)) that for A being the generator of a strongly continuous semigroup, mild solutions exist for every initial value $x \in X$ and are again given by the semigroup.

6.4 Proposition. Let $(A, D(A))$ be the generator of the strongly continuous semigroup $(T(t))_{t \geq 0}$. Then, for every $x \in X$, the orbit map

$$u : t \mapsto u(t) := T(t)x$$

is the unique mild solution of the associated abstract Cauchy problem (ACP).

PROOF. We only have to show the uniqueness of the zero solution for the initial value $x = 0$. To this end, assume u to be a mild solution of (ACP) for $x = 0$ and take $t > 0$. Then, for each $s \in (0, t)$, we obtain

$$\tfrac{d}{ds} \left(T(t-s) \int_0^s u(r) \, dr \right) = T(t-s)u(s) - T(t-s)A \int_0^s u(r) \, dr = 0.$$

Integration of this equality from 0 to t gives

$$\int_0^t u(r) \, dr = 0, \qquad \text{hence} \qquad u(t) = u(0) = 0$$

as claimed. □

The above two propositions are just reformulations of results on strongly continuous semigroups. They might suggest that the converse holds. The following example shows that this is not true.

6.5 Example. Let $(B, D(B))$ be a closed and *unbounded* operator on X. On the product space $\mathcal{X} := X \times X$, consider the operator $(\mathcal{A}, D(\mathcal{A}))$ written in matrix form as

$$\mathcal{A} := \begin{pmatrix} 0 & B \\ 0 & 0 \end{pmatrix} \qquad \text{with domain} \qquad D(\mathcal{A}) := X \times D(B).$$

Then $t \mapsto u(t) := \left(\begin{smallmatrix} x+tBy \\ y \end{smallmatrix} \right)$ is the unique solution of (ACP) associated with $\mathcal{A}$ for every $\left(\begin{smallmatrix} x \\ y \end{smallmatrix} \right) \in D(\mathcal{A})$. However, the operator $\mathcal{A}$ does not generate a strongly continuous semigroup, because for every $\lambda \in \mathbb{C}$, one has

$$(\lambda - \mathcal{A})D(\mathcal{A}) = \left\{ \left(\begin{smallmatrix} \lambda x - By \\ \lambda y \end{smallmatrix} \right) : x \in X, y \in D(B) \right\} \subset X \times D(B) \neq \mathcal{X},$$

and hence $\sigma(\mathcal{A}) = \mathbb{C}$.

We now show which properties of the solutions $u(\cdot, x)$ or of the operator $(A, D(A))$ have to be added in order to characterize semigroup generators. To this end we first consider the following existence and uniqueness condition.

(EU)

For every $x \in D(A)$, there exists a unique solution $u(\cdot, x)$ of (ACP).

To this condition we add the nonemptiness of the resolvent set $\rho(A)$ in (b) or some continuous dependence of the solutions upon the initial values in (c) below.

6.6 Theorem. *Let $A : D(A) \subset X \to X$ be a closed operator. Then for the associated abstract Cauchy problem*

(ACP)
$$\begin{cases} \dot{u}(t) = Au(t) & \text{for } t \geq 0, \\ u(0) = x \end{cases}$$

the following properties are equivalent.

(a) *A generates a strongly continuous semigroup.*

(b) *A satisfies (EU) and $\rho(A) \neq \emptyset$.*

(c) *A satisfies (EU), has dense domain, and for every sequence $(x_n)_{n \in \mathbb{N}} \subset D(A)$ satisfying $\lim_{n \to \infty} x_n = 0$, one has $\lim_{n \to \infty} u(t, x_n) = 0$ uniformly in compact intervals $[0, t_0]$.*

PROOF. From the basic properties of semigroup generators and, in particular, Proposition 6.2, it follows that (a) implies (b) and (c).

For (b) $\Rightarrow$ (c) we first show that for all $x \in X$ there exists a unique mild solution of (ACP). By assumption (EU), for each $y := R(\lambda, A)x \in D(A)$, $\lambda \in \rho(A)$, there is a classical solution $u(\cdot, y)$ with initial value y. Then it is easy to see that $v(t) := (\lambda - A)u(t, y)$ defines a mild solution for the initial value $x = (\lambda - A)y$. In order to prove uniqueness let $u(\cdot)$ be a mild solution to the initial value 0. Then $v(t) := \int_0^t u(s)\, ds$ is the classical solution for the initial value 0, hence $v = 0$ and consequently $u = 0$ as well. Because for every $x \in X$ we have, by definition of mild solutions, $\int_0^t u(s, x)\, ds \in D(A)$, we obtain from

$$\frac{1}{t} \int_0^t u(s, x)\, ds \to u(0, x) = x \qquad \text{as } t \downarrow 0$$

that $D(A)$ is dense. The uniqueness of the mild solutions implies linearity of $u(t, x)$ in x. In order to show the continuous dependence upon the initial data, we consider for fixed $t_0 > 0$ the linear map

$$\Phi : X \to \mathrm{C}([0, t_0], X), \qquad x \mapsto u(\cdot, x)$$

and show that Φ is closed. In fact, if $x_n \to x$ and $\Phi(x_n) \to y \in \mathrm{C}([0, t_0], X)$ we obtain for $t \in [0, t_0]$,

$$D(A) \ni \int_0^t u(s, x_n)\, ds \to \int_0^t y(s)\, ds$$

and

$$A \int_0^t u(s, x_n)\, ds = u(t, x_n) - x_n \to y(t) - x.$$

Hence, by the closedness of A we conclude that $\int_0^t y(s)\, ds \in D(A)$ and $A \int_0^t y(s)\, ds = y(t) - x$. Consequently $y(\cdot)$ is the unique mild solution of (ACP) with initial value x if for $t > t_0$ we define $y(t) := u(t - t_0, y(t_0))$. This shows $y(t) = u(t, x)$ for $t \in [0, t_0]$ and $\Phi(x) = y$. By the closed graph theorem Φ is continuous; hence for $x_n \to 0$ we obtain $u(t, x_n) \to 0$ in $\mathrm{C}([0, t_0], X)$, i.e., uniformly for t in the compact interval $[0, t_0]$. This proves (c).

(c) $\Rightarrow$ (a). The assumption implies the existence of bounded operators $T(t) \in \mathcal{L}(X)$ defined by

$$T(t)x := u(t, x)$$

for each $x \in D(A)$. Moreover, we claim that $\sup_{0 \le t \le 1} \|T(t)\| < \infty$. By contradiction, assume that there exists a sequence $(t_n)_{n \in \mathbb{N}} \subset [0, 1]$ such that $\|T(t_n)\| \to \infty$ as $n \to \infty$. Then we can choose $x_n \in D(A)$ such that $\lim_{n \to \infty} x_n = 0$ and $\|T(t_n)x_n\| \ge 1$. Because $u(t_n, x_n) = T(t_n)x_n$, this contradicts the assumption in (c), and therefore $\|T(t)\|$ is uniformly bounded for $t \in [0, 1]$. Now, $t \mapsto T(t)x$ is continuous for each x in the dense domain $D(A)$, and we obtain continuity for each $x \in X$ by Lemma I.1.2.

Finally, the uniqueness of the solutions implies $T(t + s)x = T(t)T(s)x$ for each $x \in D(A)$ and all $t, s \ge 0$. Thus $(T(t))_{t \ge 0}$ is a strongly continuous semigroup on X. Its generator $(B, D(B))$ certainly satisfies $A \subset B$. Moreover, the semigroup $(T(t))_{t \ge 0}$ leaves $D(A)$ invariant, which, by Proposition 1.7, is a core of B. Because A is closed, we obtain $A = B$. $\qquad\square$

Observe that (a) and (b) in the previous theorem imply that $D(A)$ is dense, whereas this property cannot be omitted in (c). Take the restriction $\tilde{A}$ of a closed operator A to the domain $D(\tilde{A}) := \{0\}$.

Intuitively, property (c) expresses what we expect for a "well-posed" problem and its solutions:

existence + uniqueness + continuous dependence on the data.

Therefore we introduce a name for this property.

6.7 Definition. *The abstract Cauchy problem*

(ACP)
$$\begin{cases} \dot{u}(t) = Au(t) & \text{for } t \ge 0, \\ u(0) = x \end{cases}$$

associated with a closed operator $A : D(A) \subset X \to X$ is called well-posed if condition (c) in Theorem 6.6 holds.

With this terminology, we can rephrase Theorem 6.6.

6.8 Corollary. *For a closed operator $A : D(A) \subset X \to X$, the associated abstract Cauchy problem (ACP) is well-posed if and only if A generates a strongly continuous semigroup on X.*

Once we agree on the well-posedness concept from Definition 6.7, strongly continuous semigroups emerge as the perfect tool for the study of abstract Cauchy problems (ACP). In addition, this explains why in this manuscript we

- Study semigroups systematically and only then
- Solve Cauchy problems.

However, we have to point out that our definition of "well-posedness" is not the only possible one. In particular, in many situations arising from physically perfectly "well-posed" problems one does not obtain a semigroup on a given Banach space. We refer to [ABHN01], [Are87], [deL94], and [Neu88] for weaker concepts of "well-posedness" and show here how to produce, for the same operator by simply varying the underlying Banach space, a series of different "well-posedness" properties.

6.9 Example. Consider the left translation group $(T(t))_{t\in\mathbb{R}}$ on $L^1(\mathbb{R})$ with generator $Af := f'$ and $D(A) := W^{1,1}(\mathbb{R})$. Decompose this space as

$$L^1(\mathbb{R}) = L^1(\mathbb{R}_-) \oplus L^1(\mathbb{R}_+),$$

and take any translation-invariant Banach space Y continuously embedded in $L^1(\mathbb{R}_-)$. Then the part $A_|$ of A in $X := Y \oplus L^1(\mathbb{R}_+)$ has domain $D(A_|) := \{f \in W^{1,1}(\mathbb{R}) : f'_{|\mathbb{R}_-} \in Y\}$. The abstract Cauchy problem

$$\dot{u}(t) = A_| u(t) \quad \text{for } t \geq 0,$$
$$u(0) = f \in D(A_|) \subset X$$

formally has the solution $t \mapsto u(t) := T(t)f$ with $\big(T(t)f\big)(s) = f(s+t)$, $s \in \mathbb{R}$. This is a classical solution if and only if $u(t) \in D(A_|)$ for all $t \geq 0$. As concrete examples, we suggest taking $Y := W^{n,1}(\mathbb{R}_-)$, or even $Y := \{0\}$, and leave the details as Exercise 6.10.

6.10 Exercise. On $X := W^{1,1}(\mathbb{R}_-) \oplus L^1(\mathbb{R}_+)$, consider the operator $Af := f'$ with $D(A) := \{(f,g) \in W^{2,1}(\mathbb{R}_-) \oplus W^{1,1}(\mathbb{R}_+) : f(0) = g(0)\}$.

  (i) Which conditions of Generation Theorem 3.8 are fulfilled by the operator $(A, D(A))$? (Hint: Use (2.1) in Section 2.b to represent $R(\lambda, A)$.)

  (ii) Show that the abstract Cauchy problem associated with $(A, D(A))$ has a classical solution only for initial values $(f, g) \in D(A)$ such that $g \in W^{2,1}(\mathbb{R}_+)$.

  (iii) Replace $W^{2,1}(\mathbb{R}_-)$ by other translation-invariant Banach function spaces on $\mathbb{R}_-$ and find the initial values for which classical solutions exist.

Chapter III

Perturbation of Semigroups

The verification of the conditions in the various generation theorems from Sections II.3–4 is not an easy task and for many important operators cannot be performed in a direct way. Therefore, one tries to build up the given operator (and its semigroup) from simpler ones. Perturbation and approximation are the standard methods for this approach and are discussed in this and the next chapter.

1. Bounded Perturbations

In many concrete situations, the evolution equation (or the associated linear operator) is given as a (formal) sum of several terms having different physical meaning and different mathematical properties. Although the mathematical analysis may be easy for each single term, it is not at all clear what happens after the formation of sums. In the context of generators of semigroups we take this as our point of departure.

1.1 Problem. Let $A : D(A) \subseteq X \to X$ be the generator of a strongly continuous semigroup $(T(t))_{t \geq 0}$ and consider a second operator $B : D(B) \subseteq X \to X$. Find conditions such that the sum $A + B$ generates a strongly continuous semigroup $(S(t))_{t \geq 0}$.

We say that the generator A is *perturbed* by the operator B or that B is a *perturbation* of A. However, before answering the above problem, we have to realize that—at this stage—the sum $A + B$ is defined as

$$(A + B)x := Ax + Bx$$

only for

$$x \in D(A + B) := D(A) \cap D(B),$$

a subspace that might be trivial in general. To emphasize this and other difficulties caused by the addition of unbounded operators, we first discuss some examples.

1.2 Examples. (i) Let $(A, D(A))$ be an unbounded generator of a strongly continuous semigroup. If we take $B := -A$, then the sum $A + B$ is the zero operator, defined on the dense subspace $D(A)$, hence not closed.

If we take $B := -2A$, then the sum is

$$A + B = -A \quad \text{with domain} \quad D(A + B) = D(A),$$

which is a generator only if A generates a strongly continuous group (see Paragraph II.3.11).

(ii) Let $A : D(A) \subseteq X \to X$ be an unbounded generator of a strongly continuous semigroup and take an isomorphism $S \in \mathcal{L}(X)$ such that $D(A) \cap S(D(A)) = \{0\}$. Then $B := SAS^{-1}$ is a generator as well (see Paragraph II.2.1), but $A + B$ is defined only on $D(A + B) = D(A) \cap D(B) = D(A) \cap S(D(A)) = \{0\}$.

A concrete example for this situation is given on $X := C_0(\mathbb{R}_+)$ by

$$Af := f' \quad \text{with its canonical domain } D(A) := C_0^1(\mathbb{R}_+)$$

and

$$Sf := q \cdot f$$

for some continuous, positive function q such that q and q^{-1} are bounded and nowhere differentiable. Defining the operator B as

$$Bf := q \cdot (q^{-1} \cdot f)' \quad \text{on} \quad D(B) := \{f \in X : q^{-1} \cdot f \in D(A)\},$$

we obtain that the sum $A + B$ is defined only on $\{0\}$.

The above examples show that the addition of unbounded operators is a delicate operation and should be studied carefully. We start with a situation in which we avoid the pitfall due to the differing domains of the operators involved. More precisely, we assume one of the two operators to be bounded.

1.3 Bounded Perturbation Theorem. *Let $(A, D(A))$ be the generator of a strongly continuous semigroup $(T(t))_{t \geq 0}$ on a Banach space X satisfying*

$$\|T(t)\| \leq Me^{wt} \text{ for all } t \geq 0$$

and some $w \in \mathbb{R}, M \geq 1$. If $B \in \mathcal{L}(X)$, then

$$C := A + B \qquad \text{with} \qquad D(C) := D(A)$$

generates a strongly continuous semigroup $(S(t))_{t \geq 0}$ satisfying

$$\|S(t)\| \leq M e^{(w + M\|B\|)t} \qquad \text{for all } t \geq 0.$$

PROOF. In the first and essential step, we assume $w = 0$ and $M = 1$. Then $\lambda \in \rho(A)$ for all $\lambda > 0$, and $\lambda - C$ can be decomposed as

$$(1.1) \qquad \lambda - C = \lambda - A - B = (I - BR(\lambda, A))(\lambda - A).$$

Because $\lambda - A$ is bijective, we conclude that $\lambda - C$ is bijective; i.e., $\lambda \in \rho(C)$, if and only if

$$I - BR(\lambda, A)$$

is invertible in $\mathcal{L}(X)$. If this is the case, we obtain

$$(1.2) \qquad R(\lambda, C) = R(\lambda, A)[I - BR(\lambda, A)]^{-1}.$$

Now choose $\operatorname{Re} \lambda > \|B\|$. Then $\|BR(\lambda, A)\| \leq \|B\|/\operatorname{Re}\lambda < 1$ by Generation Theorem II.3.5.(c), and hence $\lambda \in \rho(C)$ with

$$(1.3) \qquad R(\lambda, C) = R(\lambda, A) \sum_{n=0}^{\infty} (BR(\lambda, A))^n.$$

We now estimate

$$\|R(\lambda, C)\| \leq \frac{1}{\operatorname{Re}\lambda} \cdot \frac{1}{1 - \|B\|/\operatorname{Re}\lambda} = \frac{1}{\operatorname{Re}\lambda - \|B\|}$$

for all $\operatorname{Re}\lambda > \|B\|$ and obtain from Corollary II.3.6 that C generates a strongly continuous semigroup $(S(t))_{t \geq 0}$ satisfying

$$\|S(t)\| \leq e^{\|B\|t} \text{ for } t \geq 0.$$

For general $w \in \mathbb{R}$ and $M \geq 1$, we first do a rescaling (see Paragraph II.2.2) to obtain $w = 0$. As in Lemma II.3.10, we then introduce a new norm

$$\|\|x\|\| := \sup_{t \geq 0} \|T(t)x\|$$

on X. This norm satisfies

$$\|x\| \leq \|\|x\|\| \leq M \|x\|,$$

makes $(T(t))_{t \geq 0}$ a contraction semigroup, and yields

$$\|\|Bx\|\| \leq M \|B\| \cdot \|x\| \leq M \|B\| \cdot \|\|x\|\|$$

for all $x \in X$. By part one of this proof, the sum $C = A + B$ generates a strongly continuous semigroup $(S(t))_{t \geq 0}$ satisfying the estimate

$$\|\|S(t)\|\| \leq e^{\|\|B\|\|t} \leq e^{M\|B\|t}.$$

Hence

$$\|S(t)x\| \leq \|\|S(t)x\|\| \leq e^{M\|B\|t} \|\|x\|\| \leq M e^{M\|B\|t} \|x\|$$

for all $t \geq 0$, which is the assertion for $w = 0$. $\qquad \square$

The identities (1.1) and (1.3) are not only the basis of this proof, but are also the key to many more perturbation results. Here, we apply them to extend the above theorem to certain unbounded perturbations. For this purpose we use the terminology of Sobolev towers from Section II.2.c.

For sufficiently large λ, the generator A of a strongly continuous semigroup, and an operator $B \in \mathcal{L}(X)$, the operators

$$\left(I - BR(\lambda, A)\right) \quad \text{and} \quad \left(I - BR(\lambda, A)\right)^{-1} = \sum_{n=0}^{\infty} \left(BR(\lambda, A)\right)^n$$

are isomorphisms of the Banach space X. Therefore, for large λ, the 1-norms with respect to $\lambda - A$ and $\lambda - A - B$, i.e.,

and
$$\|x\|_1^A := \|(\lambda - A)x\|$$
$$\|x\|_1^{A+B} := \|(\lambda - A - B)x\| = \left\|(I - BR(\lambda, A))(\lambda - A)x\right\|,$$

are equivalent on $X_1 := D(A) = D(A + B)$.

Similarly, the corresponding (-1)-norms

and
$$\|x\|_{-1}^A := \|R(\lambda, A)x\|$$
$$\|x\|_{-1}^{A+B} := \|R(\lambda, A + B)x\|$$

are equivalent on X (use the identity

(1.4) $R(\lambda, A) = [I + R(\lambda, A + B)B]^{-1} R(\lambda, A + B)$

and (1.2)), and hence the Sobolev spaces X_{-1}^A for A and X_{-1}^{A+B} for $A + B$ from Definition II.2.17 coincide.

Because we know from Theorem 1.3 that $A + B$ is a generator, we obtain the following conclusion.

1.4 Corollary. *Let* $(A, D(A))$ *be the generator of a strongly continuous semigroup on a Banach space* X_0 *and take* $B \in \mathcal{L}(X_0)$. *Then the operator*

$$A + B \qquad \text{with domain } D(A + B) := D(A)$$

is a generator, and the Sobolev spaces

$$X_i^A \qquad \text{and} \qquad X_i^{A+B}$$

corresponding to A *and* $A + B$, *respectively, coincide for* $i = -1, 0, 1$.

We show in Exercise 1.13.(5) that this result is optimal in the sense that in general, only these three "floors" of the corresponding Sobolev towers coincide. Here, the above corollary immediately yields a first perturbation result for operators that are not bounded on the given Banach space.

1.5 Corollary. *Let* $(A, D(A))$ *be the generator of a strongly continuous semigroup on the Banach space* X_0. *If* B *is a bounded operator on* $X_1^A := \left(D(A), \|\cdot\|_1\right)$, *then* $A + B$ *with domain* $D(A + B) = D(A)$ *generates a strongly continuous semigroup on* X_0.

PROOF. Consider the restriction A_1 of A as a generator on X_1^A. Then $A_1 + B$ generates a strongly continuous semigroup on X_1^A by Theorem 1.3. This perturbed semigroup can be extended to its extrapolation space $(X_1^A)_{-1}^{A_1+B}$, which by Corollary 1.4 coincides with the extrapolation space $(X_1^A)_{-1}^{A_1}$. However, this is the original Banach space X_0. The generator of the extended semigroup on X_0 is the continuous extension of $A_1 + B$, hence is $A + B$. $\qquad\square$

1.6 Example. Take $Af := f'$ on $X := C_0(\mathbb{R})$ with domain $C_0^1(\mathbb{R})$. For some $h \in C_0^1(\mathbb{R})$ define the operator B by

$$Bf := f'(0) \cdot h, \ f \in C_0^1(\mathbb{R}).$$

Then B is unbounded on X but bounded on $D(A) = C_0^1(\mathbb{R})$, and hence $A + B$ is a generator on X.

Returning to Theorem 1.3, we recall that we have the series representation (1.3) for the resolvent $R(\lambda, A + B)$ of the perturbed operator $A + B$, whereas for the new semigroup $(S(t))_{t \geq 0}$ we could prove only its existence. In order to prepare for a representation formula for this new semigroup, we show first that it has to satisfy an integral equation.

1.7 Corollary. *Consider two strongly continuous semigroups* $(T(t))_{t \geq 0}$ *with generator* A *and* $(S(t))_{t \geq 0}$ *with generator* C *on the Banach space* X *and assume that*

$$C = A + B$$

for some bounded operator $B \in \mathcal{L}(X)$. *Then*

(IE) $$S(t)x = T(t)x + \int_0^t T(t - s)BS(s)x \, ds$$

holds for every $t \geq 0$ *and* $x \in X$.

PROOF. Take $x \in D(A)$ and consider the functions

$$[0, t] \ni s \mapsto \xi_x(s) := T(t - s)S(s)x \in X.$$

Because $D(A) = D(C)$ is invariant under both semigroups, it follows that $\xi_x(\cdot)$ is continuously differentiable (use Lemma A.19) with derivative

$$\tfrac{d}{ds}\xi_x(s) = T(t - s)CS(s)x - T(t - s)AS(s)x = T(t - s)BS(s)x.$$

This implies

$$S(t)x - T(t)x = \xi_x(t) - \xi_x(0) = \int_0^t \xi_x'(s) \, ds = \int_0^t T(t - s)BS(s)x \, ds.$$

Finally, the density of $D(A)$ and the boundedness of the operators involved yield that this integral equation holds for all $x \in X$. $\qquad\square$

If we replace the above functions ξ_x by

$$\eta_x(s) := S(s)T(t-s)x$$

and use the same arguments, we obtain the analogous integral equation

$$\text{(IE}^*)\qquad\qquad S(t)x = T(t)x + \int_0^t S(s)BT(t-s)x\,ds$$

for $x \in X$ and $t \geq 0$.

Both equations (IE) and (IE*) are frequently called the *variation of parameter formula* for the perturbed semigroup.

Instead of solving the integral equation (IE) by the usual fixed point method, we use an abstract and seemingly more complicated approach. However, it has the advantage of working equally well for important unbounded perturbations. For these more general perturbations we refer to [EN00, Sect. III.3].

In order to explain our method we rewrite (IE) in operator form and introduce the operator-valued function space

$$\mathfrak{X}_{t_0} := \mathrm{C}\big([0, t_0], \mathcal{L}_s(X)\big)$$

of all continuous functions from $[0, t_0]$ into $\mathcal{L}_s(X)$; i.e., $F \in \mathfrak{X}_{t_0}$ if and only if $F(t) \in \mathcal{L}(X)$ and $t \mapsto F(t)x$ is continuous for each $x \in X$. This space becomes a Banach space for the norm

$$\|F\|_\infty := \sup_{s \in [0, t_0]} \|F(s)\|, \qquad F \in \mathfrak{X}_{t_0}$$

(see Proposition A.4). We now define a "Volterra-type" operator on it.

1.8 Definition. *Let $\big(T(t)\big)_{t \geq 0}$ be a strongly continuous semigroup on X and take $B \in \mathcal{L}(X)$. For any $t_0 > 0$, we call the operator defined by*

$$VF(t)x := \int_0^t T(t-s)BF(s)x\,ds$$

for $x \in X$, $F \in \mathrm{C}\big([0, t_0], \mathcal{L}_s(X)\big)$ and $0 \leq t \leq t_0$ the associated abstract Volterra operator.

The following properties of V should be no surprise to anyone familiar with Volterra operators in the scalar-valued situation. In fact, the proof is just a repetition of the estimates there and is omitted (see Exercise 1.13.(1)).

1.9 Lemma. *The abstract Volterra operator V associated with the strongly continuous semigroup $(T(t))_{t \geq 0}$ and the bounded operator $B \in \mathcal{L}(X)$ is a bounded operator in $C([0, t_0], \mathcal{L}_s(X))$ and satisfies*

$$(1.5) \qquad \qquad \|V^n\| \leq \frac{(M \|B\| t_0)^n}{n!}$$

for all $n \in \mathbb{N}$ and with $M := \sup_{s \in [0,t_0]} \|T(s)\|$. In particular, for its spectral radius we have

$$(1.6) \qquad \qquad \mathrm{r}(V) = 0.$$

From this last assertion it follows that the resolvent of V at $\lambda = 1$ exists and is given by the Neumann series; i.e.,

$$R(1, V) = (I - V)^{-1} = \sum_{n=0}^{\infty} V^n.$$

We now turn back to our integral equation (IE), which becomes, in terms of our Volterra operator, the equation

$$T(\cdot) = (I - V)S(\cdot)$$

for the functions $T(\cdot), S(\cdot) \in C([0, t_0], \mathcal{L}_s(X))$. Therefore,

$$(1.7) \qquad \qquad S(\cdot) = R(1, V)T(\cdot) = \sum_{n=0}^{\infty} V^n T(\cdot),$$

where the series converges in the Banach space $C([0, t_0], \mathcal{L}_s(X))$. Rewriting (1.7) for each $t \geq 0$, we obtain the following representation for the semigroup $(S(t))_{t \geq 0}$. This *Dyson–Phillips series* was found by F.J. Dyson in his work [Dys49] on quantum electrodynamics and then by R.S. Phillips in his first systematic treatment [Phi53] of perturbation theory for semigroups.

1.10 Theorem. *The strongly continuous semigroup $(S(t))_{t \geq 0}$ generated by $C := A + B$, where A is the generator of $(T(t))_{t \geq 0}$ and $B \in \mathcal{L}(X)$, can be obtained as*

$$(1.8) \qquad \qquad S(t) = \sum_{n=0}^{\infty} S_n(t),$$

where $S_0(t) := T(t)$ and

$$(1.9) \qquad \qquad S_{n+1}(t) := V S_n(t) = \int_0^t T(t - s) B S_n(s) \, ds.$$

Here, the series (1.8) converges in the operator norm on $\mathcal{L}(X)$ and, because we may choose t_0 in Lemma 1.9 arbitrarily large, uniformly on compact intervals of $\mathbb{R}_+$. In contrast, the operators $S_{n+1}(t)$ in (1.9) are defined by an integral defined in the strong operator topology.

The Dyson–Phillips series and the integral equation (IE) from Corollary 1.7 are very useful when we want to compare qualitative properties of the two semigroups. Here is a simple example of such a comparison.

1.11 Corollary. *Let $(T(t))_{t\geq 0}$ and $(S(t))_{t\geq 0}$ be two strongly continuous semigroups, where the generator of $(S(t))_{t\geq 0}$ is a bounded perturbation of the generator of $(T(t))_{t\geq 0}$. Then*

$$(1.10) \qquad\qquad \|T(t) - S(t)\| \leq t\,M$$

for $t \in [0,1]$ and some constant M.

PROOF. From the integral equation (IE), we obtain

$$\|T(t)x - S(t)x\| \leq \int_0^t \|T(t-s)BS(s)x\|\,ds$$

$$\leq t \sup_{r\in[0,1]} \|T(r)\| \sup_{s\in[0,1]} \|S(s)\| \cdot \|B\| \cdot \|x\|$$

for all $x \in X$ and $t \in [0,1]$. $\qquad\qquad\qquad\qquad\qquad\qquad\qquad\qquad \square$

Conversely, one can show that an estimate such as (1.10) for the difference of two semigroups implies a close relation between their generators; see [EN00, Sect. III.3.b] for more details.

In the final result of this section we show that analyticity of a semigroup is preserved under bounded perturbation.

1.12 Proposition. *Let $(T(t))_{t\geq 0}$ be an analytic semigroup with generator A on the Banach space X and take $B \in \mathcal{L}(X)$. Then also the semigroup $(S(t))_{t\geq 0}$ generated by $A + B$ is analytic.*

PROOF. The assertion is a consequence of Theorem II.4.6.(b) and the Bounded Perturbation Theorem 1.3. $\qquad\qquad\qquad\qquad\qquad\qquad\qquad\qquad \square$

1.13 Exercises.

(1) Prove Lemma 1.9. (Hint: Show that V is a linear operator on the space $C\big([0, t_0], \mathcal{L}_s(X)\big)$. Then use induction on $n \in \mathbb{N}$ to verify (1.5). Equation (1.6) then follows from the Hadamard formula $r(V) = \lim_{n\to\infty} \|V^n\|^{1/n}$ for the spectral radius.)

(2) Let $(T(t))_{t\geq 0}$ be a strongly continuous semigroup with generator A on the Banach space X and $(S(t))_{t\geq 0}$ the semigroup with generator $A + B$ for $B \in \mathcal{L}(X)$.

(i) Show that instead of the integral equations (IE) and (IE*) we can write

$$S(t)x = T(t)x + \int_0^t T(s)BS(t-s)x\,ds$$

and

$$S(t)x = T(t)x + \int_0^t S(t-s)BT(s)x\,ds$$

for $x \in X$, $t \geq 0$.

(ii) Define a Volterra operator V^* based on the integral equation (IE*) and show that

$$S(t) = \sum_{n=0}^{\infty} S_n^*(t),$$

where $S_0^*(t) := T(t)$ and

$$S_{n+1}^*(t)x := V^*S_n^*(t)x = \int_0^t S_n^*(s)BT(t-s)x\,ds$$

for $x \in X$, $t \geq 0$.

(3) Show that the variation of parameter formulas (IE) and (IE*) also holds for perturbations $B \in \mathcal{L}(X_1)$ and $x \in D(A)$.

(4) Take the Banach space $X := C_0(\mathbb{R})$ and a function $q \in C_b(\mathbb{R})$, and define

$$T(t)f(s) := e^{\int_{s-t}^s q(\tau)\,d\tau} \cdot f(s-t)$$

for $s \in \mathbb{R}$, $t \geq 0$, and $f \in X$.

 (i) Show that $(T(t))_{t\geq 0}$ is a strongly continuous semigroup on X.

 (ii) Compute its generator.

 (iii) What happens if the function q is taken in $L^\infty(\mathbb{R})$?

 (iv) Can one allow the function q to be unbounded such that $(T(t))_{t\geq 0}$ still becomes a strongly continuous semigroup on X?

 (v) Assume that

$$\cdot u(t,s) := e^{\int_s^t q(\tau)\,d\tau}$$

 is uniformly bounded for $s, t \in \mathbb{R}$. Show that the semigroup $(T(t))_{t\geq 0}$ is similar to the left translation semigroup on X. (Hint: Use the multiplication operator $M_{u(\cdot,0)}$ as a similarity transformation.)

(5) Let $(A, D(A))$ be an unbounded generator on the Banach space X. On the product space $\mathcal{X} := X \times X$ define

$$\mathcal{A} := \begin{pmatrix} A & 0 \\ 0 & I \end{pmatrix} \quad \text{with domain} \quad D(\mathcal{A}) := D(A) \times X$$

and the bounded operator $\mathcal{B} := \begin{pmatrix} 0 & I \\ 0 & 0 \end{pmatrix}$.

 (i) Show that

$$\mathcal{X}_2^{\mathcal{A}+\mathcal{B}} = D((\mathcal{A}+\mathcal{B})^2) = \left\{ \begin{pmatrix} x \\ y \end{pmatrix} \in D(A) \times X : Ax + y \in D(A) \right\},$$

 hence is different from $\mathcal{X}_2^{\mathcal{A}} = D(A^2) \times X$.

 (ii) Prove a similar statement for the extrapolation spaces of order 2. (Hint: Consider $\mathcal{A} := \begin{pmatrix} A & 0 \\ 0 & I \end{pmatrix}$ with domain $D(\mathcal{A}) := D(A) \times X$ and $\mathcal{B} := \begin{pmatrix} 0 & 0 \\ I & 0 \end{pmatrix}$.)

This confirms the statement following Corollary 1.4.

2. Perturbations of Contractive and Analytic Semigroups

As already shown in Example 1.2, addition of two unbounded operators is a very delicate operation and can destroy many of the good properties the single operators may have. This is, in part, due to the fact that the "naive" domain

$$D(A + B) := D(A) \cap D(B)$$

for the sum $A + B$ of the operators $(A, D(A))$ and $(B, D(B))$ can be too small (see Example 1.2.(ii)). In order to avoid this pitfall, we assume in this section that the perturbing operator B behaves well with respect to the unperturbed operator A. More precisely, we assume the following property.

2.1 Definition. *Let $A : D(A) \subset X \to X$ be a linear operator on the Banach space X. An operator $B : D(B) \subset X \to X$ is called (relatively) A-bounded if $D(A) \subseteq D(B)$ and if there exist constants $a, b \in \mathbb{R}_+$ such that*

$$(2.1) \qquad \qquad \|Bx\| \leq a\,\|Ax\| + b\,\|x\|$$

for all $x \in D(A)$. The A-bound of B is

$$a_0 := \inf\{a \geq 0 : \text{there exists } b \in \mathbb{R}_+ \text{ such that } (2.1) \text{ holds}\}.$$

Before applying this notion to the perturbation problem for generators we discuss a concrete example.

2.2 Example. For an interval $I \subseteq \mathbb{R}$ we consider on $X := \mathrm{L}^p(I)$, $1 \leq p \leq \infty$, the operators

$$A := \tfrac{d^2}{dx^2}, \qquad D(A) := W^{2,p}(I),$$
$$B := \tfrac{d}{dx}, \qquad D(B) := W^{1,p}(I).$$

Proposition. *The operator B is A-bounded with A-bound $a_0 = 0$.*

PROOF. We choose an arbitrary bounded interval $J := (\alpha, \beta) \subset I$, and set $\varepsilon := \beta - \alpha$,

$$J_1 := (\alpha, \alpha + \varepsilon/3), \quad J_2 := (\alpha + \varepsilon/3, \beta - \varepsilon/3), \quad J_3 := (\beta - \varepsilon/3, \beta).$$

Then, for all $f \in D(A)$ and $s \in J_1$, $t \in J_3$ there exists, by the mean value theorem, a point $x_0 = x_0(s, t) \in J$ such that

$$f'(x_0) = \frac{f(t) - f(s)}{t - s}.$$

Using this and $t - s \geq \varepsilon/3$, we obtain

$$(2.2) \quad |f'(x)| = \left| f'(x_0) + \int_{x_0}^{x} f''(y)\,dy \right| \leq \frac{3}{\varepsilon}\big(|f(s)| + |f(t)|\big) + \int_J |f''(y)|\,dy$$

for all $x \in J$, $s \in J_1$, and $t \in J_3$. If we denote by $\|\cdot\|_{p,J}$ the p-norm in $L^p(J)$ and integrate inequality (2.2) on both sides with respect to $s \in J_1$ and $t \in J_3$, we obtain

$$\frac{\varepsilon^2}{9}|f'(x)| \leq \int_{J_1} |f(s)|\,ds + \int_{J_3} |f(t)|\,dt + \frac{\varepsilon^2}{9}\int_J |f''(y)|\,dy$$

$$\leq \|f\|_{1,J} + \frac{\varepsilon^2}{9}\|f''\|_{1,J}$$

$$\leq \varepsilon^{1/q}\|f\|_{p,J} + \frac{\varepsilon^{2+1/q}}{9}\|f''\|_{p,J},$$

where we used Hölder's inequality for $1/p + 1/q = 1$. From this estimate, it then follows that

$$\frac{\varepsilon^2}{9}\|f'\|_{p,J} \leq \varepsilon^{1/p}\varepsilon^{1/q}\|f\|_{p,J} + \varepsilon^{1/p}\frac{\varepsilon^{2+1/q}}{9}\|f''\|_{p,J}$$

$$= \varepsilon\|f\|_{p,J} + \frac{\varepsilon^3}{9}\|f''\|_{p,J};$$

i.e.,

$$\|f'\|_{p,J} \leq \frac{9}{\varepsilon}\|f\|_{p,J} + \varepsilon\|f''\|_{p,J}.$$

By splitting the interval I in finitely or countable many (depending on whether I is bounded) disjoint subintervals I_n, $n \in N \subseteq \mathbb{N}$, of length ε, we obtain by Minkowski's inequality

$$\|Bf\|_p = \Big(\sum_{n \in N} \|f'\|_{p,I_n}^p\Big)^{1/p} \leq \frac{9}{\varepsilon}\Big(\sum_{n \in N} \|f\|_{p,I_n}^p\Big)^{1/p} + \varepsilon\Big(\sum_{n \in N} \|f''\|_{p,I_n}^p\Big)^{1/p}$$

$$= \frac{9}{\varepsilon}\|f\|_p + \varepsilon\|Af\|_p.$$

Because we can choose $\varepsilon > 0$ arbitrarily small, the proof of our claim is complete. $\qquad\qquad\qquad\qquad\qquad\qquad\qquad\qquad\qquad\qquad\qquad\qquad\quad\Box$

Note that from (2.2) we immediately obtain an analogous result for the second and first derivative on $X := C_0(I)$. More precisely, if $I \subseteq \mathbb{R}$ is an arbitrary interval and

$$A := \tfrac{d^2}{dx^2}, \qquad D(A) := \big\{f \in C_0^2(I) : f', f'' \in C_0(I)\big\},$$

$$B := \tfrac{d}{dx}, \qquad D(B) := \big\{f \in C_0^1(I) : f' \in C_0(I)\big\},$$

then B is A-bounded with A-bound $a_0 = 0$.

We now return to the abstract situation and observe that for an A-bounded operator B the sum $A + B$ is defined on $D(A + B) := D(A)$. However, many desirable properties may get lost.

2.3 Examples. Take $A : D(A) \subset X \to X$ to be the generator of a strongly continuous semigroup such that $\sigma(A) = \mathbb{C}_- := \{z \in \mathbb{C} : \operatorname{Re} z \leq 0\}$ (e.g., take the generator of the translation semigroup on $C_0(\mathbb{R}_+)$; cf. Example V.1.25.(i)).

(i) If we take $B := \alpha A$ for $\alpha \in \mathbb{C}$, then $A + B$ is not a generator for $\alpha \in \mathbb{C} \setminus (-1, \infty)$, and is not even closed for $\alpha = -1$.

(ii) Consider the new operator $\mathcal{A} := \left(\begin{smallmatrix} A & 0 \\ 0 & A \end{smallmatrix} \right)$ with $D(\mathcal{A}) := D(A) \times D(A)$ on the product space $\mathcal{X} := X \times X$. If we take

$$\mathcal{B}_1 := \begin{pmatrix} 0 & \varepsilon A \\ 0 & 0 \end{pmatrix} \qquad \text{with} \qquad D(\mathcal{B}_1) := X \times D(A),$$

then $\mathcal{A} + \mathcal{B}_1$ is not a generator for every $0 \neq \varepsilon \in \mathbb{C}$ (use Exercise II.4.14.(7)). For

$$\mathcal{B}_2 := \begin{pmatrix} 0 & -A \\ A & -2A \end{pmatrix} \qquad \text{with} \qquad D(\mathcal{B}_2) := D(A) \times D(A),$$

the sum $\mathcal{A} + \mathcal{B}_2$ is not closed, and its closure is not a generator.

We now proceed with a series of lemmas showing which assumptions on the unperturbed operator A and the A-bounded perturbation B are needed such that the sum $A + B$

- Is closed,
- Has nonempty resolvent set, and, finally,
- Becomes the generator of a strongly continuous semigroup.

2.4 Lemma. If $(A, D(A))$ is closed and $(B, D(B))$ is A-bounded with A-bound $a_0 < 1$, then

$$(A + B, D(A))$$

is a closed operator.

PROOF. Because an operator is closed if and only if its domain is a Banach space for the graph norm, it suffices to show that the graph norm $\|\cdot\|_{A+B}$ of $A + B$ is equivalent to the graph norm $\|\cdot\|_A$ of A. By assumption, there exist constants $0 \leq a < 1$ and $0 < b$ such that

$$\|Bx\| \leq a \|Ax\| + b \|x\|$$

for all $x \in D(A)$. Therefore, one has

$$\|Ax\| = \|(A + B)x - Bx\| \leq \|(A + B)x\| + a \|Ax\| + b \|x\|$$

and, consequently,

$$-b \|x\| + (1 - a) \|Ax\| \leq \|(A + B)x\| \leq \|Ax\| + \|Bx\| \leq (1 + a) \|Ax\| + b \|x\|.$$

This yields the estimate

$$b \|x\| + (1 - a) \|Ax\| \leq \|(A + B)x\| + 2b \|x\| \leq (1 + a) \|Ax\| + 3b \|x\|,$$

proving the equivalence of the two graph norms. $\square$

2.5 Lemma. *Let $(A, D(A))$ be closed with $\rho(A) \neq \emptyset$ and assume $(B, D(B))$ to be A-bounded with constants $0 \leq a, b$ in estimate (2.1). If $\lambda_0 \in \rho(A)$ and*

(2.3) $$c := a\,\|AR(\lambda_0, A)\| + b\,\|R(\lambda_0, A)\| < 1,$$

then $A + B$ is closed, and one has $\lambda_0 \in \rho(A + B)$ with

(2.4). $$\|R(\lambda_0, A + B)\| \leq (1 - c)^{-1}\,\|R(\lambda_0, A)\|.$$

PROOF. As in the proof of Theorem 1.3, we decompose $\lambda_0 - A - B$ as the product

$$\lambda_0 - A - B = [I - BR(\lambda_0, A)](\lambda_0 - A)$$

and observe that $\lambda_0 - A$ is a bijection from $D(A)$ onto X, whereas $BR(\lambda_0, A)$ is bounded on X (use Exercise 2.15.(1.i)). If we show that $\|BR(\lambda_0, A)\| < 1$, we obtain that $[I - BR(\lambda_0, A)]$, hence $\lambda_0 - A - B$, is invertible with inverse

(2.5) $$R(\lambda_0, A + B) = R(\lambda_0, A) \sum_{n=0}^{\infty} (BR(\lambda_0, A))^n$$

satisfying

$$\|R(\lambda_0, A + B)\| \leq \|R(\lambda_0, A)\|\,(1 - \|BR(\lambda_0, A)\|)^{-1}.$$

To that purpose, take $x \in X$ and use (2.1) to obtain

$$\|BR(\lambda_0, A)x\| \leq a\,\|AR(\lambda_0, A)x\| + b\,\|R(\lambda_0, A)x\|$$
$$\leq (a\,\|AR(\lambda_0, A)\| + b\,\|R(\lambda_0, A)\|) \cdot \|x\|,$$

whence $\|BR(\lambda_0, A)\| \leq c < 1$ by assumption (2.3). $\qquad\square$

In the last preparatory lemma, we consider operators satisfying a Hille–Yosida type estimate for the resolvent (but not for all its powers as required in Generation Theorem II.3.8). It is shown that this class of operators remains invariant under A-bounded perturbations with *small* A-bound.

2.6 Lemma. *Let $(A, D(A))$ be an operator whose resolvent exists for all*

$$0 \neq \lambda \in \overline{\Sigma}_\delta := \{z \in \mathbb{C} : |\arg z| \leq \delta\}$$

and satisfies

$$\|R(\lambda, A)\| \leq \frac{M}{|\lambda|}$$

for some constants $\delta \geq 0$ and $M \geq 1$. Moreover, assume $(B, D(B))$ to be A-bounded with A-bound

$$a_0 < \frac{1}{M + 1}.$$

Then there exist constants $r \geq 0$ and $\widetilde{M} \geq 1$ such that

$$\overline{\Sigma}_\delta \cap \{z \in \mathbb{C} : |z| > r\} \subset \rho(A + B) \qquad \text{and} \qquad \|R(\lambda, A + B)\| \leq \frac{\widetilde{M}}{|\lambda|}$$

for all $\lambda \in \overline{\Sigma}_\delta \cap \{z \in \mathbb{C} : |z| > r\}$.

PROOF. Choose constants $0 \le a < 1/_{M+1}$ and $0 \le b$ satisfying the estimate (2.1). From this we obtain

$$c := a\,\|AR(\lambda, A)\| + b\,\|R(\lambda, A)\|$$
$$= a\,\|\lambda R(\lambda, A) - I\| + b\,\|R(\lambda, A)\|$$
$$\le a(M+1) + \frac{bM}{|\lambda|} < 1,$$

whenever $|\lambda| > r := {}^{bM}/_{(1-a(M+1))}$. Choosing $\widetilde{M} := {}^{M}/_{1-c}$ the assertion now follows from Lemma 2.5. $\qquad\square$

If we now assume the constants to be $M = \widetilde{M} = 1$, we obtain a perturbation theorem for generators of contraction semigroups. The surprising fact is that the relative bound a_0, which in Lemma 2.6 and for $M = 1$ should be smaller than $1/2$, must only satisfy $a_0 < 1$. The dissipativity (see Definition II.3.13) of the operators involved makes this possible.

2.7 Theorem. Let $(A, D(A))$ be the generator of a contraction semigroup and assume $(B, D(B))$ to be dissipative and A-bounded with A-bound $a_0 < 1$. Then $(A + B, D(A))$ generates a contraction semigroup.

PROOF. We first assume that $a_0 < 1/2$. From the criterion in Proposition II.3.23, it follows that the sum of a generator of a contraction semigroup and a dissipative operator is again dissipative. Therefore, $A + B$ is a densely defined, dissipative operator, and by Theorem II.3.15 it suffices to find $\lambda_0 > 0$ such that $\lambda_0 \in \rho(A+B)$. This, however, follows from Lemma 2.6 by choosing $\delta = 0$; i.e., $\overline{\Sigma}_\delta = [0, \infty)$.

In order to extend this to the case $0 \le a_0 < 1$, we define for $0 \le \alpha \le 1$ the operators

$$C_\alpha := A + \alpha B, \quad D(C_\alpha) := D(A).$$

Then, for $x \in D(A)$, one has

$$\|Bx\| \le a\,\|Ax\| + b\,\|x\| \le a(\|C_\alpha x\| + \alpha\,\|Bx\|) + b\,\|x\|$$
$$\le a\,\|C_\alpha x\| + a\,\|Bx\| + b\,\|x\|;$$

and hence

$$\|Bx\| \le \frac{a}{1-a}\,\|C_\alpha x\| + \frac{b}{1-a}\,\|x\| \qquad \text{for all} \qquad 0 \le \alpha \le 1.$$

Next, we choose $k \in \mathbb{N}$ such that

$$c := \frac{a}{k(1-a)} < \frac{1}{2}.$$

Then the estimate

$$\left\|\tfrac{1}{k}Bx\right\| \leq c\|C_\alpha x\| + \frac{b}{k(1-a)}\|x\|$$

shows that for each $0 \leq \alpha \leq 1$ the operator $^1\!/_k B$ is C_α-bounded with C_α-bound less than $^1\!/_2$. As observed above, this implies that

$$C_\alpha + \tfrac{1}{k}B = A + (\alpha + \tfrac{1}{k})B$$

generates a contraction semigroup whenever $C_\alpha = A + \alpha B$ does. However, A generates a contraction semigroup, hence $A + ^1\!/_k B$ does. Repeating this argument k times shows that $(A + {}^{(k-1)}\!/_k B) + {}^1\!/_k B = A + B$ generates a contraction semigroup as claimed. □

In the limit case, i.e., if one has $a = 1$ in the estimate (2.1), the result remains essentially true, provided that the adjoint of B is densely defined.

2.8 Corollary. *Let $\big(A, D(A)\big)$ be the generator of a contraction semigroup on X and assume that $\big(B, D(B)\big)$ is dissipative, A-bounded, and satisfies*

(2.6) $$\|Bx\| \leq \|Ax\| + b\|x\|$$

for all $x \in D(A)$ and some constant $b \geq 0$. If the adjoint B' is densely defined on X', then the closure of $\big(A + B, D(A)\big)$ generates a contraction semigroup on X.

PROOF. The sum $A + B$ remains dissipative and densely defined. Hence, by the Lumer–Phillips Theorem II.3.15, it suffices to show that $\mathrm{rg}(I - A - B)$ is dense in X.

Choose $y' \in X'$ satisfying $\langle z, y' \rangle = 0$ for all $z \in \mathrm{rg}(I - A - B)$ and then $y \in X$ such that $\langle y, y' \rangle = \|y'\|$. The perturbed operators $A + \varepsilon B$ with domain $D(A)$ are generators of contraction semigroups for each $0 \leq \varepsilon < 1$ by Theorem 2.7. From Generation Theorem II.3.5 we obtain $1 \in \rho(A + \varepsilon B)$, and hence there exists a unique $x_\varepsilon \in D(A)$ such that $\|x_\varepsilon\| \leq \|y\|$ and

$$x_\varepsilon - (A + \varepsilon B)x_\varepsilon = y.$$

From the estimate

$$\begin{aligned}
\|Bx_\varepsilon\| &\leq \|Ax_\varepsilon\| + b\|x_\varepsilon\| \\
&\leq \|(A + \varepsilon B)x_\varepsilon\| + \varepsilon\|Bx_\varepsilon\| + b\|x_\varepsilon\| \\
&\leq \|x_\varepsilon - y\| + \varepsilon\|Bx_\varepsilon\| + b\|x_\varepsilon\|
\end{aligned}$$

we deduce

(2.7) $$(1 - \varepsilon)\|Bx_\varepsilon\| \leq \|x_\varepsilon - y\| + b\|x_\varepsilon\| \leq (2 + b)\|y\|$$

for all $0 \leq \varepsilon < 1$.

We now use the density of $D(B')$. In fact, for $z' \in D(B')$ it follows that

$$|\langle(1-\varepsilon)Bx_\varepsilon, z'\rangle| \leq (1-\varepsilon)\|x_\varepsilon\| \cdot \|B'z'\|$$
$$\leq (1-\varepsilon)\|y\| \cdot \|B'z'\|,$$

and hence

$$\lim_{\varepsilon\uparrow 1}\langle(1-\varepsilon)Bx_\varepsilon, z'\rangle = 0.$$

Our assumption and the norm boundedness of the elements $(1-\varepsilon)Bx_\varepsilon$ (see (2.7)) then implies

$$\lim_{\varepsilon\uparrow 1}\langle(1-\varepsilon)Bx_\varepsilon, y'\rangle = 0,$$

and therefore

$$\|y'\| = \langle y, y'\rangle = \langle x_\varepsilon - (A+\varepsilon B)x_\varepsilon, y'\rangle$$
$$= \langle(1-\varepsilon)Bx_\varepsilon, y'\rangle + \langle(I - A - B)x_\varepsilon, y'\rangle$$
$$\to 0 \text{ as } \varepsilon\uparrow 1.$$

From the Hahn–Banach theorem we then conclude that $\mathrm{rg}(I - A - B)$ is dense in X. $\qquad\square$

If X is reflexive, the adjoint of every closable, densely defined operator is again densely defined on the dual space (see Proposition A.14). Because densely defined, dissipative operators are always closable (see Proposition II.3.14.(iv)), we arrive at the following result.

2.9 Corollary. *Let $(A, D(A))$ be the generator of a contraction semigroup on a reflexive Banach space X. If $(B, D(B))$ is dissipative, A-bounded, and satisfies the estimate (2.6), then the closure of $(A + B, D(A))$ generates a contraction semigroup on X.*

In order to obtain the previous perturbation results, we used Lemma 2.6 and could estimate *only* the resolvent of the perturbed operator $A + B$ and *not all* its powers. Due to the Lumer–Phillips Theorem II.3.15, this was sufficient if A was the generator of a contraction semigroup and B was dissipative. There is, however, another case where an estimate on the resolvent alone forces an operator to generate a semigroup. Such a result has been proved in Theorem II.4.6 for analytic semigroups and now easily leads to another perturbation theorem.

2.10 Theorem. *Let the operator $(A, D(A))$ generate an analytic semigroup $(T(z))_{z\in\Sigma_\delta\cup\{0\}}$ on a Banach space X. Then there exists a constant $\alpha > 0$ such that $(A + B, D(A))$ generates an analytic semigroup for every A-bounded operator B having A-bound $a_0 < \alpha$.*

PROOF. We first assume that $(T(z))_{z \in \Sigma_\delta \cup \{0\}}$ is bounded, which means, by Theorem II.4.6, that A is sectorial. Hence, there exist constants $\delta' \in (0, \pi/2]$ and $C \geq 1$ such that for every

$$0 \neq \lambda \in \overline{\Sigma}_{\pi/2+\delta'} := \left\{ z \in \mathbb{C} : |\arg z| \leq \frac{\pi}{2} + \delta' \right\}$$

we have

$$\lambda \in \rho(A) \qquad \text{and} \qquad \|R(\lambda, A)\| \leq \frac{C}{|\lambda|}.$$

If we define $\alpha := 1/C+1$, we can apply Lemma 2.6 and obtain constants $r \geq 0$ and $M \geq 1$ such that

$$\Sigma := \overline{\Sigma}_{\pi/2+\delta'} \cap \{ z \in \mathbb{C} : |z| > r \} \subseteq \rho(A+B)$$

and

$$\|R(\lambda, A+B)\| \leq \frac{M}{|\lambda|} \qquad \text{for all } \lambda \in \Sigma.$$

By Exercise II.4.14.(6), this implies that $A+B$ generates an analytic semigroup, proving the assertion in the bounded case.

In order to treat the general case, we take $w \in \mathbb{R}$ and conclude from

$$\|Bx\| \leq a \, \|Ax\| + b \, \|x\| \leq a \, \|(A-w)x\| + (aw + b) \, \|x\|$$

for all $x \in D(A)$ that B is also $A - w$ bounded with the same bound a_0. Because the semigroup generated by $A - w$ is analytic and bounded in Σ_δ for w sufficiently large, the first part of the proof implies that $A + B - w$; hence $A + B$ generates an analytic semigroup. $\qquad \square$

2.11 Examples. (i) In Example II.4.9 we showed that the second derivative

$$A := \frac{d^2}{dx^2}, \qquad D(A) := \left\{ f \in H^2[0,1] : f(0) = f(1) = 0 \right\}$$

generates an analytic semigroup on $H := L^2[0,1]$. Because by Example 2.2 the first derivative d/dx with maximal domain $H^1[0,1]$ is A-bounded with A-bound $a_0 = 0$, we conclude by Theorem 2.10 and Exercise 2.15.(1) that for all $B \in \mathcal{L}\left(H^1[0,1], L^2[0,1]\right)$ the operator

$$C := A + B, \qquad D(C) := D(A)$$

generates an analytic semigroup on H.

(ii) As in Paragraph II.2.12, we consider the diffusion semigroup on $L^1(\mathbb{R}^n)$ given by

$$\left(T(t)f\right)(s) := (4\pi t)^{-n/2} \int_{\mathbb{R}^n} e^{-|s-r|^2/4t} f(r) \, dr =: \int_{\mathbb{R}^n} K_t(s-r)f(r) \, dr.$$

It is generated by the closure of the Laplacian Δ defined on the Schwartz space $\mathscr{S}(\mathbb{R}^n)$. In Example II.4.11 we have seen that $(T(t))_{t \geq 0}$ is a bounded analytic semigroup. As a perturbation we take the multiplication operator

$$(M_q f)(s) := q(s)f(s) \quad \text{for} \quad f \in D(M_q) := \left\{ g \in L^1(\mathbb{R}^n) : qg \in L^1(\mathbb{R}^n) \right\}$$

induced by a function $q \in L^p(\mathbb{R}^n)$ for $p > \max\{1, n/2\}$.

We now show that $B := M_q$ is Δ-bounded with Δ-bound zero. To this end, we estimate for $f \in L^1(\mathbb{R}^n)$ and $\lambda > 0$

$$\|BR(\lambda, \Delta)f\|_{L^1(\mathbb{R}^n)} = \left\| B \int_0^\infty e^{-\lambda t} T(t) f \, dt \right\|_{L^1(\mathbb{R}^n)}$$

$$\leq \int_{\mathbb{R}^n} |q(s)| \int_0^\infty e^{-\lambda t} \int_{\mathbb{R}^n} K_t(s-r)|f(r)| \, dr \, dt \, ds$$

$$= \int_{\mathbb{R}^n} |f(r)| \int_0^\infty e^{-\lambda t} \int_{\mathbb{R}^n} K_t(s-r)|q(s)| \, ds \, dt \, dr$$

$$\leq \|f\|_{L^1(\mathbb{R}^n)} \sup_{r \in \mathbb{R}^n} \int_0^\infty e^{-\lambda t} \int_{\mathbb{R}^n} K_t(s-r)|q(s)| \, ds \, dt$$

$$\leq \|f\|_{L^1(\mathbb{R}^n)} \cdot \|q\|_{L^p(\mathbb{R}^n)} \int_0^\infty e^{-\lambda t} \left(\int_{\mathbb{R}^n} K_t(s)^{p'} ds \right)^{1/p'} dt$$

with $1/p + 1/p' = 1$, where we used Fubini's theorem and Hölder's inequality. It is now easy to verify that $\|K_t\|_{L^{p'}(\mathbb{R}^n)} = ct^{-n/2p}$ for a constant $c > 0$. Hence, we conclude that $D(\Delta) \subset D(B)$ and

$$\|Bf\|_{L^1(\mathbb{R}^n)} \leq c\|q\|_{L^p(\mathbb{R}^n)} \int_0^\infty e^{-\lambda t} t^{-n/2p} dt \, \|(\lambda - \Delta)f\|_{L^1(\mathbb{R}^n)}$$

$$=: a_\lambda \|(\lambda - \Delta)f\|_{L^1(\mathbb{R}^n)} \leq \lambda a_\lambda \|f\|_{L^1(\mathbb{R}^n)} + a_\lambda \|\Delta f\|_{L^1(\mathbb{R}^n)}$$

for all $f \in D(\Delta)$. Because $a_\lambda := c\|q\|_{L^p(\mathbb{R}^n)} \int_0^\infty e^{-\lambda t} t^{-n/2p} dt$ converges to zero as $\lambda \to \infty$, this proves our claim. Thus, by Theorem 2.10, the operator $(\Delta + M_q, D(\Delta))$ generates an analytic semigroup for every $q \in L^p(\mathbb{R}^n)$ with $p > \max\{1, n/2\}$.

We now introduce a class of operators always having A-bound zero with respect to a given operator A.

2.12 Definition. *Let $(A, D(A))$ be a closed operator on a Banach space X. An operator $(B, D(B))$ is called (relatively) A-compact if $D(A) \subseteq D(B)$ and $B : X_1 \to X$ is compact, where X_1 denotes the domain $D(A)$ equipped with the graph norm $\|\cdot\|_A$.*

If $\rho(A)$ is nonempty, one can show that an A-bounded operator B is A-compact if and only if $BR(\lambda, A) \in \mathcal{L}(X)$ is compact for some/all $\lambda \in \rho(A)$, see Exercise 2.15.(1). Because compact operators are "small" in some sense, one might hope that an A-compact operator is A-bounded with bound 0. This is, however, not true in general (see [Hes70]), and we need some additional conditions to ensure it.

2.13 Lemma. *Let $(A, D(A))$ be a closed operator on a Banach space X and assume $(B, D(B))$ to be A-compact. If*

  (i) *A is a generator and X is reflexive, or if*

  (ii) *$(B, D(B))$ is closable in X,*

then B is A-bounded with A-bound $a_0 = 0$.

PROOF. (i) For $0 < \mu$ sufficiently large and $x \in D(A)$, we write

$$
\begin{aligned}
Bx &= BR(\mu, A)(\mu - A)x \\
&= \mu BR(\mu, A)x - BR(\mu, A)AR(\lambda, A)(\lambda - A)x \\
&= \mu BR(\mu, A)x - BR(\mu, A)A\lambda R(\lambda, A)x + BR(\mu, A)AR(\lambda, A)Ax
\end{aligned}
$$

for all $\lambda > \mu$. Because the operators appearing in the first two terms are bounded, it suffices to show that for each $\varepsilon > 0$ there exist $\lambda > \mu$ such that

$$
\begin{aligned}
\varepsilon &> \|BR(\mu, A)AR(\lambda, A)\| = \|BR(\mu, A)(\lambda R(\lambda, A) - I)\| \\
&= \|(\lambda R(\lambda, A') - I)(BR(\mu, A))'\|.
\end{aligned}
$$

If X is reflexive, then the adjoint operator A' is again a generator (see Paragraph I.1.13). Therefore, by Lemma II.3.4, $\lambda R(\lambda, A')$ converges strongly to I as $\lambda \to \infty$. Moreover, $BR(\mu, A)$ and therefore its adjoint $(BR(\mu, A))'$ are compact operators. Combining these two properties and applying Proposition A.3 yields

$$
\lim_{\lambda \to \infty} \|(\lambda R(\lambda, A') - I)(BR(\mu, A))'\| = 0.
$$

(ii) Assume the assertion to be false. Then there exists $\varepsilon > 0$ and a sequence $(x_n)_{n \in \mathbb{N}} \subset D(A)$ such that

$$
(2.8) \qquad \|Bx_n\| > \varepsilon\|Ax_n\| + n\|x_n\| \quad \text{for all } n \in \mathbb{N}.
$$

For $y_n := {}^{x_n}/_{\|x_n\|_A}$ this means

$$
(2.9) \qquad \|By_n\| > \varepsilon\|Ay_n\| + n\|y_n\|.
$$

Because $\|y_n\|_A = 1$ for all $n \in \mathbb{N}$ and because B is A-compact, there exists a subsequence $(z_n)_{n \in \mathbb{N}}$ of $(y_n)_{n \in \mathbb{N}}$ such that $(Bz_n)_{n \in \mathbb{N}}$ converges in X. Moreover, $\|z_n\| < \|Bz_n\|/n$ and $(Bz_n)_{n \in \mathbb{N}}$ is bounded in X; hence $\lim_{n \to \infty} \|z_n\| = 0$. Using the assumption that B is closable, this implies $\lim_{n \to \infty} \|Bz_n\| = 0$ and therefore $\lim_{n \to \infty} \|Az_n\| = 0$ by (2.9). This, however, yields a contradiction, in as much as

$$
1 = \|z_n\|_A = \|z_n\| + \|Az_n\| \qquad \text{for all } n \in \mathbb{N}.
$$

$\square$

We again combine this lemma with our previous perturbation results.

2.14 Corollary. Let $(A, D(A))$ be the generator of a strongly continuous semigroup on a Banach space X and assume the operator $(B, D(B))$ to be A-compact. If X is reflexive or if B is closable, then the following assertions are true.

(i) If A and B are dissipative, then $(A + cB, D(A))$ generates a contraction semigroup on X for all $c \in \mathbb{R}_+$.

(ii) If the semigroup generated by A is analytic, then $(A + cB, D(A))$ generates an analytic semigroup on X for all $c \in \mathbb{C}$.

One can show that Corollary 2.14.(ii) holds without the extra assumptions that B is closable or that X is reflexive (see [DS88]).

2.15 Exercises. (1) Let A be an operator on a Banach space X having nonempty resolvent set $\rho(A)$. Show that for a linear operator $B : D(A) \to X$ the following assertions are true.

    (i) B is A-bounded if and only if $B \in \mathcal{L}(X_1, X)$ if and only if $BR(\lambda, A) \in \mathcal{L}(X)$ for some/all $\lambda \in \rho(A)$.

    (ii) B is A-compact if and only if $BR(\lambda, A)$ is compact for some/all $\lambda \in \rho(A)$.

(2) Let $(A, D(A))$ be the generator of a strongly continuous semigroup $(T(t))_{t \geq 0}$ on a Banach space X and let $(B, D(B))$ be a closed operator on X. If there exists

    (i) A $(T(t))_{t \geq 0}$-invariant dense subspace $D \subset D(A) \cap D(B)$ such that the map $t \mapsto BT(t)x$ is continuous for all $x \in D$ and

    (ii) Constants $t_0 > 0$ and $q \geq 0$ such that

$$\int_0^{t_0} \|BT(t)x\| \, dt \leq q\|x\| \qquad \text{for all } x \in D,$$

then B is A-bounded with A-bound less than or equal to q. (Hint: Use the formula

$$(2.10) \quad BR(\lambda, A)x = \sum_{n=0}^{\infty} e^{-\lambda n t_0} \int_0^{t_0} e^{-\lambda r} BT(r)T(nt_0)x \, dr, \qquad x \in D,$$

in order to show that $BR(\lambda, A)$ is bounded on D. Then it follows from Proposition A.6.(i) and Theorem A.10 that $D(A) \subseteq D(B)$. Finally, take in (2.10) the limit as $\lambda \to \infty$ to estimate the A-bound of B.)

(3) Assume $(A, D(A))$ to generate an analytic semigroup of angle $\delta \in (0, \pi]$. Show that in the situation of Theorem 2.10 the semigroup generated by $A + B$ is analytic of angle at least δ.

(4) Take the operators $Af := f''$ and $Bf := f'$ with maximal domains in $X := C_0(\mathbb{R})$. Show that $A + \alpha B - \beta$ generates a contraction semigroup for $\alpha \in \mathbb{R}, \beta \geq 0$. Can one replace the constants α and β by certain functions?

(5) Let $(A, D(A))$ be the generator of a contraction semigroup on the Banach space X.

    (i) If $(B, D(B))$ is dissipative, then $(A + B, D(A) \cap D(B))$ is again dissipative.

    (ii) If B is dissipative and bounded, then $(A + B, D(A))$ generates a contraction semigroup.

(6) Take $X := c_0$ and define $A(x_n) := (inx_n)$ with domain $D(A)$ consisting of all finite sequences.

(i) Show that the closure $\overline{A}$ of A generates a group of isometries on X.

(ii) Construct a different semigroup generator $(B, D(B))$ on X such that A and B coincide on $D(A)$.

(7) Let B be an operator on a Banach space X such that there exists a sequence $(\lambda_n)_{n \in \mathbb{N}} \subset \rho(B)$ satisfying $\lim_{n \to \infty} \|R(\lambda_n, B)\| = 0$. Show that B is $A := B^2$-bounded with A-bound $a_0 = 0$. (Hint: Compute $B^2 R(\lambda, B)$ using the formula $BR(\lambda, B) = \lambda R(\lambda, B) - I$.)

Chapter IV

Approximation of Semigroups

1. Trotter–Kato Approximation Theorems

Approximation, besides perturbation, is the other main method used to study a complicated operator and the semigroup it generates. We already encountered an example for such an approximation procedure in our proof of the Generation Theorem II.3.5. For an operator $(A, D(A))$ on X satisfying the Hille–Yosida conditions, we defined the (bounded) *Yosida approximants*[1]

$$A_n := nAR(n, A), \ n \in \mathbb{N}$$

(see Chapter II, (3.7)) generating the (uniformly continuous) semigroups $\left(e^{tA_n}\right)_{t \geq 0}$. Using the fact that $A_n \to A$ pointwise on $D(A)$ as $n \to \infty$ (see Lemma II.3.4.(ii)), we could show that the semigroups converge as well; that is,

$$e^{tA_n} \to T(t) \quad \text{as} \quad n \to \infty.$$

In this section we study this situation systematically and consider the three objects *semigroup*, *generator*, and *resolvent*, visualized by the triangle

---

[1] In this context, this notation should not cause any confusion with the operators A_n induced on the abstract Sobolev spaces from Section II.2.c.

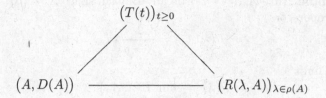

from Chapter II. We then try to show that the convergence at one "vertex" implies convergence in the two other "vertices." That the truth is not as simple is shown by the following example.

1.1 Example. On the Banach space $X := c_0$, we take the multiplication operator

$$A(x_k) := (\mathrm{i}kx_k)$$

with domain

$$D(A) := \big\{(x_k) \in c_0 : (\mathrm{i}kx_k) \in c_0\big\}.$$

As we know from Example I.3.7.(iii), it generates the strongly continuous semigroup $(T(t))_{t\geq 0}$ given by

$$T(t)(x_k) = (\mathrm{e}^{\mathrm{i}kt}x_k), \quad t \geq 0.$$

Perturbing A by the bounded operators

$$P_n(x_k) := (0, \ldots, nx_n, 0, \ldots),$$

we obtain new operators

$$A_n := A + P_n.$$

Each A_n is the generator of a strongly continuous semigroup $(T_n(t))_{t\geq 0}$ (use Theorem III.1.3), and for each $x = (x_k) \in D(A)$, we have

$$\|A_n x - Ax\| = \|P_n x\| = n|x_n| \to 0.$$

However, the semigroups $(T_n(t))_{t\geq 0}$ do not converge. In fact, one has

$$T_n(t)x = (\mathrm{e}^{\mathrm{i}t}x_1, \mathrm{e}^{2\mathrm{i}t}x_2, \ldots, \mathrm{e}^{(\mathrm{i}n+n)t}x_n, \mathrm{e}^{(n+1)\mathrm{i}t}x_{n+1}, \ldots)$$

and therefore

$$\|T_n(t)\| \geq \mathrm{e}^{nt} \quad \text{for } n \in \mathbb{N} \text{ and } t \geq 0.$$

By the uniform boundedness principle, this implies that there exists $x \in X$ such that $(T_n(t)x)_{n\in\mathbb{N}}$ does not converge.

The example shows that the convergence of the generators (pointwise on the domain of the limit operator) does not imply convergence of the corresponding semigroups. Another unpleasant phenomenon may happen for a converging sequence of resolvent operators.

1.2 Example. Take $A_n := -n \cdot I$ on any Banach space $X \neq \{0\}$. Then the resolvent operators

$$R(\lambda, A_n) = \frac{1}{\lambda + n} \cdot I$$

and their limit

$$R(\lambda) := \lim_{n \to \infty} R(\lambda, A_n)$$

exist for all $\operatorname{Re} \lambda > 0$. However, the limit $R(\lambda)$ is equal to zero, hence cannot be the resolvent of an operator on X.

For our purposes we must exclude such a phenomenon. In order to do so, we need a new concept.

a. A Technical Tool: Pseudoresolvents

In this subsection we consider bounded operators on a Banach space X that depend on a complex parameter and satisfy the resolvent equation (see Paragraph V.1.2, (1.2)). Here is the formal definition.

1.3 Definition. *Let $\Lambda \subset \mathbb{C}$ and consider operators $\mathcal{J}(\lambda) \in \mathcal{L}(X)$ for each $\lambda \in \Lambda$. The family $\{\mathcal{J}(\lambda) : \lambda \in \Lambda\}$ is called a pseudoresolvent if*

$$(1.1) \qquad \mathcal{J}(\lambda) - \mathcal{J}(\mu) = (\mu - \lambda)\mathcal{J}(\lambda)\mathcal{J}(\mu)$$

holds for all $\lambda, \mu \in \Lambda$.

The limit operators $R(\lambda)$ from Example 1.2 form a (trivial) pseudoresolvent for $\operatorname{Re} \lambda > 0$. However, they are not injective, and therefore they cannot be the resolvent operators $R(\lambda, A)$ of an operator A. It is our goal, and crucial for the proofs in Section 1.b, to find conditions implying that a pseudoresolvent is indeed a resolvent. Before doing so, we discuss the typical situation in which we encounter pseudoresolvents.

1.4 Proposition. *For each $n \in \mathbb{N}$, let A_n be the generator of a semigroup $(T_n(t))_{t \geq 0}$ on X satisfying $\|T_n(t)\| \leq M$ for all $n \in \mathbb{N}$, $t \geq 0$ and some fixed $M \geq 1$. Moreover, assume that for some $\lambda_0 > 0$*

$$\lim_{n \to \infty} R(\lambda_0, A_n)x$$

exists for all $x \in X$. Then, the limit

$$R(\lambda)x := \lim_{n \to \infty} R(\lambda, A_n)x, \quad x \in X,$$

exists for all $\operatorname{Re} \lambda > 0$ and defines a pseudoresolvent $\{R(\lambda) : \operatorname{Re} \lambda > 0\}$.

PROOF. Consider the set

$$\Omega := \left\{\lambda \in \mathbb{C} : \operatorname{Re}\lambda > 0, \lim_{n \to \infty} R(\lambda, A_n)x \quad \text{exists for all } x \in X\right\},$$

which is nonempty by assumption. As in Proposition V.1.3, one shows that for given $\mu \in \Omega$ one has

$$R(\lambda, A_n) = \sum_{k=0}^{\infty} (\mu - \lambda)^k R(\mu, A_n)^{k+1}$$

as long as $|\mu - \lambda| < \operatorname{Re}\mu$ (use (3.17) from Chapter II). The convergence is with respect to the operator norm and uniform in $\{\lambda \in \mathbb{C} : |\mu - \lambda| \leq \alpha \operatorname{Re}\mu\}$ for each $0 < \alpha < 1$. Because the series $M/\operatorname{Re}\mu \sum_{k=0}^{\infty} \alpha^k$ majorizes all the series $\sum_{k=0}^{\infty} |\mu - \lambda|^k \left\|R(\mu, A_n)^{k+1}\right\|$, we can conclude that $R(\lambda, A_n)x$ converges as $n \to \infty$ for all λ satisfying $|\mu - \lambda| \leq \alpha \operatorname{Re}\mu$. Therefore, the set Ω is open.

On the other hand, take an accumulation point λ of Ω with $\operatorname{Re}\lambda > 0$. For $0 < \alpha < 1$, we can find $\mu \in \Omega$ such that $|\mu - \lambda| \leq \alpha \operatorname{Re}\mu$. Hence, by the above considerations, λ must belong to Ω; i.e., Ω is relatively closed in $S := \{\lambda \in \mathbb{C} : \operatorname{Re}\lambda > 0\}$. The only set satisfying both properties is S itself; hence we obtain the existence of the operators $R(\lambda)$ for $\operatorname{Re}\lambda > 0$.

Evidently, the resolvent equation (1.1) remains valid for the limit operators.   □

In the subsequent lemma, we state the basic properties of pseudoresolvents.

1.5 Lemma. *Let $\{\mathcal{J}(\lambda) : \lambda \in \Lambda\}$ be a pseudoresolvent on X. Then the following properties hold for all $\lambda, \mu \in \Lambda$.*

(i) $\mathcal{J}(\lambda)\mathcal{J}(\mu) = \mathcal{J}(\mu)\mathcal{J}(\lambda)$.

(ii) $\ker \mathcal{J}(\lambda) = \ker \mathcal{J}(\mu)$.

(iii) $\operatorname{rg}\mathcal{J}(\lambda) = \operatorname{rg}\mathcal{J}(\mu)$.

PROOF. The commutativity (i) follows from the resolvent equation (1.1). If we rewrite it in the form

$$\mathcal{J}(\lambda) = \mathcal{J}(\mu)\left[I + (\mu - \lambda)\mathcal{J}(\lambda)\right] = \left[I + (\mu - \lambda)\mathcal{J}(\lambda)\right]\mathcal{J}(\mu),$$

we see that $\operatorname{rg}\mathcal{J}(\lambda) \subseteq \operatorname{rg}\mathcal{J}(\mu)$ and $\ker \mathcal{J}(\mu) \subseteq \ker \mathcal{J}(\lambda)$. By symmetry, the assertions (ii) and (iii) follow.   □

If we now require that $\ker \mathcal{J}(\lambda) = \{0\}$ and $\operatorname{rg}\mathcal{J}(\lambda)$ is dense, then the pseudoresolvent $\{\mathcal{J}(\lambda) : \lambda \in \Lambda\}$ becomes the resolvent of a closed, densely defined operator.

1.6 Proposition. *For a pseudoresolvent* $\{\mathcal{J}(\lambda) : \lambda \in \Lambda\}$ *on* X, *the following assertions are equivalent.*

(a) *There exists a densely defined, closed operator* $(A, D(A))$ *such that* $\Lambda \subset \rho(A)$ *and* $\mathcal{J}(\lambda) = R(\lambda, A)$ *for all* $\lambda \in \Lambda$.

(b) $\ker \mathcal{J}(\lambda) = \{0\}$, *and* $\operatorname{rg} \mathcal{J}(\lambda)$ *is dense in* X *for some/all* $\lambda \in \Lambda$.

PROOF. We have only to show that (b) implies (a). Because $\mathcal{J}(\lambda)$ is injective, we can define

$$A := \lambda_0 - \mathcal{J}(\lambda_0)^{-1}$$

for some $\lambda_0 \in \Lambda$. This yields a closed operator with dense domain $D(A) := \operatorname{rg} \mathcal{J}(\lambda_0)$. From the definition of A, it follows that

$$(\lambda_0 - A)\mathcal{J}(\lambda_0) = \mathcal{J}(\lambda_0)(\lambda_0 - A) = I;$$

hence $\mathcal{J}(\lambda_0) = R(\lambda_0, A)$. For arbitrary $\lambda \in \Lambda$, we have

$$
\begin{aligned}
(\lambda - A)\mathcal{J}(\lambda) &= \big[(\lambda - \lambda_0) + (\lambda_0 - A)\big]\mathcal{J}(\lambda) \\
&= \big[(\lambda - \lambda_0) + (\lambda_0 - A)\big]\mathcal{J}(\lambda_0)\big[I - (\lambda - \lambda_0)\mathcal{J}(\lambda)\big] \\
&= I + (\lambda - \lambda_0)\big[\mathcal{J}(\lambda_0) - \mathcal{J}(\lambda) - (\lambda - \lambda_0)\mathcal{J}(\lambda)\mathcal{J}(\lambda_0)\big] \\
&= I,
\end{aligned}
$$

and similarly, $\mathcal{J}(\lambda)(\lambda - A) = I$. This shows that $\mathcal{J}(\lambda) = R(\lambda, A)$ for all $\lambda \in \Lambda$ and, in particular, that A does not depend on the choice of λ_0.   □

We conclude these considerations with some useful sufficient conditions that make a pseudoresolvent a resolvent.

1.7 Corollary. *Let* $\{\mathcal{J}(\lambda) : \lambda \in \Lambda\}$ *be a pseudoresolvent on* X *and assume that* Λ *contains an unbounded sequence* $(\lambda_n)_{n \in \mathbb{N}}$. *If*

$$(1.2) \qquad \lim_{n \to \infty} \lambda_n \mathcal{J}(\lambda_n)x = x \quad \text{for all } x \in X,$$

then $\{\mathcal{J}(\lambda) : \lambda \in \Lambda\}$ *is the resolvent of a densely defined operator. In particular,* (1.2) *holds if* $\operatorname{rg} \mathcal{J}(\lambda)$ *is dense and*

$$(1.3) \qquad \|\lambda_n \mathcal{J}(\lambda_n)\| \leq M$$

for some constant M *and all* $n \in \mathbb{N}$.

PROOF. If (1.2) holds, we have $X = \overline{\bigcup_{n \in \mathbb{N}} \operatorname{rg} \mathcal{J}(\lambda_n)} = \overline{\operatorname{rg} \mathcal{J}(\lambda)}$, and hence $\mathcal{J}(\lambda)$ has dense range for each $\lambda \in \Lambda$. If $x \in \ker \mathcal{J}(\lambda)$, we obtain $x = \lim \lambda_n \mathcal{J}(\lambda_n)x = 0$; hence $\ker \mathcal{J}(\lambda) = \{0\}$. The first assertion now follows from Proposition 1.6.(b).

From the estimate $\|\mathcal{J}(\lambda_n)\| \le \frac{M}{|\lambda_n|}$, $n \in \mathbb{N}$, and the resolvent equation, we obtain

$$\lim_{n\to\infty} \|(\lambda_n \mathcal{J}(\lambda_n) - I)\mathcal{J}(\mu)\| = 0$$

for fixed $\mu \in \Lambda$. Therefore, it follows that

$$\lim_{n\to\infty} \lambda_n \mathcal{J}(\lambda_n)x = x$$

for $x \in \mathrm{rg}\,\mathcal{J}(\mu)$. Because this is a dense subspace of X, the norm boundedness in (1.3) allows us to conclude that (1.2) holds. $\qquad\square$

b. The Approximation Theorems

We now turn our attention to the approximation problem stated above; i.e., we study the relation among convergence of semigroups, generators, and resolvents. The adequate type of convergence for strongly continuous semigroups (and unbounded operators) is pointwise convergence.

If we *assume* that the limit operator is known to be a generator, we obtain our first main result. However, we need a uniform bound on the semigroups involved.

1.8 First Trotter–Kato Approximation Theorem. (TROTTER 1958, KATO 1959). *Let $(T(t))_{t\ge 0}$ and $(T_n(t))_{t\ge 0}$, $n \in \mathbb{N}$, be strongly continuous semigroups on X with generators A and A_n, respectively, and assume that they satisfy the estimate*

$$\|T(t)\|,\ \|T_n(t)\| \le Me^{wt} \qquad \text{for all } t \ge 0,\ n \in \mathbb{N},$$

and some constants $M \ge 1$, $w \in \mathbb{R}$. Take D to be a core for A and consider the following assertions.

(a) *$D \subset D(A_n)$ for all $n \in \mathbb{N}$ and $A_n x \to Ax$ as $n \to \infty$ for all $x \in D$.*

(b) *For each $x \in D$, there exists $x_n \in D(A_n)$ such that*

$$x_n \to x \qquad \text{and} \qquad A_n x_n \to Ax \qquad \text{as } n \to \infty.$$

(c) *$R(\lambda, A_n)x \to R(\lambda, A)x$ as $n \to \infty$ for all $x \in X$ and some/all $\lambda > w$.*

(d) *$T_n(t)x \to T(t)x$ as $n \to \infty$ for all $x \in X$, uniformly for t in compact intervals.*

Then the implications

$$\text{(a)} \implies \text{(b)} \iff \text{(c)} \iff \text{(d)}$$

hold, and (b) does not imply (a).

PROOF. Before starting, we perform a rescaling and assume without loss of generality that $w = 0$; i.e.,

$$\|T(t)\|,\ \|T_n(t)\| \le M \qquad \text{for all } t \ge 0,\ n \in \mathbb{N}.$$

Because the implication (a) $\Rightarrow$ (b) is trivial, we start by showing (b) $\Rightarrow$ (c).

Let $\lambda > 0$. Because $\|R(\lambda, A_n)\| \leq M/\lambda$ for all $n \in \mathbb{N}$, it suffices to show that

$$\lim_{n \to \infty} R(\lambda, A_n)y = R(\lambda, A)y$$

for y in the dense subspace $(\lambda - A)D$. Take $x \in D$ and define $y := (\lambda - A)x$. By assumption, there exists $x_n \in D(A_n)$ such that

$$x_n \to x \quad \text{and} \quad A_n x_n \to Ax;$$

hence

$$y_n := (\lambda - A_n)x_n \to y.$$

Therefore, the estimate

$$\begin{aligned}
\|R(\lambda, A_n)y - R(\lambda, A)y\| &\leq \|R(\lambda, A_n)y - R(\lambda, A_n)y_n\| \\
&\quad + \|R(\lambda, A_n)y_n - R(\lambda, A)y\| \\
&\leq \|R(\lambda, A_n)\| \cdot \|y - y_n\| + \|x_n - x\|
\end{aligned}$$

implies the assertion.

The implication (c) $\Rightarrow$ (b) follows if we take $x := R(\lambda, A)y$, and $x_n := R(\lambda, A_n)y$ for fixed $\lambda > 0$ and then observe that

$$A_n x_n = A_n R(\lambda, A_n)y = \lambda R(\lambda, A_n)y - y$$

converges to

$$\lambda R(\lambda, A)y - y = Ax.$$

(d) $\Rightarrow$ (c). The integral representation of the resolvent yields, for each $\lambda > 0$ and $x \in X$, that

$$\|R(\lambda, A)x - R(\lambda, A_n)x\| \leq \int_0^\infty e^{-\lambda t} \|T(t)x - T_n(t)x\|\, dt.$$

The desired convergence is now a consequence of Lebesgue's dominated convergence theorem.

To prove the final implication (c) $\Rightarrow$ (d), we fix some $t_0 > 0$ and assume that $R(\lambda, A_n)x \to R(\lambda, A)x$ as $n \to \infty$ for some $\lambda > 0$ and all $x \in X$. Then for all $t \in [0, t_0]$ we obtain

$$
\begin{aligned}
\big\|[T_n(t) - T(t)]R(\lambda, A)x\big\| &\leq \big\|T_n(t)[R(\lambda, A) - R(\lambda, A_n)]x\big\| \\
&\quad + \big\|R(\lambda, A_n)[T_n(t) - T(t)]x\big\| \\
&\quad + \big\|[R(\lambda, A_n) - R(\lambda, A)]T(t)x\big\| \\
&=: D_1(n, x) + D_2(n, x) + D_3(n, x),
\end{aligned}
$$

(1.4)

where we used the fact that a semigroup commutes with the resolvent of its generator. Because $\|T_n(t)\| \leq M$ for all $n \in \mathbb{N}$ and $t \in [0, t_0]$, the first term $D_1(n, x) \to 0$ as $n \to \infty$ uniformly on $[0, t_0]$. Moreover, because $(T(t))_{t \geq 0}$ is strongly continuous, the set $\{T(t)x : t \in [0, t_0]\} \subset X$ is compact and hence, by Exercise I.1.8.(1), also $D_3(n, x) \to 0$ as $n \to \infty$ uniformly for $t \in [0, t_0]$.

In order to estimate $D_2(n, x)$ we first show that

(1.5)
$$R(\lambda, A_n)\big[T(t) - T_n(t)\big]R(\lambda, A)z = \int_0^t T_n(t-s)\big[R(\lambda, A) - R(\lambda, A_n)\big]T(s)z\,ds$$

for every $z \in X$ and $t > 0$. To this end we first observe that, by Lemma A.19, the function $[0, t] \ni s \mapsto T_n(t - s)R(\lambda, A_n)T(s)R(\lambda, A)\dot{z} \in X$ is differentiable. Using also Lemma II.1.3.(ii) and (1.1) in Chapter V we obtain

$$\frac{d}{ds}\big[T_n(t - s)R(\lambda, A_n)T(s)R(\lambda, A)z\big]$$
$$= T_n(t - s)\big[-A_n R(\lambda, A_n)T(s) + R(\lambda, A_n)T(s)A\big]R(\lambda, A)z$$
$$= T_n(t - s)\big[R(\lambda, A) - R(\lambda, A_n)\big]T(s)z.$$

Integrating the last equation with respect to s from 0 to t then gives (1.5). This implies that

$$\big\|R(\lambda, A_n)\big[T(t) - T_n(t)\big]R(\lambda, A)z\big\|$$
$$\leq \int_0^t \|T_n(t - s)\| \cdot \big\|\big[R(\lambda, A) - R(\lambda, A_n)\big]T(s)z\big\|\,ds$$
$$\leq t_0 M \cdot \sup_{s \in [0, t_0]} \big\|\big[R(\lambda, A) - R(\lambda, A_n)\big]T(s)z\big\|.$$

Using the same reasoning as above to prove that $D_3(n, x) \to 0$, we conclude that

$$\big\|R(\lambda, A_n)\big[T(t) - T_n(t)\big]R(\lambda, A)z\big\| = \big\|D_2\big(n, R(\lambda, A)z\big)\big\| \to 0 \quad \text{as } n \to \infty$$

uniformly for $t \in [0, t_0]$. Because every $x \in D(A)$ can be written as $x = R(\lambda, A)z$ for $z = (\lambda - A)x \in X$, this shows that for $x \in D(A)$ the term $D_2(n, x) \to 0$ as $n \to \infty$ uniformly for $t \in [0, t_0]$.

Summing up, we conclude that by inequality (1.4),

$$\|T_n(t)x - T(t)x\| \to 0 \quad \text{as } n \to \infty$$

for all $x \in D(A^2)$, uniformly on $[0, t_0]$. Because $\|T_n(t) - T(t)\| \leq 2M$ and $D(A^2)$ is dense in X by Proposition II.1.8, from Proposition A.3 we finally obtain (d).

That (b) does not imply (a) in general can be seen from Counterexample 2.8 below. $\qquad\square$

For the above result we had to *assume* that the limit operator A is already known to be a generator. This is a major defect, because in the applications one wants to approximate the operator A by (simple) operators A_n and then *conclude* that A becomes a generator. Moreover, the semigroup generated by A should be obtained as the limit of the known semigroups generated by the operators A_n. In fact, we encountered this problem already in the proof (of the nontrivial implication) of Generation Theorem II.3.5. Therefore, the following result can be viewed as a generalization of the Hille–Yosida theorem.

1.9 Second Trotter–Kato Approximation Theorem. (TROTTER 1958, KATO 1959). *Let* $(T_n(t))_{t\geq 0}$, $n \in \mathbb{N}$, *be strongly continuous semigroups on* X *with generators* $(A_n, D(A_n))$ *satisfying the stability condition*

$$(1.6) \qquad \qquad \|T_n(t)\| \leq Me^{wt}$$

for constants $M \geq 1$, $w \in \mathbb{R}$ *and all* $t \geq 0$, $n \in \mathbb{N}$. *For some* $\lambda_0 > w$ *consider the following assertions.*

(a) *There exists a densely defined operator* $(A, D(A))$ *such that* $A_n x \to Ax$ *as* $n \to \infty$ *for all* x *in a core* D *of* A *and such that the range* $\mathrm{rg}(\lambda_0 - A)$ *is dense in* X.

(b) *The operators* $R(\lambda_0, A_n)$, $n \in \mathbb{N}$, *converge strongly as* $n \to \infty$ *to an operator* $R \in \mathcal{L}(X)$ *with dense range* $\mathrm{rg}\,R$.

(c) *The semigroups* $(T_n(t))_{t\geq 0}$, $n \in \mathbb{N}$, *converge strongly (and uniformly for* $t \in [0, t_0]$) *as* $n \to \infty$ *to a strongly continuous semigroup* $(T(t))_{t\geq 0}$ *with generator* B *such that* $R = R(\lambda_0, B)$.

Then the implications (a) $\Rightarrow$ (b) $\Longleftrightarrow$ (c) *hold. In particular, if* (a) *holds, then* $B = \overline{A}$.

PROOF. Without loss of generality, and after the usual rescaling, it suffices to consider uniformly bounded semigroups only, i.e., the case $w = 0$.

(a) $\Rightarrow$ (b). As in the above proof, it suffices to show convergence of the sequence $(R(\lambda_0, A_n)y)_{n\in\mathbb{N}}$ for $y := (\lambda_0 - A)x$, $x \in D$, only. This follows, because

$$R(\lambda_0, A_n)y = R(\lambda_0, A_n)[(\lambda_0 - A_n)x - (\lambda_0 - A_n)x + (\lambda_0 - A)x]$$
$$= x + R(\lambda_0, A_n)(A_n x - Ax) \to x = Ry$$

as $n \to \infty$. Moreover, $\mathrm{rg}\,R$ contains D, hence is dense in X.

Because the implication (c) $\Rightarrow$ (b) holds by the above theorem, it remains to prove that (b) $\Rightarrow$ (c). By Proposition 1.4, we obtain a pseudoresolvent $\{R(\lambda) : \lambda > 0\}$ by defining

$$R(\lambda)x := \lim_{n\to\infty} R(\lambda, A_n)x, \quad x \in X.$$

This pseudoresolvent satisfies, for all $\lambda > 0$,

$$\|\lambda R(\lambda)\| \leq M,$$

and, because $R(\lambda)^k = \lim_{n\to\infty} R(\lambda, A_n)^k$,

$$\|\lambda^k R(\lambda)^k\| \leq M \qquad \text{for all } k \in \mathbb{N}.$$

Moreover, it has dense range $\mathrm{rg}\,R(\lambda) = \mathrm{rg}\,R$. Therefore, Corollary 1.7 yields the existence of a densely defined operator $(B, D(B))$ such that

$R(\lambda) = R(\lambda, B)$ for $\lambda > 0$. Moreover, this operator satisfies the Hille–Yosida estimate

$$\left\| \lambda^k R(\lambda, B)^k \right\| \leq M \qquad \text{for all } k \in \mathbb{N},$$

hence generates a bounded strongly continuous semigroup $\big(T(t)\big)_{t\geq 0}$. We can now apply the implication (c) $\Rightarrow$ (d) from the First Trotter–Kato Approximation Theorem 1.8 in order to conclude that the semigroups $\big(T_n(t)\big)_{n\geq 0}$ converge—in the desired way—to the semigroup $\big(T(t)\big)_{t\geq 0}$.

In the final step, we show that (a) implies $\overline{A} = B$. Because $R(\lambda_0, B) = R$, we have

$$R(\lambda_0, B)(\lambda_0 - A)x = x$$

for all $x \in D$. However, D is a core for $\overline{A}$, and therefore

$$R(\lambda_0, B)(\lambda_0 - \overline{A})x = x$$

for all $x \in D(\overline{A})$. From this it follows that λ_0 is not an approximate eigenvalue of $\overline{A}$. Moreover, $\mathrm{rg}(\lambda_0 - A)$ is dense in X by assumption; hence λ_0 does not belong to the residual spectrum of $\overline{A}$. Therefore, $\lambda_0 \in \rho(\overline{A})$, and we obtain $R(\lambda_0, \overline{A}) = R(\lambda_0, B)$; i.e., $\overline{A} = B$ as claimed. $\qquad\square$

The importance of the above theorems cannot be overestimated. In fact, they yield the theoretical background for many approximation schemes in abstract operator theory and applied numerical analysis. However, we restrict ourselves to rather abstract examples and applications.

c. Examples

The Hille–Yosida Generation Theorem II.3.8 was the main tool in our proof of the Trotter–Kato approximation theorems. Conversely, this theorem was proved using an approximation argument. It is enlightening to start our series of examples by reformulating this part of the proof.

1.10 Yosida Approximants. Let $\big(A, D(A)\big)$ be an operator on X satisfying the conditions (in the contractive case, for simplicity) from Generation Theorem II.3.5.(b). For each $n \in \mathbb{N}$, define the *Yosida approximant*

$$A_n := nAR(n, A) \in \mathcal{L}(X).$$

By Lemma II.3.4, the sequence $(A_n)_{n\in\mathbb{N}}$ converges pointwise on $D(A)$ to A. Because $\lambda - A$ is already supposed to be surjective, we can apply the Second Trotter–Kato Approximation Theorem 1.9 to conclude the existence of the limit semigroup $\big(T(t)\big)_{t\geq 0}$ with

$$T(t)x := \lim_{n\to\infty} e^{tA_n}x, \quad x \in X,$$

and generator $\big(A, D(A)\big)$.

Clearly, in a logical sense, these arguments do not re-prove the Hille–Yosida generation theorem, which we already used for the proof of the Trotter–Kato approximation theorem. However, it might be helpful for the beginner to observe that the above approximating sequence enjoys a special feature: The operators A_n, $n \in \mathbb{N}$, mutually commute. This property allows a simple and direct proof of the essential step in Approximation Theorem 1.8.

Lemma. *Let $(T(t))_{t \geq 0}$ and $(T_n(t))_{t \geq 0}$, $n \in \mathbb{N}$, be strongly continuous semigroups on X with generator $(A, D(A))$ and bounded generators A_n, respectively. In addition, suppose that $(T(t))_{t \geq 0}$ and $(T_n(t))_{t \geq 0}$ satisfy the stability condition (1.6) and that*

$$A_n T(t) = T(t) A_n$$

for all $n \in \mathbb{N}$ and $t > 0$. If

$$A_n x \to A x$$

for all x in a core D of A, then

$$T_n(t) x \to T(t) x$$

for all $x \in X$ uniformly for $t \in [0, t_0]$.

PROOF. For $x \in D$ and $n \in \mathbb{N}$, we have

$$
\begin{aligned}
T_n(t)x - T(t)x &= -\int_0^t \tfrac{d}{ds}[T_n(t-s)T(s)x]\,ds \\
&= \int_0^t T_n(t-s)(A_n - A)T(s)x\,ds \\
&= \int_0^t T_n(t-s)T(s)(A_n x - A x)\,ds;
\end{aligned}
$$

hence

$$\|T_n(t)x - T(t)x\| \leq t M^2 e^{wt} \|A_n x - A x\|. \qquad \square$$

We encounter this situation in our next example, by which we re-prove a classical theorem.

1.11 Weierstrass Approximation Theorem. Take the function space $X := C_0(\mathbb{R})$ (or $C_{ub}(\mathbb{R})$) and the (left) translation group $(T(t))_{t \in \mathbb{R}}$ with

$$T(t)f(s) := f(s+t) \qquad \text{for } s, t \in \mathbb{R}$$

and generator

$$A f := f', \quad D(A) := \{f \in X : f' \in X\}$$

(see Paragraph II.2.9). The bounded operators

$$A_n := \frac{T(1/n) - I}{1/n}, \quad n \in \mathbb{N},$$

- Commute with all operators $T(t)$,
- Generate contraction semigroups, because

$$(1.7) \qquad \left\| e^{tA_n} \right\| = \left\| e^{nt(T(1/n) - I)} \right\| \le e^{-nt} e^{nt\|T(1/n)\|} = 1,$$

and

- Satisfy, by definition of the derivative,

$$A_n f \to A f$$

for each $f \in D(A)$.

Therefore, the (First) Trotter–Kato Approximation Theorem 1.8 (or the lemma in 1.10) can be applied and yields

$$(1.8) \qquad f(s + t) = \lim_{n \to \infty} \sum_{k=0}^{\infty} \frac{t^k}{k!} (A_n^k f)(s)$$

for all $f \in X$ and uniformly for $s \in \mathbb{R}$, $t \in [0, 1]$. If we now take $s = 0$, choose an appropriate sequence $(m_n)_{n \in \mathbb{N}}$ of natural numbers, and observe that $\sum_{k=0}^{m_n} t^k/k! (A_n^k f)(0)$ is a polynomial, we obtain the Weierstrass approximation theorem as a consequence.

Proposition. *For every $f \in X$ there exists a sequence $(m_n)_{n \in \mathbb{N}} \subset \mathbb{N}$ such that*

$$(1.9) \qquad f(t) = \lim_{n \to \infty} \sum_{k=0}^{m_n} \frac{t^k}{k!} (A_n^k f)(0)$$

uniformly for $t \in [0, 1]$.

It is very instructive to observe how convergence breaks down if we reverse the order of the limit and of the series summation in (1.9). See the illuminating remarks in [Gol85, Sect. I.8.3].

1.12 A First Approximation Formula. The idea employed in Paragraph 1.11 is very simple and can be formulated in a general context. Let $(T(t))_{t \ge 0}$ be a strongly continuous contraction semigroup on X with generator $(A, D(A))$. Then the bounded operators

$$A_n := \frac{T(1/n) - I}{1/n}, \quad n \in \mathbb{N},$$

approximate A on $D(A)$ and generate contraction semigroups $(e^{tA_n})_{t \ge 0}$ (see (1.7)). Therefore, we obtain the following approximation formula.

Proposition. *With the above definitions, one has*

(1.10)
$$T(t)x = \lim_{n \to \infty} e^{-nt} e^{ntT(1/n)} x$$

for all $x \in X$ and uniformly in $t \in [0, t_0]$.

This formula might seem useless, because it assumes that the operators $T(t)$ are already known, at least for small $t > 0$. However, it is the first step towards more interesting approximation formulas to be developed in the next section.

1.13 Exercises. (1) Consider the operator $Af := f''$ with maximal domain on $X := C_0(\mathbb{R})$. For each $n \in \mathbb{N}$, we define bounded difference operators

$$A_n f(s) := n^2 \big[f(s + 1/n) - 2f(s) + f(s - 1/n) \big], \quad s \in \mathbb{R}, f \in X.$$

Prove the following statements.

   (i) $\big(A, D(A)\big)$ is a closed, densely defined operator.

  (ii) $\big\| e^{tA_n} \big\| \leq 1$ for each $n \in \mathbb{N}$, and $A_n f \to Af$ for $f \in D(A)$.

 (iii) For each $g \in X$, there exists a unique $f \in D(A)$ such that $f - f'' = g$. (Hint: Use the formal identity $(I - (d/ds)^2)^{-1} = (I - d/ds)^{-1}(I + d/ds)^{-1}$ and the resolvent formula for d/ds from Proposition 2 in Paragraph II.2.9. Check that this yields the correct solution.)

 (iv) $\big(A, D(A)\big)$ generates the strongly continuous semigroup $\big(T(t)\big)_{t \geq 0}$ given by

$$T(t)f(s) = \lim_{n \to \infty} e^{-2n^2 t} \sum_{k=0}^{\infty} \frac{(n^2 t)^k}{k!} \sum_{l=0}^{k} \binom{k}{l} f\big(s + (k-2l)/n\big)$$

for $s \in \mathbb{R}$, $f \in X$.

(2) What happens in Exercise I.2.15.(1) as $\alpha \downarrow 0$?

2. The Chernoff Product Formula

As announced in the previous section, it is now our aim to obtain more or less explicit formulas for the semigroup operators $T(t)$. These formulas are based on some knowledge of the generator (and its resolvent) and the Trotter–Kato approximation theorem.

Our first approach is via the Chernoff product formula, from which many approximation formulas can be derived. For its proof the following estimate is essential.

2.1 Lemma. *Let* $S \in \mathcal{L}(X)$ *satisfy* $\|S^m\| \leq M$ *for some* $M \geq 1$ *and all* $m \in \mathbb{N}$*. Then we have*

$$(2.1) \qquad \left\| e^{n(S-I)}x - S^n x \right\| \leq \sqrt{n} M \|Sx - x\|$$

for every $n \in \mathbb{N}$ *and* $x \in X$.

PROOF. Let $n \in \mathbb{N}$ be fixed and observe that

$$e^{n(S-I)} - S^n = e^{-n}\left(e^{nS} - e^n S^n\right) = e^{-n} \sum_{k=0}^{\infty} \frac{n^k}{k!}\left(S^k - S^n\right).$$

For $k > n$, we write

$$S^k - S^n = \sum_{j=n}^{k-1}\left(S^{j+1} - S^j\right) = \sum_{j=n}^{k-1} S^j(S - I),$$

and similarly for $k < n$. Therefore, and because $\|S^m\| \leq M$, we obtain

$$\left\| S^k x - S^n x \right\| \leq |n - k| \cdot M \|Sx - x\|$$

for all $k \in \mathbb{N}$, $x \in X$. This allows the estimate

$$\left\| e^{n(S-I)}x - S^n x \right\| \leq e^{-n} M \|Sx - x\| \cdot \sum_{k=0}^{\infty} \left(\frac{n^k}{k!}\right)^{1/2} \left(\frac{n^k}{k!}\right)^{1/2} |n - k|$$

$$\leq e^{-n} M \|Sx - x\| \cdot \left(\sum_{k=0}^{\infty} \frac{n^k}{k!}\right)^{1/2} \left(\sum_{k=0}^{\infty} \frac{n^k}{k!}(n-k)^2\right)^{1/2}$$

$$= e^{-n} M \|Sx - x\| \cdot \left(e^n\right)^{1/2} \left(ne^n\right)^{1/2}$$

$$= \sqrt{n} M \|Sx - x\|,$$

where we used the Cauchy–Schwarz inequality and the identity

$$\sum_{k=0}^{\infty} \frac{n^k}{k!}(n-k)^2 = ne^n.$$

$\square$

This lemma, combined with Approximation Theorem 1.9, yields the main result of this section.

2.2 Theorem. (*Chernoff Product Formula*). *Consider a function*

$$V : \mathbb{R}_+ \to \mathcal{L}(X)$$

satisfying $V(0) = I$ *and* $\left\| [V(t)]^m \right\| \leq M$ *for all* $t \geq 0$, $m \in \mathbb{N}$, *and some* $M \geq 1$. *Assume that*

$$Ax := \lim_{h \downarrow 0} \frac{V(h)x - x}{h}$$

exists for all $x \in D \subset X$, *where* D *and* $(\lambda_0 - A)D$ *are dense subspaces in* X *for some* $\lambda_0 > 0$. *Then the closure* $\overline{A}$ *of* A *generates a bounded strongly continuous semigroup* $(T(t))_{t \geq 0}$, *which is given by*

$$(2.2) \qquad T(t)x = \lim_{n \to \infty} [V(t/n)]^n \, x$$

for all $x \in X$ *and uniformly for* $t \in [0, t_0]$.

PROOF. For $s > 0$, define

$$A_s := \frac{V(s) - I}{s} \in \mathcal{L}(X),$$

and observe that $A_s x \to Ax$ for all $x \in D$ as $s \downarrow 0$. The semigroups $(e^{t A_s})_{t \geq 0}$ all satisfy

$$\left\| e^{t A_s} \right\| \leq e^{-t/s} \left\| e^{t V(s)/s} \right\| \leq e^{-t/s} \sum_{m=0}^{\infty} \frac{t^m \left\| [V(s)]^m \right\|}{s^m m!} \leq M \quad \text{for every } t \geq 0.$$

This shows that the assumptions of the Second Trotter–Kato Approximation Theorem 1.9 are fulfilled (with the discrete parameter $n \in \mathbb{N}$ replaced by the continuous parameter $s > 0$). Hence, the closure $\overline{A}$ of A generates a strongly continuous semigroup $(T(t))_{t \geq 0}$ satisfying

$$\left\| T(t)x - e^{t A_s} x \right\| \to 0 \qquad \text{for all } x \in X \text{ as } s \downarrow 0$$

uniformly for $t \in [0, t_0]$, and therefore

$$(2.3) \qquad \left\| T(t)x - e^{t A_{t/n}} x \right\| \to 0 \qquad \text{for all } x \in X \text{ as } n \to \infty$$

uniformly for $t \in [0, t_0]$.

On the other hand, we have by Lemma 2.1 that

$$\left\| e^{t A_{t/n}} x - [V(t/n)]^n \, x \right\| = \left\| e^{n(V(t/n) - I)} x - [V(t/n)]^n \, x \right\|$$

$$(2.4) \qquad\qquad\qquad \leq \sqrt{n} M \left\| V(t/n)x - x \right\|$$

$$= \frac{tM}{\sqrt{n}} \left\| A_{t/n} x \right\| \to 0$$

for all $x \in D$ as $n \to \infty$, uniformly on $(0, t_0]$. Because $\left\| e^{t A_{t/n}} - [V(t/n)]^n \right\| \leq 2M$, the combination of (2.3), (2.4), and Proposition A.3 yields (2.2). $\square$

As before, we pass to the unbounded case by a rescaling procedure.

2.3 Corollary. *Consider a function*

$$V : \mathbb{R}_+ \to \mathcal{L}(X)$$

satisfying $V(0) = I$ and

$$\left\| [V(t)]^k \right\| \le M e^{kwt} \qquad \text{for all } t \ge 0, k \in \mathbb{N}$$

and some constants $M \ge 1$, $w \in \mathbb{R}$. Assume that

$$Ax := \lim_{t\downarrow 0} \frac{V(t)x - x}{t}$$

exists for all $x \in D \subset X$, where D and $(\lambda_0 - A)D$ are dense subspaces in X for some $\lambda_0 > w$. Then the closure $\overline{A}$ of A generates a strongly continuous semigroup $(T(t))_{t \ge 0}$ given by

$$(2.5) \qquad T(t)x = \lim_{n \to \infty} [V(t/n)]^n x$$

for all $x \in X$ and uniformly for $t \in [0, t_0]$. Moreover, $(T(t))_{t \ge 0}$ satisfies the estimate

$$\|T(t)\| \le M e^{wt} \qquad \text{for all } t \ge 0.$$

PROOF. From the function $V(\cdot)$ we pass to

$$\widetilde{V}(t) := e^{-wt} V(t),$$

which then satisfies

$$\left\| \widetilde{V}(t)^k \right\| \le M \qquad \text{for all } k \in \mathbb{N} \text{ and } t \ge 0$$

and whose derivative in zero is the operator $A - w$. The assertions then follow from Theorem 2.2. □

Next, we substitute the "time steps" of size "t/n" in the definition of the approximating operators $V(t/n)$ by an arbitrary null sequence $(t_n)_{n \in \mathbb{N}}$.

2.4 Corollary. *Let $V : \mathbb{R}_+ \to \mathcal{L}(X)$ satisfy the assumptions in Corollary 2.3. If for fixed $t > 0$ we take a positive null sequence $(t_n)_{n \in \mathbb{N}} \in c_0$ and a strictly increasing sequence of integers k_n such that*

$$k_n t_n \to t,$$

then

$$(2.6) \qquad T(t)x = \lim_{n \to \infty} [V(t_n)]^{k_n} x$$

for all $x \in X$.

PROOF. Using the function

$$\xi(s) := \begin{cases} s \cdot {(t_n k_n)}/{t} & \text{for } s \in ({t}/{k_{n+1}}, {t}/{k_n}], \\ 0 & \text{for } s = 0 \quad \text{or } s > {t}/{k_1}, \end{cases}$$

we introduce a new operator-valued function $W : \mathbb{R}_+ \to \mathcal{L}(X)$ by

$$W(t) := V\big(\xi(t)\big), \quad t \ge 0.$$

This function still satisfies $W(0) = I$ and $\big\|W(t)^k\big\| \le M e^{kwt}$ for all $t \ge 0$, $k \in \mathbb{N}$. For $x \in D$, we show that

$$\lim_{t \downarrow 0} \frac{W(t)x - x}{t} = Ax.$$

Let $(t_n)_{n\in\mathbb{N}} \in c_0$ be an arbitrary null sequence and for each t_m choose $n_m \in \mathbb{N}$ such that $t_m \in ({t}/{k_{n_m+1}}, {t}/{k_{n_m}}]$. Then

$$\frac{W(t_m)x - x}{t_m} = \frac{V\big(\xi(t_m)\big)x - x}{\xi(t_m)} \cdot \frac{\xi(t_m)}{t_m}$$

$$= \frac{V\big(\xi(t_m)\big)x - x}{\xi(t_m)} \cdot \frac{t_{n_m} k_{n_m} \cdot t_m}{t \cdot t_m};$$

hence

$$\lim_{m\to\infty} \frac{W(t_m)x - x}{t_m} = Ax \cdot \lim_{m\to\infty} \frac{t_{n_m} k_{n_m}}{t} = Ax.$$

By Corollary 2.3, we conclude that $\overline{A}$ generates the semigroup $(T(t))_{t\ge 0}$ given by

$$T(t)x = \lim_{n\to\infty} [W({t}/{n})]^n x$$

uniformly for $t \in [0, t_0]$. In particular, we obtain for the subsequence $({t}/{k_n})_{n\in\mathbb{N}}$ that

$$T(t)x = \lim_{n\to\infty} [W({t}/{k_n})]^{k_n} x$$

$$= \lim_{n\to\infty} \big[V\big(\xi({t}/{k_n})\big)\big]^{k_n} x$$

$$= \lim_{n\to\infty} \big[V(t_n)\big]^{k_n} x \qquad \text{for all } x \in X.$$

$\square$

The following application of the Chernoff Product Formula Theorem 2.2 (or of Corollary 2.3) finally gives us an explicit formula, called the *Post–Widder Inversion Formula*, for the semigroup in terms of the resolvent of its generator. This adds a missing arrow to the "triangle" from Diagram II.1.14, and, at the same time, corresponds to Hille's original proof of Generation Theorem II.3.5.

2.5 Corollary. *For every strongly continuous semigroup* $(T(t))_{t\geq 0}$ *on* X *with generator* $(A, D(A))$, *one has*

$$(2.7) \qquad T(t)x = \lim_{n\to\infty} \left[{}^{n}/_{t} R({}^{n}/_{t}, A) \right]^{n} x = \lim_{n\to\infty} \left[I - {}^{t}/_{n} A \right]^{-n} x, \quad x \in X,$$

uniformly for t *in compact intervals.*

PROOF. Assume that $\|T(t)\| \leq M e^{wt}$ for constants $M \geq 1$, $w > 0$ and define

$$V(t) := \begin{cases} I & \text{for } t = 0, \\ {}^{1}/_{t} R({}^{1}/_{t}, A) & \text{for } t \in (0, \delta), \\ 0 & \text{for } t \geq \delta, \end{cases}$$

for some $\delta \in (0, {}^{1}/_{w})$. In this way we obtain a function $V : \mathbb{R}_+ \to \mathcal{L}(X)$ satisfying

$$\left\| V(t)^k \right\| \leq {}^{1}/_{t^k} \left\| R({}^{1}/_{t}, A)^k \right\| \leq \frac{M}{t^k ({}^{1}/_{t} - w)^k} = \frac{M}{(1 - wt)^k} \leq M e^{k(w+1)t}$$

for all $t \in (0, \delta)$, provided that we choose $\delta > 0$ sufficiently small. Moreover, by Lemma II.3.4, we have

$$\lim_{t\downarrow 0} \frac{V(t)x - x}{t} = \lim_{t\downarrow 0} {}^{1}/_{t} R({}^{1}/_{t}, A) A x = A x \qquad \text{if } x \in D(A).$$

Therefore, the Chernoff product formula as stated in Corollary 2.3 applies, and (2.5) becomes the above formula. □

For the sake of completeness, we add this new relation to the diagram relating the semigroup, its generator, and its resolvent operators.

2.6 Diagram.

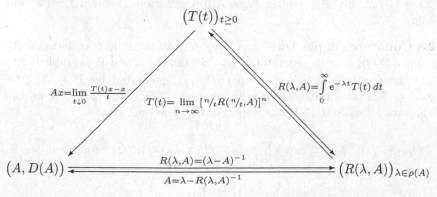

We now apply the Chernoff product formula from Theorem 2.2 to perturbation theory yielding another important formula, called the *Trotter product formula*, for the perturbed semigroup. In contrast to the situation studied in Sections III.1 and 2, we obtain a result that is symmetric in the operators A and B.

2.7 Corollary. *Let $(T(t))_{t\geq 0}$ and $(S(t))_{t\geq 0}$ be strongly continuous semigroups on X satisfying the* stability condition

$$(2.8) \qquad \|[T(t/n)S(t/n)]^n\| \leq Me^{wt} \qquad \text{for all } t \geq 0, n \in \mathbb{N},$$

and for constants $M \geq 1$, $w \in \mathbb{R}$. Consider the "sum" $A + B$ on $D := D(A) \cap D(B)$ of the generators $(A, D(A))$ of $(T(t))_{t\geq 0}$ and $(B, D(B))$ of $(S(t))_{t\geq 0}$, and assume that D and $(\lambda_0 - A - B)D$ are dense in X for some $\lambda_0 > w$. Then $C := \overline{A + B}$ generates a strongly continuous semigroup $(U(t))_{t\geq 0}$ given by the Trotter product formula

$$(2.9) \qquad U(t)x = \lim_{n\to\infty} [T(t/n)S(t/n)]^n x, \quad x \in X,$$

with uniform convergence for t in compact intervals.

PROOF. In order to apply the Chernoff product formula from Corollary 2.3, it suffices to define
$$V(t) := T(t)S(t), \quad t \geq 0,$$
and observe that

$$\lim_{t\downarrow 0} \frac{T(t)S(t)y - y}{t} = \lim_{t\downarrow 0} T(t)\frac{S(t)y - y}{t} + \lim_{t\downarrow 0}\frac{T(t)y - y}{t}$$
$$= By + Ay$$

for all $y \in D$. □

We now show first that the density of $D(A) \cap D(B)$ is not necessary for the convergence (to a strongly continuous semigroup) of the Trotter Product Formula (2.9) and second that the converse of the implication (a) $\Rightarrow$ (b) in the First Trotter–Kato Approximation Theorem 1.8 does not hold.

2.8 Counterexample. On $X := L^2(\mathbb{R})$ we take the right translation semigroup $(T(t))_{t\geq 0}$ with generator A (see Section I.3.c and Paragraph II.2.9) and the multiplication semigroup $(S(t))_{t\geq 0}$ generated by $B = M_{iq}$ for $q : \mathbb{R} \to \mathbb{R}$ a measurable and locally integrable function. For $f \in X$ we can compute (cf. [EN00, (5.16) in Expl. III.5.9]) the products

$$[T(t/n)S(t/n)]^n f(s) = \exp\left(i\sum_{k=1}^n q\left(s - kt/n\right) t/n\right) \cdot f(s-t) \quad \text{for } t \geq 0, s \in \mathbb{R}.$$

They converge in L^2-norm to $U(t)f$ with

$$U(t)f(s) := e^{i\int_{s-t}^s q(\tau)\,d\tau} \cdot f(s - t).$$

These operators $U(t)$ form a strongly continuous semigroup (of isometries) on X. Observe that no assumption on $D(A) \cap D(B)$ was made.

In fact, this intersection can be $\{0\}$. Take $\mathbb{Q} = \{\alpha_k : k \in \mathbb{N}\}$ and define

$$q(s) := \sum_{k=1}^{\infty} \frac{1}{k!} |s - \alpha_k|^{-1/2} \quad \text{for } s \in \mathbb{R}.$$

Then $q \in \mathrm{L}^1_{\mathrm{loc}}(\mathbb{R})$. However, $q \notin \mathrm{L}^2[a,b]$ for any $a < b$. Therefore, no continuous function $f \neq 0$ belongs to $D(B)$; hence $D(A) \cap D(B) = 0$. However, at least formally, the generator C of $(U(t))_{t \geq 0}$ is the "sum" $A + B$.

In fact, one can show that the domain of C is

$$D(C) = \{f \in \mathrm{L}^2(\mathbb{R}) : f \text{ is absolutely continuous and } -f' + qf \in \mathrm{L}^2(\mathbb{R})\}$$

and

$$Cf = -f' + qf \quad \text{for } f \in D(C).$$

Using the same q, we now define semigroups on X by

$$U_n(t)f(s) := e^{i/n \int_{s-}^{s} t(n+1)/n \, q(\tau) \, d\tau} \cdot f\left(s - \frac{t(n+1)}{n}\right)$$

for every $n \in \mathbb{N}$. Then $\lim_{n \to \infty} U_n(t)f = T(t)f$ for every $f \in X$, and the semigroups $(U_n(t))_{t \geq 0}$ and the right translation semigroup $(T(t))_{t \geq 0}$ satisfy the equivalent conditions (b), (c), and (d) in the First Trotter–Kato Approximation Theorem 1.8. However, the intersections of the respective domains are trivial; hence condition (a) does not hold.

2.9 Exercise. Let $(T(t))_{t \geq 0}$ be a strongly continuous semigroup with generator A on a Banach space X. If $B \in \mathcal{L}(X)$, then the semigroup $(U(t))_{t \geq 0}$ generated by $A + B$ is given by the Trotter product formula

$$U(t)x = \lim_{n \to \infty} \left[T(t/n) e^{tB/n} \right]^n x$$

for all $t \geq 0$ and $x \in X$. (Hint: By renorming X as in Chapter II, (3.18) (or by Lemma II.3.10) one may assume that $(T(t))_{t \geq 0}$ is a contraction semigroup. To verify the stability condition (2.8) observe that $\|e^{tB}\| \leq e^{t\|B\|}$.)

Chapter V

Spectral Theory and Asymptotics for Semigroups

Up to now, our main concern was to show that strongly continuous semigroups have a generator (with nice properties) and, conversely, that certain operators generate strongly continuous semigroups (with nice properties). In the perspective of Section II.6 this means that certain evolution equations have unique solutions, hence are well-posed.

Having established this kind of well-posedness, that is, the existence of a strongly continuous semigroup, we now turn our attention to the qualitative behavior of these solutions, i.e., of these semigroups. Our main tool for this investigation is provided by *spectral theory*.

This is already evident from the Hille–Yosida theorem (and its variants), where generators were characterized by the location of their spectrum and by norm estimates of the resolvent. Moreover, the classical Liapunov Stability Theorem for matrix semigroups $(e^{tA})_{t \geq 0}$ characterizes the stability, i.e., $\lim_{t \to \infty} \|e^{tA}\| = 0$, by the location of the eigenvalues of A (see Theorem 3.6 below).

In order to continue in this direction, we first develop a spectral theory for semigroups and their generators. The importance of these techniques becomes evident in Section 3, where we apply it to the study of the asymptotic behavior of strongly continuous semigroups.

We start with an introductory section, in which we explain the basic spectral theoretic notions and results for general closed operators. Because many of these notions have already been used in the preceding chapters, the reader may skip (most of) this section.

1. Spectrum of Semigroups and Generators

a. Spectral Theory for Closed Operators

The guiding idea of spectral theory is to associate numbers with linear operators in the hope of recovering properties of the operator from these numbers. So let

$$A : D(A) \subset X \to X$$

be a closed linear operator on some Banach space X. Note that we do not assume a dense domain, whereas the closedness is essential for a reasonable spectral theory.

1.1 Definition. *We call*

$$\rho(A) := \{\lambda \in \mathbb{C} : \lambda - A : D(A) \to X \text{ is bijective}\}$$

the resolvent set and its complement $\sigma(A) := \mathbb{C} \setminus \rho(A)$ *the spectrum of* A. *For* $\lambda \in \rho(A)$, *the inverse*

$$R(\lambda, A) := (\lambda - A)^{-1}$$

is, by the closed graph theorem, a bounded operator on X *and is called the resolvent (of* A *at the point* λ).

It follows immediately from the definition that the identity

$$(1.1) \qquad\qquad AR(\lambda, A) = \lambda R(\lambda, A) - I$$

holds for every $\lambda \in \rho(A)$. The next identity is the reason for many of the nice properties of the resolvent set $\rho(A)$ and the *resolvent map*

$$\rho(A) \ni \lambda \mapsto R(\lambda, A) \in \mathcal{L}(X).$$

1.2 Resolvent Equation. *For* $\lambda, \mu \in \rho(A)$, *one has*

$$(1.2) \qquad\qquad R(\lambda, A) - R(\mu, A) = (\mu - \lambda)R(\lambda, A)R(\mu, A).$$

PROOF. The definition of the resolvent implies

$$[\lambda R(\lambda, A) - AR(\lambda, A)]R(\mu, A) = R(\mu, A)$$
and
$$R(\lambda, A)[\mu R(\mu, A) - AR(\mu, A)] = R(\lambda, A).$$

If we subtract these equations and use the fact that $R(\lambda, A)$ and $R(\mu, A)$ commute, we obtain (1.2).　　　　　　　　　　　　　　　□

The basic properties of the resolvent set and the resolvent map are now collected in the following proposition.

1.3 Proposition. For a closed operator $A : D(A) \subset X \to X$, the following properties hold.

(i) The resolvent set $\rho(A)$ is open in $\mathbb{C}$, and for $\mu \in \rho(A)$ one has

$$(1.3) \qquad R(\lambda, A) = \sum_{n=0}^{\infty} (\mu - \lambda)^n R(\mu, A)^{n+1}$$

for all $\lambda \in \mathbb{C}$ satisfying $|\mu - \lambda| < {}^1/_{\|R(\mu,A)\|}$.

(ii) The resolvent map $\lambda \mapsto R(\lambda, A)$ is locally analytic with

$$(1.4) \qquad \frac{d^n}{d\lambda^n} R(\lambda, A) = (-1)^n n! \, R(\lambda, A)^{n+1} \qquad \text{for all } n \in \mathbb{N}.$$

(iii) Let $\lambda_n \in \rho(A)$ with $\lim_{n \to \infty} \lambda_n = \lambda_0$. Then $\lambda_0 \in \sigma(A)$ if and only if $\lim_{n \to \infty} \|R(\lambda_n, A)\| = \infty$.

PROOF. (i) For $\lambda \in \mathbb{C}$ write

$$\lambda - A = \mu - A + \lambda - \mu = [I - (\mu - \lambda)R(\mu, A)](\mu - A).$$

This operator is bijective if $[I - (\mu - \lambda)R(\mu, A)]$ is invertible, which is the case for $|\mu - \lambda| < {}^1/_{\|R(\mu,A)\|}$. The inverse is then obtained as

$$R(\lambda, A) = R(\mu, A)[I - (\mu - \lambda)R(\mu, A)]^{-1} = \sum_{n=0}^{\infty} (\mu - \lambda)^n R(\mu, A)^{n+1}.$$

Assertion (ii) follows immediately from the series representation (1.3) for the resolvent.

To show (iii) we use (i), which implies $\|R(\mu, A)\| \geq {}^1/_{\text{dist}(\mu, \sigma(A))}$ for all $\mu \in \rho(A)$. This already proves one implication. For the converse, assume that $\lambda_0 \in \rho(A)$. Then the continuous resolvent map remains bounded on the compact set $\{\lambda_n : n \geq 0\}$. This contradicts the assumption that $\lim_{n \to \infty} \|R(\lambda_n, A)\| = \infty$; hence $\lambda_0 \in \sigma(A)$. $\square$

As an immediate consequence, we have that the spectrum $\sigma(A)$ is a closed subset of $\mathbb{C}$. Nothing more can be said in general (see the examples below). However, if A is bounded, it follows that

$$\sigma(A) \subset \{\lambda \in \mathbb{C} : |\lambda| \leq \|A\|\},$$

because

$$R(\lambda, A) = \frac{1}{\lambda}\left(1 - \frac{A}{\lambda}\right)^{-1} = \sum_{n=0}^{\infty} \frac{A^n}{\lambda^{n+1}}$$

exists for all $|\lambda| > \|A\|$. In addition, an application of Liouville's theorem to the resolvent map implies $\sigma(A) \neq \emptyset$ (see [TL80, Chap. V, Thm. 3.2]).

1.4 Corollary. For a bounded operator A on a Banach space X, the spectrum $\sigma(A)$ is always compact and nonempty; hence its spectral radius

$$\mathrm{r}(A) := \sup\{|\lambda| : \lambda \in \sigma(A)\} = \lim_{n \to \infty} \|A^n\|^{1/n}$$

is finite and satisfies $\mathrm{r}(A) \leq \|A\|$.

The above formula for the spectral radius is called the *Hadamard formula* because it resembles the well-known Hadamard formula for the radius of convergence of a power series. For its proof we refer to [TL80, Chap. V, Thm. 3.5] or [Yos65, XIII.2, Thm. 3].

Before proceeding with a more detailed analysis of $\sigma(A)$, we show by some simple examples that for unbounded operators $\sigma(A)$ can be any closed subset of $\mathbb{C}$.

1.5 Examples. (i) On $X := C[0,1]$ take the differential operators

$$A_i f := f' \qquad \text{for } i = 1, 2$$

with domain

$$D(A_1) := C^1[0,1] \quad \text{and}$$
$$D(A_2) := \{ f \in C^1[0,1] : f(1) = 0 \}.$$

Then

$$\sigma(A_1) = \mathbb{C},$$

because for each $\lambda \in \mathbb{C}$ one has $(\lambda - A_1)\varepsilon_\lambda = 0$ for $\varepsilon_\lambda := e^{\lambda s}$, $0 \le s \le 1$. On the other hand,

$$\sigma(A_2) = \emptyset,$$

because

$$R_\lambda f(s) := \int_s^1 e^{\lambda(s-t)} f(t)\, dt, \quad 0 \le s \le 1, \, f \in X,$$

yields the inverse of $(\lambda - A_2)$ for every $\lambda \in \mathbb{C}$.

(ii) Take any nonempty, closed subset $\Omega \subset \mathbb{C}$. On the space $X := C_0(\Omega)$ consider the multiplication operator

$$M f(\lambda) := \lambda \cdot f(\lambda)$$

for $\lambda \in \Omega$, $f \in X$. From Proposition I.3.2 we obtain that

$$\sigma(M) = \Omega.$$

As a next step, we look at the fine structure of the spectrum. We start with a particularly important subset of $\sigma(A)$.

1.6 Definition. *For a closed operator* $A : D(A) \subseteq X \to X$, *we call*

$$P\sigma(A) := \{ \lambda \in \mathbb{C} : \lambda - A \text{ is not injective} \}$$

the point spectrum *of* A. *Moreover, each* $\lambda \in P\sigma(A)$ *is called an* eigenvalue, *and each* $0 \ne x \in D(A)$ *satisfying* $(\lambda - A)x = 0$ *is an* eigenvector *of* A *(corresponding to* λ).

In most cases, the eigenvalues of an operator are simpler to determine than arbitrary spectral values. However, they do not, in general, exhaust the entire spectrum.

1.7 Examples. (i) For the operator A_1 in Example 1.5.(i), one has

$$\sigma(A_1) = P\sigma(A_1) = \mathbb{C}.$$

(ii) In contrast, for the multiplication operator M in Example 1.5.(ii) one has

$$\sigma(M) = \Omega, \quad \text{but } P\sigma(M) = \{\lambda \in \mathbb{C} : \lambda \text{ is isolated in } \Omega\}.$$

As a variant of the point spectrum, we introduce the following larger subset of $\sigma(A)$.

1.8 Definition. *For a closed operator* $A : D(A) \subseteq X \to X$, *we call*

$$A\sigma(A) := \left\{ \lambda \in \mathbb{C} : \begin{array}{l} \lambda - A \text{ is not injective or} \\ \text{rg}(\lambda - A) \text{ is not closed in } X \end{array} \right\}$$

the approximate point spectrum *of* A.

The inclusion $P\sigma(A) \subset A\sigma(A)$ is evident from the definition, but the reason for calling it "approximate point spectrum" is not. This is explained by the next lemma.

1.9 Lemma. *For a closed operator* $A : D(A) \subset X \to X$ *and a number* $\lambda \in \mathbb{C}$ *one has* $\lambda \in A\sigma(A)$; *i.e.,* λ *is an* approximate eigenvalue, *if and only if there exists a sequence* $(x_n)_{n \in \mathbb{N}} \subset D(A)$, *called an* approximate eigenvector, *such that* $\|x_n\| = 1$ *and* $\lim_{n \to \infty} \|Ax_n - \lambda x_n\| = 0$.

PROOF. We only have to consider the case in which $\lambda - A$ is injective. As usual, we denote by $X_1 := \big(D(A), \|\cdot\|_A\big)$ the first Sobolev space for A; cf. Section II.2.c. Then the inverse $(\lambda - A)^{-1} : \text{rg}(\lambda - A) \to X_1$ exists and, by the closed graph theorem, is unbounded if and only if $\text{rg}(\lambda - A)$ is not closed. On the other hand, if $(\lambda - A)^{-1} : \text{rg}(\lambda - A) \to X$ is bounded, the closedness of A implies the closedness of $\text{rg}(\lambda - A)$. Hence $(\lambda - A)^{-1} : X \to X_1$ is unbounded if and only if $(\lambda - A)^{-1} : X \to X$ is unbounded, and this property can be expressed by the condition above. $\qquad\square$

The approximate point spectrum generalizes the point spectrum. However, as we show in the following corollary, it has the advantage that it can be empty only if $\sigma(A) = \emptyset$ or $\sigma(A) = \mathbb{C}$.

1.10 Proposition. *For a closed operator* $A : D(A) \subset X \to X$, *the topological boundary* $\partial\sigma(A)$ *of the spectrum* $\sigma(A)$ *is contained in the approximate point spectrum* $A\sigma(A)$.

PROOF. For each $\lambda_0 \in \partial\sigma(A) \subseteq \sigma(A)$ we can find a sequence $(\lambda_n)_{n\in\mathbb{N}} \subseteq \rho(A)$ such that $\lambda_n \to \lambda_0$. By Proposition 1.3.(iii), using the uniform boundedness principle and passing to a subsequence, we find $x \in X$ such that $\lim_{n\to\infty} \|R(\lambda_n, A)x\| = \infty$. Define $y_n \in D(A)$ by

$$y_n := \frac{R(\lambda_n, A)x}{\|R(\lambda_n, A)x\|}.$$

The identity

$$(\lambda_0 - A)y_n = (\lambda_0 - \lambda_n)y_n + (\lambda_n - A)y_n$$

shows that (y_n) is an approximate eigenvector corresponding to λ_0. $\square$

The remaining part of the spectrum is now taken care of by the following definition.

1.11 Definition. *For a closed operator* $A : D(A) \subseteq X \to X$, *we call*

$$R\sigma(A) := \{\lambda \in \mathbb{C} : \text{rg}(\lambda - A) \text{ is not dense in } X\}$$

the residual spectrum of A.

All possibilities for $\lambda - A$ *not* being bijective are now covered by Definitions 1.8 and 1.11, and hence

$$\sigma(A) = A\sigma(A) \cup R\sigma(A).$$

However, there is no reason for the union to be disjoint. It is easy to find examples by applying the following very useful dual characterization of $R\sigma(A)$. Note that we now need a dense domain in order to define the adjoint operator (see Definition A.12).

1.12 Proposition. *For a closed, densely defined operator* A, *the residual spectrum* $R\sigma(A)$ *coincides with the point spectrum* $P\sigma(A')$ *of* A'.

PROOF. The closure of $\text{rg}(\lambda - A)$ is different from X if and only if there exists a linear form $0 \neq x' \in X'$ vanishing on $\text{rg}(\lambda - A)$. By the definition of A', this means $x' \in D(A')$ and $(\lambda - A')x' = 0$. $\square$

In the next theorem we show that for each $\lambda_0 \in \rho(A)$ there is a canonical relation, called the *spectral mapping theorem*, between the spectrum of the unbounded operator A and the spectrum of the bounded operator $R(\lambda_0, A)$. This allows us to transfer results from the spectral theory of bounded operators to the unbounded case.

1.13 Spectral Mapping Theorem for the Resolvent. *Let* $A : D(A) \subseteq X \to X$ *be a closed operator with nonempty resolvent set* $\rho(A)$.

(i) $\sigma\big(R(\lambda_0, A)\big) \setminus \{0\} = (\lambda_0 - \sigma(A))^{-1} := \big\{\frac{1}{\lambda_0 - \mu} : \mu \in \sigma(A)\big\}$ *for each* $\lambda_0 \in \rho(A)$.

(ii) *Analogous statements hold for the point, approximate point, and residual spectra of* A *and* $R(\lambda_0, A)$.

PROOF. For $0 \neq \mu \in \mathbb{C}$ and $\lambda_0 \in \rho(A)$ we have

$$(\mu - R(\lambda_0, A))x = \mu\big[(\lambda_0 - \tfrac{1}{\mu}) - A\big]R(\lambda_0, A)x \qquad \text{for } x \in X,$$
$$= \mu R(\lambda_0, A)\big[(\lambda_0 - \tfrac{1}{\mu}) - A\big]x \qquad \text{for } x \in D(A).$$

This identity shows that

$$\ker\big(\mu - R(\lambda_0, A)\big) = \ker\big[(\lambda_0 - \tfrac{1}{\mu}) - A\big]$$

and

$$\mathrm{rg}\big(\mu - R(\lambda_0, A)\big) = \mathrm{rg}\big[(\lambda_0 - \tfrac{1}{\mu}) - A\big].$$

Recalling Definitions 1.6, 1.8, and 1.11 for the various parts of the spectrum, we see that $\mu \in P\sigma\big(R(\lambda_0, A)\big)$ if and only if $(\lambda_0 - 1/\mu) \in P\sigma(A)$ and similarly for the approximate point spectrum and the residual spectrum. This proves assertion (ii), and hence (i). $\qquad\qquad\square$

This relation between $\sigma(A)$ and $\sigma\big(R(\lambda_0, A)\big)$ determines the spectral radius of $R(\lambda_0, A)$.

1.14 Corollary. *For each $\lambda_0 \in \rho(A)$ one has*

$$(1.5) \qquad \mathrm{dist}\big(\lambda_0, \sigma(A)\big) = \frac{1}{\mathrm{r}\big(R(\lambda_0, A)\big)} \geq \frac{1}{\|R(\lambda_0, A)\|}.$$

The Spectral Mapping Theorem for the Resolvent combined with the Riesz–Schauder theory for compact operators, cf. [TL80, Sect. V.7], [Yos65, X.5], or [Lan93, Chap. XVII], gives the following result. It states that, as in finite dimensions, for resolvent compact operators (cf. Definition II.5.7) the spectrum and the point spectrum coincide.

1.15 Corollary. *If the operator A has compact resolvent, then*

$$\sigma(A) = P\sigma(A).$$

We now study so-called *spectral decompositions*, which are one of the most important features of spectral theory. First, we recall briefly their construction in the bounded case (see, e.g., [DS58, Sect. VII.3], [GGK90, I.2], or [TL80, Sect. V.9]).

Let $T \in \mathcal{L}(X)$ be a bounded operator and assume that the spectrum $\sigma(T)$ can be decomposed as

$$(1.6) \qquad\qquad \sigma(T) = \sigma_c \cup \sigma_u,$$

where σ_c, σ_u are closed and disjoint sets. From the functional calculus (already used in Section I.2.b) one obtains the associated *spectral projection*

$$(1.7) \qquad\qquad P := P_c := \frac{1}{2\pi \mathrm{i}} \int_\gamma R(\lambda, T)\, d\lambda,$$

where γ is a Jordan path in the complement of σ_u and enclosing σ_c. This projection commutes with T and yields the *spectral decomposition*

$$X = X_c \oplus X_u$$

with the T-invariant spaces $X_c := \operatorname{rg} P$, $X_u := \ker P$. The restrictions $T_c \in \mathcal{L}(X_c)$ and $T_u \in \mathcal{L}(X_u)$ of T satisfy

$$(1.8) \qquad \sigma(T_c) = \sigma_c \ \text{ and } \ \sigma(T_u) = \sigma_u,$$

a property that characterizes the above decomposition of X and T in a unique way.

For unbounded operators A and an arbitrary decomposition of the spectrum $\sigma(A)$ into closed sets it is not always possible to find an associated spectral decomposition (for counterexamples see [EN00, Exer. IV.2.30] or [Nag86, A-III, Expl. 3.2]). However, if one of these sets is compact, the spectral mapping theorem for the resolvent allows us to deduce the result from the bounded case. To prove this, we first need the following lemma.

1.16 Lemma. *Let Y be a Banach space continuously embedded in X. If $\lambda \in \rho(A)$ such that $R(\lambda, A)Y \subset Y$, then $\lambda \in \rho(A_|)$ and $R(\lambda, A_|) = R(\lambda, A)_|$.*

PROOF. By the definition of $D(A_|)$ and because $R(\lambda, A)Y \subseteq Y$, we already know that $R(\lambda, A)_|$ maps Y onto $D(A_|)$ and therefore is the algebraic inverse of $\lambda - A_|$. To show that it is bounded in Y, it suffices to observe that it is a closed, everywhere defined operator. □

1.17 Proposition. *Let $A : D(A) \subset X \to X$ be a closed operator such that its spectrum $\sigma(A)$ can be decomposed into the disjoint union of two closed subsets σ_c and σ_u; i.e.,*

$$\sigma(A) = \sigma_c \cup \sigma_u.$$

If σ_c is compact, then there exists a unique spectral decomposition $X = X_c \oplus X_u$ for A in the following sense.

(i) *X_c and X_u are A-invariant.*

(ii) *The restriction $A_c := A_{|X_c}$ is bounded on the Banach space X_c.*

(iii) *$X_1^A = X_c \oplus (X_u)_1^{A_u}$, where $A_u := A_{|X_u}$ (and X_1^A denotes the first Sobolev space with respect to A as introduced in Exercise II.2.22.(1)).*

(iv) *$A = A_c \oplus A_u$.*

(v) *$\sigma(A_c) = \sigma_c$ and $\sigma(A_u) = \sigma_u$.*

(vi) *If $X = X_1 \oplus X_2$ for two A-invariant closed subspaces X_1 and X_2 of X such that $A_{|X_1}$ is bounded, $\sigma(A_{|X_1}) = \sigma_c$ and $\sigma(A_{|X_2}) = \sigma_u$, then $X_1 = X_c$ and $X_2 = X_u$.*

PROOF. When A is bounded, we have already indicated a proof based on Formula (1.7). Therefore, we may assume A to be unbounded and fix some $\lambda \in \rho(A)$. Then $0 \in \sigma(R(\lambda, A))$. Hence, by Theorem 1.13, we obtain

$$(1.9) \qquad \sigma\big(R(\lambda, A)\big) = (\lambda - \sigma_c)^{-1} \bigcup \big((\lambda - \sigma_u)^{-1} \cup \{0\}\big)$$
$$=: \tau_c \cup \tau_u,$$

where τ_c, τ_u are compact and disjoint subsets of $\mathbb{C}$. Now let P be the spectral projection for $R(\lambda, A)$ associated with the decomposition (1.9) and put $X_c := \operatorname{rg} P$, $X_u := \ker P$. Because $R(\lambda, A)$ and P commute, we have $R(\lambda, A)X_c \subseteq X_c$, and Lemma 1.16 implies

$$(1.10) \qquad \lambda \in \rho(A_c) \qquad \text{and} \qquad R(\lambda, A_c) = R(\lambda, A)_{|X_c}.$$

Moreover, we know that $\sigma\big(R(\lambda, A_c)\big) = \tau_c \not\ni 0$. Therefore, the operator $A_c = \lambda - R(\lambda, A_c)^{-1}$ is bounded on X_c and we obtain (ii).

To show (i) we observe that $X_c \subseteq D(A)$ and $AX_c = A_c X_c \subseteq X_c$; i.e., X_c is A-invariant. Because $A(I - P)x = (I - P)Ax$ for $x \in D(A)$, also $X_u = \operatorname{rg}(I - P)$ is A-invariant.

To verify (iii), observe that by similar arguments as above we obtain

$$(1.11) \qquad \lambda \in \rho(A_u) \qquad \text{and} \qquad R(\lambda, A_u) = R(\lambda, A)_{|X_u}.$$

Combining this with (1.10) yields

$$\begin{aligned} X_c + D(A_u) &= R(\lambda, A_c)X_c + R(\lambda, A_u)X_u \\ &\subseteq D(A) = R(\lambda, A)(X_c + X_u) \\ &\subseteq R(\lambda, A_c)X_c + R(\lambda, A_u)X_u \\ &= X_c + D(A_u); \end{aligned}$$

i.e., $X_1^A = X_c + D(A_u)$. Because $P \in \mathcal{L}(X)$, the restriction $P_{|X_1^A} : X_1^A \to X_1^A$ is closed and therefore bounded by the closed graph theorem. This proves (iii), and assertion (iv) then follows from (ii) and (iii).

Finally, (v) is a consequence of the Spectral Mapping Theorem 1.13 and (1.9), (1.10), (1.11), and (vi) follows from Theorem 1.13 and the uniqueness of the spectral decomposition for bounded operators; see [GGK90, Prop. I.2.4]. $\qquad\qquad\square$

1.18 Isolated Singularities. We now sketch a particularly important case of the above decomposition that occurs when $\sigma_c = \{\mu\}$ consists of a single point only. This means that μ is isolated in $\sigma(A)$ and therefore the holomorphic function $\lambda \mapsto R(\lambda, A)$ can be expanded as a Laurent series

$$R(\lambda, A) = \sum_{n=-\infty}^{\infty} (\lambda - \mu)^n U_n$$

for $0 < |\lambda - \mu| < \delta$ and some sufficiently small $\delta > 0$. The coefficients U_n of this series are bounded operators given by the formulas

$$(1.12) \qquad U_n = \frac{1}{2\pi i} \int_\gamma \frac{R(\lambda, A)}{(\lambda - \mu)^{n+1}} \, d\lambda, \qquad n \in \mathbb{Z},$$

where γ is, for example, the positively oriented boundary of the disc with radius $\delta/2$ centered at μ. The coefficient U_{-1} is exactly the spectral projection P corresponding to the decomposition $\sigma(A) = \{\mu\} \cup (\sigma(A) \setminus \{\mu\})$ of the spectrum of A (cf. (1.7)). It is called the *residue* of $R(\cdot, A)$ at μ. From (1.12) (or using the multiplicativity of the functional calculus in [TL80, Thm. V.8.1]), one deduces the identities

$$(1.13) \qquad \begin{aligned} U_{-(n+1)} &= (A - \mu)^n P \qquad \text{and} \\ U_{-(n+1)} \cdot U_{-(m+1)} &= U_{-(n+m+1)} \end{aligned}$$

for $n, m \geq 0$. If there exists $k > 0$ such that $U_{-k} \neq 0$ and $U_{-n} = 0$ for all $n > k$, then the spectral value μ is called a *pole of $R(\cdot, A)$ of order k*. In view of (1.13), this is true if and only if $U_{-k} \neq 0$ and $U_{-(k+1)} = 0$. Moreover, we can obtain U_{-k} as

$$U_{-k} = \lim_{\lambda \to \mu} (\lambda - \mu)^k R(\lambda, A).$$

The dimension of the spectral subspace $\operatorname{rg} P$ is called the *algebraic multiplicity* m_a of μ, and $m_g := \dim \ker(\mu - A)$ is the *geometric multiplicity*. In the case $m_a = 1$, we call μ an *algebraically simple* (or *first-order*) *pole*.

If k is the order of the pole, where we set $k = \infty$ if $R(\cdot, A)$ has an essential singularity at μ, one can show the inequalities

$$(1.14) \qquad m_g + k - 1 \leq m_a \leq m_g \cdot k$$

if we put $\infty \cdot 0 := \infty$. This implies that

(i) $m_a < \infty$ if and only if μ is a pole with $m_g < \infty$, and

(ii) if μ is a pole of order k, then $\mu \in P\sigma(A)$ and $\operatorname{rg} P = \ker(\mu - A)^k$.

For proofs of these facts we refer to [GGK90, Chap. II], [Kat80, III.5], [TL80, V.10], or [Yos65, VIII.8].

1.19 The Essential Spectrum. As we already mentioned above, spectral decomposition is a powerful method to split an operator on a Banach space into two, it is hoped simpler, parts acting on invariant subspaces. In this paragraph we present the tools for a decomposition in which one of these subspaces is finite-dimensional. The results are used in Sections 4, VI.3, and VI.4. We start with the following notion.

An operator $S \in \mathcal{L}(X)$ on a Banach space X is called a *Fredholm operator* if

$$\dim \ker S < \infty \qquad \text{and} \qquad \dim {}^X/_{\mathrm{rg}\,S} < \infty.$$

For $T \in \mathcal{L}(X)$, we then define its *Fredholm domain* $\rho_{\mathrm{F}}(T)$ by

$$\rho_{\mathrm{F}}(T) := \{\lambda \in \mathbb{C} : \lambda - T \text{ is a Fredholm operator}\},$$

and call its complement

$$\sigma_{\mathrm{ess}}(T) := \mathbb{C} \setminus \rho_{\mathrm{F}}(T)$$

the *essential spectrum* of the operator T. One can show (see for instance [GGK90, Chap. XI, Thm. 5.1]) that
(1.15)
$$S \text{ is a Fredholm operator} \iff \begin{cases} \text{there exists } T \in \mathcal{L}(X) \text{ such that} \\ I - TS \text{ and } I - ST \text{ are compact.} \end{cases}$$

Using this fact, an equivalent characterization of $\sigma_{\mathrm{ess}}(T)$ is obtained through the *Calkin algebra* $\mathcal{C}(X) := {}^{\mathcal{L}(X)}/_{\mathcal{K}(X)}$, where $\mathcal{K}(X)$ stands for the two-sided closed ideal in $\mathcal{L}(X)$ of all compact operators. In fact, $\mathcal{C}(X)$ equipped with the quotient norm

$$\|\widehat{T}\| := \mathrm{dist}(T, \mathcal{K}(X)) = \inf\{\|T - K\| : K \in \mathcal{K}(X)\}$$

for $\widehat{T} := T + \mathcal{K}(X) \in \mathcal{C}(X)$ is a Banach algebra with unit. Then, by the equivalence in (1.15), we have

$$\rho_{\mathrm{F}}(T) = \rho(\widehat{T})$$

and

$$\sigma_{\mathrm{ess}}(T) = \sigma(\widehat{T})$$

for all $T \in \mathcal{L}(X)$, where the spectrum of $\widehat{T}$ is defined in the Banach algebra $\mathcal{C}(X)$ (see [CPY74, Chap. 1]). In particular, this implies that $\sigma_{\mathrm{ess}}(T)$ is closed and, if X is infinite-dimensional, nonempty.

In the sequel, we also use the notation

$$\|T\|_{\mathrm{ess}} := \|\widehat{T}\|$$

and

$$\mathrm{r}_{\mathrm{ess}}(T) := \mathrm{r}(\widehat{T}) = \sup\{|\lambda| : \lambda \in \sigma_{\mathrm{ess}}(T)\}$$

for the *essential norm* and the *essential spectral radius*, respectively, of the operator T. Because $\|T\|_{\mathrm{ess}} = \|T + K\|_{\mathrm{ess}}$ for every compact operator K on X, we have

$$\mathrm{r}_{\mathrm{ess}}(T + K) = \mathrm{r}_{\mathrm{ess}}(T)$$

for all $K \in \mathcal{K}(X)$. Moreover, using the Hadamard formula for the spectral radius of $\widehat{T}$, cf. Corollary 1.4, we obtain the equality

$$r_{ess}(T) = \lim_{n \to \infty} \|T^n\|_{ess}^{1/n}.$$

For a detailed analysis of the essential spectrum of an operator, we refer to [Kat80, Sect. IV.5.6], [GGK90, Chap. XVII], or [Gol66, Sect. IV.2]. Here, we only recall that the poles of $R(\cdot, T)$ with finite algebraic multiplicity belong to $\rho_F(T)$. Conversely, an element of the unbounded connected component of $\rho_F(T)$ either belongs to $\rho(T)$ or is a pole of finite algebraic multiplicity. Thus $r_{ess}(T)$ can be characterized by

$$(1.16) \quad r_{ess}(T) = \inf \left\{ r > 0 : \begin{array}{l} \text{each } \lambda \in \sigma(T) \text{ satisfying } |\lambda| > r \text{ is a pole} \\ \text{of } R(\cdot, T) \text{ of finite algebraic multiplicity} \end{array} \right\}.$$

1.20 Exercises. (1) Let A be a complex $n \times n$ matrix. Show that for $\lambda \in \sigma(A)$

   (i) The pole order of $R(\cdot, A)$ in λ,

   (ii) The size of the largest Jordan block of A corresponding to λ,

   (iii) The multiplicity of λ as zero of the minimal polynomial m_A of A
coincide.

(2) Compute the spectrum $\sigma(A)$ for the following operators on the Banach space $X := C[0,1]$.

   (i) $Af := \frac{1}{s(1-s)} \cdot f(s)$, $D(A) := \{f \in X : Af \in X\}$.

   (ii) $Bf(s) := is^2 \cdot f(s)$, $D(B) := X$.

   (iii) $Cf(s) := f'(s)$, $D(C) := \{C^1[0,1] : f(0) = 0\}$.

   (iv) $Df(s) := f'(s)$, $D(D) := \{f \in C^1[0,1] : f'(1) = 0\}$.

   (v) $Ef(s) := f'(s)$, $D(E) := \{f \in C^1[0,1] : f(0) = f(1)\}$.

   (vi) $Ff(s) := f''(s)$, $D(G) := C^2[0,1]$.

   (vii) $Gf(s) := f''(s)$, $D(H) := \{f \in C^2[0,1] : f(0) = f(1) = 0\}$.

(viii) $Hf(s) := f''(s)$, $D(J) := \{f \in C^2[0,1] : f''(0) = 0\}$.

Which of these operators are generators on X?

(3) Consider $X := C_0(\mathbb{R}, \mathbb{C}^2)$ and

$$Af(s) := f'(s) + Mf(s), \quad s \in \mathbb{R},$$

where $M := \left(\begin{smallmatrix} 0 & 1 \\ 1 & 0 \end{smallmatrix} \right)$ and $D(A) := C_0^1(\mathbb{R}, \mathbb{C}^2)$. Show that $\sigma(A)$ decomposes into $-1 + i\mathbb{R}$ and $1 + i\mathbb{R}$ and that there exists a corresponding spectral decomposition. (Hint: Transform M into a diagonal matrix.)

(4) Let A be an operator on a Banach space X and let B be a restriction of A. If B is surjective and A is injective, then $A = B$. In particular, $A = B$ if $B \subset A$ and $\rho(A) \cap \rho(B) \neq \emptyset$.

b. Spectral Theory for Generators

The Hille–Yosida theorem already ensures that the spectrum of the generator of a strongly continuous semigroup always lies in a proper left half-plane and thus satisfies a property not shared by arbitrary closed operators. In this section we study the spectrum of generators and its relation to the spectrum of the semigroup operators more closely.

For (unbounded) semigroup generators, the role played by the spectral radius in the case of bounded operators is taken over by the following quantity.

1.21 Definition. *Let $A : D(A) \subset X \to X$ be a closed operator. Then*

$$s(A) := \sup\{\operatorname{Re}\lambda : \lambda \in \sigma(A)\}$$

is called the spectral bound *of A.*

Note that $s(A)$ can be any real number including $-\infty$ (if $\sigma(A) = \emptyset$) and $+\infty$. For the generator A of a strongly continuous semigroup $\mathcal{T} = (T(t))_{t\geq 0}$, however, the spectral bound $s(A)$ is always dominated by the *growth bound*

$$\omega_0 := \omega_0(\mathcal{T}) := \inf\left\{w \in \mathbb{R} : \begin{array}{l} \text{there exists } M_w \geq 1 \text{ such that} \\ \|T(t)\| \leq M_w e^{wt} \text{ for all } t \geq 0 \end{array}\right\}$$

of the semigroup[1] (see Definition I.1.5 and Corollary II.1.13).

We now show that ω_0 is related to the spectrum (more precisely, to the spectral radius) of the operators $T(t)$.

1.22 Proposition. *For the spectral bound $s(A)$ of a generator A and for the growth bound ω_0 of the generated semigroup $(T(t))_{t\geq 0}$, one has*

(1.17)
$$-\infty \leq s(A) \leq \omega_0 = \inf_{t>0} \frac{1}{t} \log \|T(t)\| = \lim_{t\to\infty} \frac{1}{t} \log \|T(t)\|$$
$$= \frac{1}{t_0} \log \operatorname{r}(T(t_0)) < \infty$$

for each $t_0 > 0$. In particular, the spectral radius of the semigroup operator $T(t)$ is given by

(1.18)
$$\operatorname{r}(T(t)) = e^{\omega_0 t} \qquad \text{for all } t \geq 0.$$

For the proof we need the following elementary fact.

---

[1] Occasionally, we write "$\omega_0(A)$" instead of "$\omega_0(\mathcal{T})$," because by Theorem II.1.4 the semigroup $\mathcal{T}$ is uniquely determined by its generator A.

1.23 Lemma. *Let $\xi : \mathbb{R}_+ \to \mathbb{R}$ be bounded on compact intervals and subadditive; i.e., $\xi(s + t) \le \xi(s) + \xi(t)$ for all $s, t \ge 0$. Then*

$$\inf_{t>0} \frac{\xi(t)}{t} = \lim_{t \to \infty} \frac{\xi(t)}{t}$$

exists.

PROOF. Fix $t_0 > 0$ and write $t = kt_0 + s$ with $k \in \mathbb{N}$, $s \in [0, t_0)$. The subadditivity implies

$$\frac{\xi(t)}{t} \le \frac{1}{kt_0} \left(\xi(kt_0) + \xi(s) \right) \le \frac{\xi(t_0)}{t_0} + \frac{\xi(s)}{kt_0}.$$

Because $k \to \infty$ if $t \to \infty$, we obtain

$$\varlimsup_{t \to \infty} \frac{\xi(t)}{t} \le \frac{\xi(t_0)}{t_0}$$

for each $t_0 > 0$ and therefore

$$\varlimsup_{t \to \infty} \frac{\xi(t)}{t} \le \inf_{t>0} \frac{\xi(t)}{t} \le \varliminf_{t \to \infty} \frac{\xi(t)}{t},$$

which proves the assertion. $\qquad\qquad\qquad\qquad\qquad\qquad\qquad\qquad\square$

PROOF OF PROPOSITION 1.22. Because the function

$$t \mapsto \xi(t) := \log \|T(t)\|$$

satisfies the assumptions of Lemma 1.23, we can define

$$v := \inf_{t>0} \frac{1}{t} \log \|T(t)\| = \lim_{t \to \infty} \frac{1}{t} \log \|T(t)\|.$$

From this identity, it follows that

$$e^{vt} \le \|T(t)\|$$

for all $t \ge 0$; hence $v \le \omega_0$ by the definition of ω_0. Now choose $w > v$. Then there exists $t_0 > 0$ such that

$$\frac{1}{t} \log \|T(t)\| \le w$$

for all $t \ge t_0$; hence $\|T(t)\| \le e^{wt}$ for $t \ge t_0$. On $[0, t_0]$, the norm of $T(t)$ remains bounded, so we find $M \ge 1$ such that

$$\|T(t)\| \le Me^{wt}$$

for all $t \ge 0$; i.e., $\omega_0 \le w$. In as much as we have already proved that $v \le \omega_0$, this implies $\omega_0 = v$.

To prove the identity $\omega_0 = {}^{1}\!/_{t_0} \log r\big(T(t_0)\big)$, we use the Hadamard formula for the spectral radius in Corollary 1.4; i.e.,

$$r\big(T(t)\big) = \lim_{n\to\infty} \|T(nt)\|^{1/n} = \lim_{n\to\infty} e^{t \cdot {}^{1}\!/_{nt} \log \|T(nt)\|}$$
$$= e^{t \cdot \lim_{n\to\infty} ({}^{1}\!/_{nt} \log \|T(nt)\|)} = e^{t\,\omega_0}.$$

The remaining inequalities have already been proved in Corollary II.1.13.

$\square$

We now state a simple consequence of this proposition.

1.24 Corollary. *For the generator A of a strongly continuous semigroup $(T(t))_{t\geq 0}$ with growth bound $\omega_0 = -\infty$ (e.g., for a nilpotent semigroup) one has*

$$r\big(T(t)\big) = 0 \quad \text{for all } t > 0 \qquad \text{and} \qquad \sigma(A) = \emptyset.$$

The inequalities in (1.17) establish an interesting relation between spectral properties of the generator A, expressed by the spectral bound $s(A)$, and the qualitative behavior of the semigroup $(T(t))_{t\geq 0}$, expressed by the growth bound ω_0. In particular, if spectral and growth bound coincide, we obtain infinite-dimensional versions of the Liapunov Stability Theorem 3.6 below. For general strongly continuous semigroups, however, the situation is more complex, as shown by the following examples and counterexamples.

1.25 Examples. We first discuss (left) translation semigroups on various function spaces (see Section I.3.c and Paragraph II.2.9) and show that the spectra heavily depend on the choice of the Banach space. Before starting the discussion, it is useful to observe that the exponential functions

$$\varepsilon_\lambda(s) := e^{\lambda s}, \quad s \in \mathbb{R},$$

satisfy

$$\tfrac{d}{ds}\varepsilon_\lambda = \lambda\varepsilon_\lambda \qquad \text{for each } \lambda \in \mathbb{C}.$$

Because the generator A of a translation semigroup is the first derivative with appropriate domain (see Paragraph II.2.9), it follows that λ is an eigenvalue of A if and only if ε_λ belongs to the domain $D(A)$.
(i) Consider the (left) translation semigroup $(T(t))_{t\geq 0}$ on the space $X := C_0(\mathbb{R}_+)$. Its generator is

$$Af = f'$$

with domain

$$D(A) = \{f \in C_0(\mathbb{R}_+) \cap C^1(\mathbb{R}_+) : f' \in C_0(\mathbb{R}_+)\}.$$

Therefore, we have $\varepsilon_\lambda \in D(A)$ if and only if $\lambda \in \mathbb{C}$ satisfies $\operatorname{Re}\lambda < 0$. This shows that

$$P\sigma(A) = \{\lambda \in \mathbb{C} : \operatorname{Re}\lambda < 0\}.$$

We have that $s(A) \le \omega_0 \le 0$, because $(T(t))_{t\ge0}$ is a contraction semigroup. This implies, because the spectrum is closed, that

$$\sigma(A) = \{\lambda \in \mathbb{C} : \operatorname{Re}\lambda \le 0\}.$$

The same eigenfunctions ε_λ yield eigenvalues $e^{\lambda t}$ for the operators $T(t)$. Again by the contractivity of $T(t)$ we obtain that

and
$$P\sigma(T(t)) = \{z \in \mathbb{C} : |z| < 1\}$$
$$\sigma(T(t)) = \{z \in \mathbb{C} : |z| \le 1\} \quad \text{for } t > 0.$$

(ii) Next, we consider the (left) translation group $(T(t))_{t\in\mathbb{R}}$ on $X :=$ $C_0(\mathbb{R})$. Then $P\sigma(A) = \emptyset$, because no ε_λ belongs to $D(A)$. However, for each $\alpha \in \mathbb{R}$, the functions

$$f_n(s) := e^{i\alpha s} \cdot e^{-s^2/n}, \quad n \in \mathbb{N},$$

form an approximate eigenvector of A for the approximate eigenvalue $i\alpha$. This shows that

$$A\sigma(A) = \sigma(A) = i\mathbb{R},$$

and analogously

$$\sigma(T(t)) = \{z \in \mathbb{C} : |z| = 1\}.$$

(iii) The nilpotent right translation semigroup $(T(t))_{t\ge0}$ on $X := C_0(0,1]$ satisfies $\omega_0 = -\infty$ (see Example II.3.19), hence it follows from Corollary 1.24 that

$$\sigma(T(t)) = \{0\} \qquad \text{and} \qquad \sigma(A) = \emptyset.$$

In addition, for each $\lambda \in \mathbb{C}$, the resolvent is given by

$$(1.19) \qquad (R(\lambda, A)f)(s) = \int_0^s e^{-\lambda(s-\tau)} f(\tau)\, d\tau, \qquad s \in (0,1], f \in X.$$

(iv) For the periodic translation group on, e.g., $X = C_{2\pi}(\mathbb{R})$ (see Paragraph I.3.15), the functions ε_λ belong to $D(A)$ if and only if $\lambda \in i\mathbb{Z}$. Because A has compact resolvent (use Example II.5.9), we obtain from Corollary 1.15,

$$\sigma(A) = P\sigma(A) = i\mathbb{Z}.$$

The spectra of the operators $T(t)$ are always contained in $\Gamma := \{z \in \mathbb{C} : |z| = 1\}$ and contain the eigenvalues e^{ikt} for $k \in \mathbb{Z}$. Because $\sigma(T(t))$ is closed, it follows that

$$\sigma(T(t)) = \begin{cases} \Gamma & \text{if } t/2\pi \notin \mathbb{Q}, \\ \Gamma_q & \text{if } t/2\pi = p/q \in \mathbb{Q} \text{ with } p \text{ and } q \text{ coprime}, \end{cases}$$

where $\Gamma_q := \{z \in \mathbb{C} : z^q = 1\}$.

In each of these examples there is a close relationship between the spectrum $\sigma(A)$ and the spectra $\sigma(T(t))$ implying $\omega_0 = s(A)$. As we show next this is not always the case.

1.26 Counterexample. Consider the Banach space

$$X := \mathrm{C}_0(\mathbb{R}_+) \cap \mathrm{L}^1(\mathbb{R}_+, e^s ds)$$

of all continuous functions on $\mathbb{R}_+$ that vanish at infinity and are integrable for $e^s\, ds$ endowed with the norm

$$\|f\| := \|f\|_\infty + \|f\|_1 = \sup_{s \geq 0} |f(s)| + \int_0^\infty |f(s)| e^s\, ds.$$

The (left) translations define a strongly continuous semigroup $(T(t))_{t \geq 0}$ on X whose generator is

$$Af = f',$$
$$D(A) = \{f \in X : f \in \mathrm{C}^1(\mathbb{R}_+), f' \in X\}$$

(use Proposition II.2.3). As a first observation, we note that $\|T(t)\| = 1$ for all $t \geq 0$. Thus, we have $\omega_0 = 0$, and hence $s(A) \leq 0$. On the other hand, $\varepsilon_\lambda \in D(A)$ only if $\mathrm{Re}\,\lambda < -1$. Hence, we obtain for the point spectrum

$$P\sigma(A) = \{\lambda \in \mathbb{C} : \mathrm{Re}\,\lambda < -1\}$$

and for the spectral bound $s(A) \geq -1$.

We now show that $\lambda \in \rho(A)$ if $\mathrm{Re}\,\lambda > -1$. In fact, for every $f \in X$ we have that

$$\|\cdot\|_1\text{-}\lim_{t \to \infty} \int_0^t e^{-\lambda s} T(s) f\, ds$$

exists, because $\|T(s)f\|_1 \leq e^{-s} \|f\|_1$ for all $s \geq 0$. Moreover, the limit

$$\|\cdot\|_\infty\text{-}\lim_{t \to \infty} \int_0^t e^{-\lambda s} T(s) f\, ds$$

exists, because $\int_0^\infty e^s |f(s)|\, ds < \infty$. Consequently, the improper integral

$$(1.20) \qquad \int_0^\infty e^{-\lambda s} T(s) f\, ds$$

exists in X for every $f \in X$ and yields the inverse of $\lambda - A$ (see Theorem II.1.10.(i)). We conclude that

$$\sigma(A) = \{\lambda \in \mathbb{C} : \mathrm{Re}\,\lambda \leq -1\}, \quad \text{whence} \quad s(A) = -1,$$

whereas $\omega_0 = 0$ and $\mathrm{r}(T(t)) = 1$ by (1.18). In particular, for $t > 0$, $T(t)$ has spectral values that are not the exponential of a spectral value of A.

The above phenomenon makes the spectral theory of semigroups interesting and nontrivial. Before analyzing carefully what we call the "spectral mapping theorem" for semigroups in Section 2, we first discuss an example showing spectral theory at work.

1.27 Delay Differential Operators. We return to the delay differential operator from Paragraph II.3.29 defined as

$$Af := f' \quad \text{on} \quad D(A) := \{f \in C^1[-1,0] : f'(0) = Lf\}$$

on the Banach space $X := C[-1,0]$ for some linear form $L \in X'$ and try to compute its point spectrum $P\sigma(A)$. As for the above translation semigroups, we see that a function $f \in C[-1,0]$ is an eigenfunction of A only if it is (up to a scalar factor) of the form $f = \varepsilon_\lambda$, where

$$\varepsilon_\lambda(s) := e^{\lambda s}, \quad s \in [-1,0],$$

for some $\lambda \in \mathbb{C}$. However, such a function ε_λ belongs to $D(A)$ if and only if it satisfies the *boundary condition*

which becomes
$$\varepsilon'_\lambda(0) = L\varepsilon_\lambda,$$
$$\lambda = L\varepsilon_\lambda.$$

Therefore, if we define $\xi(\lambda) := \lambda - L\varepsilon_\lambda$, we obtain the point spectrum $P\sigma(A)$ as

$$P\sigma(A) = \{\lambda \in \mathbb{C} : \xi(\lambda) = 0\}.$$

Because $\xi(\cdot)$ is an analytic function on $\mathbb{C}$, its zeros are isolated, and therefore $P\sigma(A)$ is a discrete subset of $\mathbb{C}$.

In order to identify the entire spectrum $\sigma(A)$, we observe that $X_1 := (D(A), \|\cdot\|_A)$ is a closed subspace of $C^1[-1,0]$ and that the canonical injection

$$i : C^1[-1,0] \to C[-1,0]$$

is compact by the Arzelà–Ascoli theorem. Therefore, it follows from Proposition II.5.8 that $R(\lambda, A)$ is a compact operator, and by Corollary 1.15, we obtain

$$\sigma(A) = P\sigma(A).$$

Proposition. *The spectrum of the above delay differential operator consists of isolated eigenvalues only. More precisely, we call*

$$\lambda \mapsto \xi(\lambda) := \lambda - L\varepsilon_\lambda$$

the corresponding characteristic function and obtain

$$\sigma(A) = \{\lambda \in \mathbb{C} : \xi(\lambda) = 0\}.$$

In other words, the spectrum of A consists of the zeros of the *characteristic equation*

$$\xi(\lambda) = 0.$$

For arbitrary $L \in C[-1,0]'$, it is still difficult to determine all complex zeros of the analytic function $\xi(\cdot)$. However, for many applications as in Section 3, it suffices to know the spectral bound $s(A)$. To determine it, we now assume that the linear form L is decomposed as

$$L = L_0 + a\delta_0,$$

where L_0 is a *positive* linear form on $C[-1,0]$ having no *atomic part* in 0. This means that $\lim_{n\to\infty} L_0(f_n) = 0$ whenever $(f_n)_{n\in\mathbb{N}}$ is a bounded sequence in X satisfying $\lim_{n\to\infty} f_n(s) = 0$ for all $-1 \le s < 0$. As usual, δ_0 denotes the point evaluation at 0, and we take $a \in \mathbb{R}$. In this case, we can determine $s(A)$ by discussing the characteristic equation as an equation on $\mathbb{R}$ only.

Corollary. *Consider the above delay differential operator* $(A, D(A))$ *on* $X := C[-1,0]$ *and assume that* $L \in X'$ *is of the form*

$$L = L_0 + a\delta_0$$

for some $a \in \mathbb{R}$ *and some positive* $L_0 \in X'$ *having no atomic part in* 0. *Then the spectral bound* $s(A)$ *is given by*

$$s(A) = \sup\{\lambda \in \mathbb{R} : \lambda = L_0\varepsilon_\lambda + a\},$$

and one has the equivalence

$$s(A) < 0 \iff \|L_0\| + a < 0.$$

PROOF. The characteristic function $\lambda \mapsto \xi(\lambda) := \lambda - L_0\varepsilon_\lambda - a$, considered as a function on $\mathbb{R}$, is continuous and strictly increasing from $-\infty$ to $+\infty$. This holds, because we assumed L_0 to be positive having no atomic part in 0, hence satisfying

$$L_0\varepsilon_\lambda \downarrow 0 \qquad \text{as} \qquad \lambda \to \infty.$$

Therefore, ξ has a unique real zero λ_0 satisfying

$$\lambda_0 < 0 \iff 0 < \xi(0).$$

It remains to show that $\lambda_0 = s(A)$. Take $\lambda = \mu + i\nu \in \sigma(A)$. Using the above characteristic equation, this can be restated as

$$\mu + i\nu = L_0(\varepsilon_\mu\varepsilon_{i\nu}) + a.$$

By taking the real parts in this identity and using the positivity of L_0, we obtain

$$\mu = \text{Re}(L_0(\varepsilon_\mu\varepsilon_{i\nu}) + a) \le |L_0(\varepsilon_\mu\varepsilon_{i\nu})| + a \le L_0(\varepsilon_\mu) + a,$$

which, by the above properties of the characteristic function ξ on $\mathbb{R}$, implies $\mu \le \lambda_0$. Therefore, we conclude that

$$\mu = \text{Re}\,\lambda \le \lambda_0 = s(A)$$

for all $\lambda \in \sigma(A)$. $\qquad\qquad\qquad\qquad\qquad\qquad\qquad\qquad\qquad\qquad\square$

We recommend restating the above results for

$$L_1 f := af(0) + bf(-1)$$

or

$$L_2 f := af(0) + \int_{-1}^{0} k(s)f(s)\,ds$$

with $a \in \mathbb{R}$, $0 \le b$, and $0 \le k \in L^{\infty}[-1,0]$.

1.28 Exercises. (1) Use the rescaling procedure and Counterexample 1.26 to show that for arbitrary real numbers $\alpha < \beta$, there exists a strongly continuous semigroup $(T(t))_{t \ge 0}$ with generator A such that

$$s(A) = \alpha \quad \text{and} \quad \omega_0 = \beta.$$

(2) Let $(T(t))_{t \in \mathbb{R}}$ be a strongly continuous group on X with generator A. Then there exist constants m, $M \ge 1$, v, $w \in \mathbb{R}$ such that

$$\frac{1}{m}\,e^{-vt}\|x\| \le \|T(t)x\| \le Me^{wt}\|x\| \qquad \text{for all } t \ge 0,\ x \in X.$$

Show that

$$-v \le -s(-A) \le s(A) \le w.$$

(3) Let $(T(t))_{t \ge 0}$ be the semigroup from Counterexample 1.26. Find an approximate eigenvector $(f_n)_{n \in \mathbb{N}}$ corresponding to the approximative eigenvalue $\lambda = 1$ of $T(t)$ for $t > 0$.

(4) Modify Counterexample 1.26 to obtain $s(A) = -\infty$, $\omega_0 = 0$. (Hint: Consider $X := C_0(\mathbb{R}_+) \cap L^1(\mathbb{R}_+, e^{x^2}\,dx)$.)

(5*) Consider the translations on

$$X := \left\{ f \in C(\mathbb{R}) : \lim_{s \to \infty} f(s) = \lim_{s \to -\infty} e^{3s}f(s) = 0 \text{ and } \int_{-\infty}^{\infty} e^{2s}|f(s)|\,ds < \infty \right\}$$

endowed with the norm

$$\|f\| := \sup_{s \ge 0} |f(s)| + \sup_{s \le 0} e^{3s}|f(s)| + \int_{-\infty}^{\infty} e^{2s}|f(s)|\,ds.$$

Show that this yields a strongly continuous group on X with growth bound $\omega_0 = 0$, but spectral bound $s(A) < -1$. (Hint: See [Wol81].)

2. Spectral Mapping Theorems

It is our ultimate goal to describe the semigroup $(T(t))_{t \geq 0}$ by the spectrum $\sigma(A)$ of its generator A. However, as we have already seen in Counterexample 1.26, the general case is much more complex. As a first, but essential, step, we now study in detail the relation between the spectrum $\sigma(A)$ of the generator A and the spectrum $\sigma(T(t))$ of the semigroup operators $T(t)$. The intuitive interpretation of $T(t)$ as the exponential "e^{tA}" of A leads us to the following principle.

2.1 Leitmotif. *The spectra $\sigma(T(t))$ of the semigroup operators $T(t)$ should be obtained from the spectrum $\sigma(A)$ of the generator A by a relation of the form*

$$(2.1) \qquad "\sigma(T(t)) = e^{t\sigma(A)} := \{e^{t\lambda} : \lambda \in \sigma(A)\}."$$

a. Examples and Counterexamples

If (2.1), or a similar relation, holds, we say that the semigroup $(T(t))_{t \geq 0}$ and its generator A satisfy a *spectral mapping theorem*. However, before proving results in this direction, we explain in a series of examples and counterexamples what might go wrong.

2.2 Examples. (i) Take a strongly continuous semigroup $(T(t))_{t \geq 0}$ that cannot be extended to a group (e.g., the left translation semigroup on $C_0(\mathbb{R}_+)$; see Paragraph I.3.16). Then $0 \in \sigma(T(t))$ for all $t > 0$, although evidently 0 is never contained in $e^{t\sigma(A)}$.

Therefore, we are led to modify (2.1) and call a *spectral mapping theorem* the relation

$$(\text{SMT}) \qquad \sigma(T(t)) \setminus \{0\} = e^{t\sigma(A)} \quad \text{for } t \geq 0.$$

(ii) For the periodic translation group in Example 1.25.(iv) we have $\sigma(A) = i\mathbb{Z}$ and $\sigma(T(t)) = \Gamma$ if $t/2\pi$ is irrational, hence (SMT) does not hold.

The phenomenon appearing in this example is referred to as a *weak spectral mapping theorem*, meaning that only

$$(\text{WSMT}) \qquad \sigma(T(t)) \setminus \{0\} = \overline{e^{t\sigma(A)}} \setminus \{0\} \quad \text{for } t \geq 0$$

holds.

The above modifications of the spectral mapping theorem are simply caused by properties of the complex exponential map $z \mapsto e^z$ and have no serious consequences for our applications in Section 3. Much more problematic is the failure of (SMT) or (WSMT) due to the particular form of the operator A and the semigroup $(T(t))_{t \geq 0}$.

Such a breakdown always occurs for generators A for which the so-called *spectral bound equal growth bound condition*

(SBeGB) $\mathrm{s}(A) = \omega_0$

does *not* hold. In fact, if $\mathrm{s}(A) < \omega_0$, then

$$e^{t\sigma(A)} \subseteq \left\{ \lambda \in \mathbb{C} : |\lambda| \le e^{t\,\mathrm{s}(A)} \right\},$$

and $\mathrm{r}\big(T(t)\big) = e^{t\,\omega_0} > e^{t\,\mathrm{s}(A)}$ (use Proposition 1.22).

For later reference, it is useful to state this fact explicitly.

2.3 Proposition. *For a strongly continuous semigroup $(T(t))_{t\ge 0}$ with generator A one always has the implications*

$$(\text{SMT}) \implies (\text{WSMT}) \implies (\text{SBeGB}).$$

Therefore, the generator and the semigroup in Counterexample 1.26 do not satisfy (WSMT). Whereas the semigroup in this example was the well-known translation semigroup, the chosen Banach space seems to be artificial. Therefore, we present more examples for a drastic failure of (WSMT) on more natural spaces.

Even for semigroups on Hilbert spaces the spectral mapping theorem may fail.

2.4 Counterexample (on Hilbert Spaces). We start by considering the n-dimensional Hilbert space $X_n := \mathbb{C}^n$ (with the $\| \cdot \|_2$-norm) and the $n \times n$ matrix

$$A_n := \begin{pmatrix} 0 & 1 & 0 & 0 \\ \vdots & \ddots & \ddots & 0 \\ \vdots & & \ddots & 1 \\ 0 & \cdots & \cdots & 0 \end{pmatrix}.$$

Because A_n is nilpotent, we obtain $\sigma(A_n) = \{0\}$. Moreover, the semigroups $\big(e^{tA_n}\big)_{t\ge 0}$ generated by A_n satisfy

$$\big\| e^{tA_n} \big\| \le e^t$$

for $t \ge 0$. We now collect some elementary facts about these matrices.

Lemma. *For the elements $x_n := n^{-1/2}(1, \ldots, 1) \in X_n$ we have $\|x_n\| = 1$ and*

(i) $\|A_n x_n - x_n\| \le n^{-1/2}$,

(ii) $\big\| e^{tA_n} x_n - e^t x_n \big\| \le te^t n^{-1/2}$ *for $t \ge 0$ and $n \in \mathbb{N}$.*

PROOF. Assertion (i) follows directly from the definition, whereas (ii) is obtained from

$$e^{tA_n}x_n - e^t x_n = \int_0^t e^{t-s} e^{sA_n}(A_n x_n - x_n)\, ds$$

(see (1.10) in Lemma II.1.9) and the estimate $\|e^{tA_n}\| \le e^t$.   □

Consider now the Hilbert space

$$X := \bigoplus_{n\in\mathbb{N}}^2 X_n := \Big\{ (x_n)_{n\in\mathbb{N}} : x_n \in X_n \text{ and } \sum_{n\in\mathbb{N}} \|x_n\|^2 < \infty \Big\},$$

with inner product

$$((x_n)\,|\,(y_n)) := \sum_{n\in\mathbb{N}} (x_n\,|\,y_n)$$

on which we define $A := \oplus_{n\in\mathbb{N}}(A_n + in)$ with maximal domain $D(A)$ in X. This operator generates the strongly continuous semigroup $(T(t))_{t\ge 0}$ given by

$$T(t) := \bigoplus_{n\in\mathbb{N}} (e^{int} e^{tA_n})$$

and satisfying ,

$$\|T(t)\| \le \sup_{n\in\mathbb{N}} \|e^{int} e^{tA_n}\| \le e^t$$

for $t \ge 0$. This implies that its growth bound satisfies

$$\omega_0 \le 1.$$

We now show that $s(A) = 0$. For $\lambda \in \mathbb{C}$ with $\operatorname{Re}\lambda > 0$, we have

$$R(\lambda, A_n + in) = R(\lambda - in, A_n) = \sum_{k=0}^{n-1} \frac{A_n^k}{(\lambda - in)^{k+1}}.$$

Because $\|A_n\| = 1$, we conclude that

$$\|R(\lambda, A_n + in)\| \le \sum_{k=0}^{n-1} \frac{1}{|\lambda - in|^{k+1}} \le \frac{1}{|\lambda - in| - 1}$$

for $n \in \mathbb{N}$ sufficiently large. This implies $\sup_{n\in\mathbb{N}} \|R(\lambda, A_n + in)\| < \infty$, and therefore

$$\bigoplus_{n\in\mathbb{N}} (R(\lambda, A_n + in))$$

is a bounded operator on X, which evidently gives the inverse of $(\lambda - A)$. Hence, $s(A) \le 0$, whereas $s(A) \ge 0$ follows from the fact that each in is an eigenvalue of A.

To prove $\omega_0 \geq 1$, we show that $r(T(t_0)) \geq e^{t_0}$ for $t_0 = 2\pi$. Take x_n as in the lemma, identify it with the element $(0, \ldots, x_n, 0, \ldots) \in X$, and consider the sequence $(x_n)_{n \in \mathbb{N}}$ in X. Then $(x_n)_{n \in \mathbb{N}}$ is an approximate eigenvector of $T(2\pi)$ with eigenvalue $e^{2\pi}$. So we have proved the following.

Proposition. *For the strongly continuous semigroup* $(T(t))_{t \geq 0}$ *with*

$$T(t) := \bigoplus_{n \in \mathbb{N}} (e^{int} e^{tA_n})$$

and its generator

$$A := \bigoplus_{n \in \mathbb{N}} (A_n + in)$$

on the Hilbert space $X := \oplus_{n \in \mathbb{N}}^2 X_n$, *one has*

$$s(A) = 0 < \omega_0 = 1.$$

For still more examples we refer to Exercises 2.13.

b. Spectral Mapping Theorems for Semigroups

After having seen so many failures of our Leitmotif 2.1, it is now time to present some positive results. Surprisingly, "most" of (SMT) still holds.

2.5 Spectral Inclusion Theorem. *For the generator* $(A, D(A))$ *of a strongly continuous semigroup* $(T(t))_{t \geq 0}$ *on a Banach space* X, *we have the inclusions*

$$(2.2) \qquad \sigma(T(t)) \supset e^{t\sigma(A)} \qquad \text{for} \qquad t \geq 0.$$

More precisely, for the point, approximate point, and residual spectra the inclusions

$$(2.3) \qquad P\sigma(T(t)) \supset e^{tP\sigma(A)},$$

$$(2.4) \qquad A\sigma(T(t)) \supset e^{tA\sigma(A)},$$

$$(2.5) \qquad R\sigma(T(t)) \supset e^{tR\sigma(A)}$$

hold for all $t \geq 0$.

PROOF. Recalling the identities

$$(2.6) \qquad \begin{aligned} e^{\lambda t}x - T(t)x &= (\lambda - A) \int_0^t e^{\lambda(t-s)} T(s)x\, ds \quad \text{for } x \in X, \\ &= \int_0^t e^{\lambda(t-s)} T(s)(\lambda - A)x\, ds \quad \text{for } x \in D(A) \end{aligned}$$

from Lemma II.1.9, we see that $(e^{\lambda t} - T(t))$ is not bijective if $(\lambda - A)$ fails to be bijective. This proves (2.2).

We now prove (2.4) and, by the same arguments, (2.3). Take $\lambda \in A\sigma(A)$ and a corresponding approximate eigenvector $(x_n)_{n\in\mathbb{N}} \subset D(A)$. Define a new sequence $(y_n)_{n\in\mathbb{N}}$ by

$$y_n := e^{\lambda t}x_n - T(t)x_n = \int_0^t e^{\lambda(t-s)}T(s)(\lambda - A)x_n \, ds.$$

These vectors satisfy for some constant $c > 0$ the estimate

$$\|y_n\| \leq \int_0^t \left\|e^{\lambda(t-s)}T(s)(\lambda - A)x_n\right\| ds \leq c\,\|(\lambda - A)x_n\| \to 0 \text{ as } n \to \infty.$$

Hence, $e^{\lambda t}$ is an approximate eigenvalue of $T(t)$, and $(x_n)_{n\in\mathbb{N}}$ serves as the same approximate eigenvector for all $t \geq 0$.

Next, take $\lambda \in R\sigma(A)$ and use (2.6) to obtain that

$$\mathrm{rg}\big(e^{\lambda t} - T(t)\big) \subset \mathrm{rg}(\lambda - A)$$

is not dense in X. Hence (2.5) holds. $\qquad\square$

It follows from the above examples and counterexamples that not all converse inclusions can hold in general. In fact, we show that it is only the approximate point spectrum that is responsible for the failure of (SMT). For the point spectrum and the residual spectrum, however, we are able to prove a spectral mapping formula.

2.6 Spectral Mapping Theorem for Point and Residual Spectrum. *For the generator $\big(A, D(A)\big)$ of a strongly continuous semigroup $(T(t))_{t\geq 0}$ on a Banach space X, we have the identities*

(2.7) $$P\sigma\big(T(t)\big) \setminus \{0\} = e^{tP\sigma(A)},$$

(2.8) $$R\sigma\big(T(t)\big) \setminus \{0\} = e^{tR\sigma(A)}$$

for all $t \geq 0$.

PROOF. Take $t_0 > 0$ and $0 \neq \lambda \in P\sigma\big(T(t_0)\big)$. According to Paragraphs I.1.10 and II.2.2, we can pass from the semigroup $(T(t))_{t\geq 0}$ to the rescaled semigroup $\big(S(t)\big)_{t\geq 0} := \big(e^{-t\log\lambda}T(t_0 t)\big)_{t\geq 0}$ having the generator $B = t_0 A - \log\lambda$. Because for this rescaled semigroup 1 is an eigenvalue of $S(1)$, we can assume that $t_0 = 1$ and $\lambda = 1$ from the beginning.

Take $0 \neq x \in X$ satisfying $T(1)x = x$. Then the function $t \mapsto T(t)x \neq 0$ is periodic, hence there exists at least one $k \in \mathbb{Z}$ such that the Fourier coefficient

$$y_k := \int_0^1 e^{2\pi i k(1-s)}T(s)x \, ds$$

is nonzero. However, by Lemma II.1.9, $y_k \in D(A)$ and

$$(A - 2\pi i k)y_k = T(1)x - e^{2\pi i k}x = 0.$$

Therefore, $2\pi i k \in P\sigma(A)$ satisfying $e^{2\pi i k} = 1 \in P\sigma\big(T(1)\big)$. This and (2.3) prove (2.7).

The identity for the residual spectrum follows from (2.7) if we consider the sun dual semigroup $(T(t)^{\odot})_{t \geq 0}$ and use that $R\sigma(A) = P\sigma(A^{\odot})$ and $R\sigma(T(t)) = P\sigma(T(t)^{\odot})$; cf. [EN00, IV.2.18]. $\qquad\qquad\square$

Because we have proved spectral mapping theorems for the point as well as for the residual spectrum, it follows that in Counterexample 2.4 there must be approximate eigenvalues μ of $T(t)$ that do not stem from some $\lambda \in \sigma(A)$ via the exponential map. In order to overcome this failure and to obtain a spectral mapping theorem for the entire spectrum, we could exclude the existence of such approximate eigenvalues and assume

$$\sigma(T(t)) = P\sigma(T(t)) \cup R\sigma(T(t))$$

(e.g., if $(T(t))_{t \geq 0}$ is eventually compact). A more interesting and useful way to save the validity of (SMT), however, is to look for additional properties of the semigroup that guarantee even

(2.9)
$$A\sigma(T(t)) \setminus \{0\} = e^{tA\sigma(A)}.$$

Eventual norm continuity seems to be the most general hypothesis doing this job.

However, we first characterize those approximate eigenvalues that satisfy the spectral mapping property.

2.7 Lemma. *For an approximate eigenvalue $\lambda \neq 0$ of the operator $T(t_0)$ the following statements are equivalent.*

(a) *There exists a sequence $(x_n)_{n \in \mathbb{N}} \subset X$ satisfying $\|x_n\| = 1$ and $\|T(t_0)x_n - \lambda x_n\| \to 0$ such that $\lim_{t \downarrow 0} \sup_{n \in \mathbb{N}} \|T(t)x_n - x_n\| = 0$.*

(b) *There exists $\mu \in A\sigma(A)$ such that $\lambda = e^{\mu t_0}$.*

PROOF. The implication (b) $\Rightarrow$ (a) follows from identity (2.6).

To show the converse implication it suffices, as in the proof of Theorem 2.6, to consider the case $\lambda = 1$ and $t_0 = 1$ only. To this end we take an approximate eigenvector $(x_n)_{n \in \mathbb{N}}$ as in (a). The uniform continuity of $(T(t))_{t \geq 0}$ on the vectors x_n implies that the maps $[0,1] \ni t \mapsto T(t)x_n$, $n \in \mathbb{N}$, are equicontinuous. Choose now $x_n' \in X'$, $\|x_n'\| \leq 1$, satisfying $\langle x_n, x_n' \rangle \geq 1/2$ for all $n \in \mathbb{N}$. Then the functions

$$[0,1] \ni s \mapsto \xi_n(s) := \langle T(s)x_n, x_n' \rangle$$

are uniformly bounded and equicontinuous. Hence, there exists, by the Arzelà–Ascoli theorem, a convergent subsequence, still denoted by $(\xi_n)_{n \in \mathbb{N}}$, such that $\lim_{n \to \infty} \xi_n =: \xi \in C[0,1]$. From $\xi(0) = \lim_{n \to \infty} \xi_n(0) \geq 1/2$ we obtain that $\xi \neq 0$. Therefore, this function has a nonzero Fourier coefficient; i.e., there exists $\mu_m := 2\pi i m$, $m \in \mathbb{Z}$, such that

$$\int_0^1 e^{-\mu_m s} \xi(s) \, ds \neq 0.$$

If we set

$$z_n := \int_0^1 e^{-\mu_m s} T(s) x_n \, ds,$$

we have $z_n \in D(A)$ by Lemma II.1.3. In addition, the elements z_n satisfy

$$(\mu_m - A) z_n = \left(1 - e^{-\mu_m} T(1)\right) x_n = \left(1 - T(1)\right) x_n \to 0$$

and

$$\lim_{n \to \infty} \|z_n\| \geq \lim_{n \to \infty} |\langle z_n, x_n' \rangle|$$

$$\geq \lim_{n \to \infty} \left| \int_0^1 e^{-\mu_m s} \langle T(s) x_n, x_n' \rangle \, ds \right|$$

$$\geq \left| \int_0^1 e^{-\mu_m s} \xi(s) \, ds \right| > 0.$$

This shows that $\left(z_n / \|z_n\| \right)_{n \in \mathbb{N}}$ is an approximate eigenvector of A corresponding to the approximate eigenvalue $\mu_m = 2\pi i m$. $\qquad \square$

For eventually norm-continuous semigroups we can always construct approximate eigenvectors satisfying condition (a) of the previous lemma. Therefore, we obtain (SMT).

2.8 Spectral Mapping Theorem for Eventually Norm-Continuous Semigroups. *Let $\left(T(t)\right)_{t \geq 0}$ be an eventually norm-continuous semigroup with generator $\left(A, D(A)\right)$ on the Banach space X. Then the spectral mapping theorem*

(SMT) $$\qquad\qquad \sigma\left(T(t)\right) \setminus \{0\} = e^{t\sigma(A)}, \quad t \geq 0,$$

holds.

PROOF. Taking into account all our previous theorems such as 2.5 and 2.6 and using the rescaling technique, we have to show the following.

If $1 \in A\sigma\left(T(1)\right)$, then there exists $m \in \mathbb{Z}$ such that $\mu_m := 2\pi i m \in A\sigma(A)$.

To prove this claim, we take an approximate eigenvector $(x_n)_{n \in \mathbb{N}}$ of $T(1)$; i.e., we assume $\|x_n\| = 1$ and $\|T(1) x_n - x_n\| \to 0$. Moreover, we assume that $t \mapsto T(t)$ is norm-continuous for $t \geq t_0$. Now choose $t_0 < k \in \mathbb{N}$ and observe that

$$\|T(k) x_n - x_n\| = \|T(k) x_n - T(k-1) x_n + T(k-1) x_n - \cdots - x_n\| \to 0$$

as $n \to \infty$. The semigroup $\left(T(t)\right)_{t \geq 0}$ is then uniformly continuous on $\left(T(k) x_n\right)_{n \in \mathbb{N}}$ by assumption and on $\left(T(k) x_n - x_n\right)_{n \in \mathbb{N}}$, because this is a null sequence (use Proposition A.3). Therefore, $\left(T(t)\right)_{t \geq 0}$ is uniformly continuous on $(x_n)_{n \in \mathbb{N}} = \left(T(k) x_n\right)_{n \in \mathbb{N}} - \left(T(k) x_n - x_n\right)_{n \in \mathbb{N}}$, and the assertion follows from Lemma 2.7. $\qquad \square$

Combining the previous result with Proposition 2.3 yields the following.

2.9 Corollary. *For an eventually norm-continuous semigroup* $(T(t))_{t \geq 0}$ *with generator* $(A, D(A))$ *on a Banach space* X, *we have*

(SBeGB) $$s(A) = \omega_0.$$

Finally, we know from Section II.5 that many important regularity properties of semigroups imply eventual norm continuity. We state the spectral mapping theorem for these semigroups.

2.10 Corollary. *The spectral mapping theorem*

(SMT) $$e^{t\sigma(A)} = \sigma(T(t)) \setminus \{0\}, \quad t \geq 0,$$

and the spectral bound equal growth bound condition

(SBeGB) $$s(A) = \omega_0$$

hold for the following classes of strongly continuous semigroups:

   (i) *Eventually compact semigroups,*

  (ii) *Eventually differentiable semigroups,*

 (iii) *Analytic semigroups, and*

 (iv) *Uniformly continuous semigroups.*

It is the above condition (SBeGB) that is used in Section 3 (e.g., in Theorem 3.7) to characterize stability of semigroups. However, not all of (SMT) is needed to derive (SBeGB). The weaker property (WSMT), already encountered in Example 2.2.(ii), is sufficient. Therefore, the following simple result on multiplication operators (see Section I.3.a and Paragraph II.2.8) is a useful addition to the above corollaries.

2.11 Proposition. *Let* M_q *be the generator of a multiplication semigroup* $(T_q(t))_{t \geq 0}$ *on* $X := C_0(\Omega)$ *(or* $X := L^p(\Omega, \mu)$*) defined by an appropriate function* $q : \Omega \to \mathbb{C}$. *Then*

(WSMT) $$\sigma(T_q(t)) = \overline{e^{t\sigma(M_q)}} \qquad \text{for } t \geq 0,$$

hence (SBeGB) *hold.*

PROOF. In Proposition I.3.2.(iv), we stated that the spectrum of a multiplication operator is the closed (essential) range of the corresponding function. Therefore, we obtain

$$\sigma(T_q(t)) = \overline{e^{tq_{(ess)}(\Omega)}} = \overline{e^{t\overline{q_{(ess)}(\Omega)}}} = \overline{e^{t\sigma(M_q)}}$$

for all $t \geq 0$. □

A simple, but typical, example is given by the multiplication operator

$$M_q(x_n)_{n \in \mathbb{Z}} := (inx_n)_{n \in \mathbb{Z}}$$

for $(x_n)_{n \in \mathbb{Z}} \in \ell^p(\mathbb{Z})$. Then $\sigma(M_q) = i\mathbb{Z}$ and $\sigma(T_q(t)) = \Gamma$ whenever $t/2\pi \notin \mathbb{Q}$. Therefore, only (WSMT) but not (SMT) holds. See also Example 2.2.(ii).

Most important, the above proposition can be applied to semigroups of normal operators on Hilbert spaces. In fact, due to the Spectral Theorem I.3.9, these semigroups are always isomorphic to multiplication semigroups on L^2-spaces; hence (WSMT) holds.

2.12 Corollary. *Let $(T(t))_{t\geq0}$ be a strongly continuous semigroup of normal operators on a Hilbert space and denote its generator by $(A, D(A))$. Then*

(WSMT) $\sigma(T(t)) = \overline{e^{t\sigma(A)}}$ *for $t \geq 0$,*

hence (SBeGB) *hold.*

2.13 Exercises. (1) Show that the semigroup in Counterexample 2.4 is in fact a group whose generator has compact resolvent.

(2) (Counterexample on reflexive Banach spaces). Take $1 < p < q < \infty$ and the (reflexive) Banach space $X := L^p[1,\infty) \cap L^q[1,\infty)$ with norm $\|f\| := \|f\|_p + \|f\|_q$. Then the following hold.

(i) The operator family $(T(t))_{t\geq0}$ given by $T(t)f(s) := f(se^t)$ for $s \geq 1$, $t \geq 0$, and $f \in X$, defines a strongly continuous semigroup on X.

(ii) The generator A of $(T(t))_{t\geq0}$ is given by $Af(s) = sf'(s)$, $s \geq 1$, with domain

$$D(A) = \left\{ f \in X : \begin{array}{l} f \text{ is absolutely continuous} \\ \text{and } s \mapsto sf'(s) \text{ belongs to } X \end{array} \right\}.$$

(iii) Spectral and growth bound of A are given by $s(A) = -\frac{1}{p} < -\frac{1}{q} = \omega_0$. (Hint: See Exercise I.1.8.(3) and [EN00, IV.3.3].)

(3) On the space $L^2_{2\pi}$ of all 2π-periodic functions on $\mathbb{R}^2$ that are square integrable on $[0, 2\pi]^2$ consider the second-order partial differential equation

(2.10) $$\begin{cases} \dfrac{\partial^2 u(t,x,y)}{\partial t^2} = \dfrac{\partial^2 u(t,x,y)}{\partial x^2} + \dfrac{\partial^2 u(t,x,y)}{\partial y^2} + e^{iy}\dfrac{\partial u(t,x,y)}{\partial x}, \\ u(0,x,y) = u_0(x,y), \quad \dfrac{\partial u(0,x,y)}{\partial t} = u_1(x,y) \end{cases}$$

for $(x,y) \in [0, 2\pi]^2$ and $t \geq 0$.

(i) Show that (2.10) is equivalent to the abstract Cauchy problem (ACP) for the operator $(A, D(A))$ defined by

$$A(u,v) := \left(v, \tfrac{d^2}{dx^2}u + \tfrac{d^2}{dy^2}u + e^i\tfrac{d}{dx}u \right), \qquad D(A) := H^2_{2\pi} \times H^1_{2\pi}$$

on $X := H^1_{2\pi} \times L^2_{2\pi}$ and for the initial value (u_0, u_1).

(ii) Show that A generates a strongly continuous semigroup on X.

(iii*) Show that $s(A) = 0$, whereas $\omega_0 \geq 1/2$. (Hint: See [HW03], [BLX05], [Ren94].)

(4) Assume that for some $t_0 > 0$ the spectral radius $\mathrm{r}(T(t_0))$ is an eigenvalue of $T(t_0)$ (or of its adjoint $T(t_0)'$). Show that in this case one has (SBeGB); i.e., $\mathrm{s}(A) = \omega_0$.

(5) Let $(T(t))_{t \geq 0}$ be a strongly continuous semigroup on some $\mathrm{L}^1(\Omega, \mu)$ and assume that $0 \leq T(t)f$ for all $0 \leq f \in \mathrm{L}^1(\Omega, \mu)$ and all $t \geq 0$. Show that (SBeGB) holds; that is, $\mathrm{s}(A) = \omega_0$. (Hint: Use Lemma VI.2.1.)

(6*) A strongly continuous semigroup $(T(t))_{t \geq 0}$ with growth bound ω_0 is called *asymptotically norm-continuous* if

$$\lim_{t \to \infty} \left(\overline{\lim_{h \downarrow 0}} \, \mathrm{e}^{-\omega_0 t} \| T(t+h) - T(t) \| \right) = 0.$$

(i) Show that a semigroup $(T(t))_{t \geq 0}$ is asymptotically norm-continuous if it can be written as $T(t) = U_0(t) + U_1(t)$ for operator families $(U_0(t))_{t \geq 0}$ and $(U_1(t))_{t \geq 0}$ where $(U_0(t))_{t \geq 0}$ is eventually norm-continuous and $\lim_{t \to \infty} \mathrm{e}^{-\omega_0 t} \| U_1(t) \| = 0$.

(ii) Construct an example of such a decomposition using Theorem III.1.10.

(iii) For a semigroup $(T(t))_{t \geq 0}$ that is norm-continuous at infinity, the spectral mapping theorem holds for the boundary spectrum; i.e.,

$$\sigma(T(t)) \cap \{ \lambda \in \mathbb{C} : |\lambda| = \mathrm{r}(T(t)) \} = \mathrm{e}^{t(\sigma(A) \cap (\mathrm{s}(A) + i\mathbb{R}))}$$

for $t \geq 0$ and $\mathrm{r}(T(t)) > 0$. See [MM96], [Bla01], and [NP00].

3. Stability and Hyperbolicity of Semigroups

We now come to one of the most interesting aspects of semigroup theory. After having established generation, perturbation, and approximation theorems in the previous chapters, we investigate the qualitative behavior of a given semigroup. We already dealt with this problem when we classified strongly continuous semigroups according to their regularity properties in Section II.5, but we now concentrate on their "asymptotic" behavior. By this we mean the behavior of the semigroup $(T(t))_{t \geq 0}$ for large $t > 0$ or, more precisely, the existence (or nonexistence) of

$$\lim_{t \to \infty} T(t),$$

where the limit is understood in various ways and for different topologies. If we recall that the function $t \mapsto T(t)x$ yields the (mild) solutions of the corresponding abstract Cauchy problem

(ACP) $\qquad \begin{cases} \dot{x}(t) = Ax(t), \quad t \geq 0, \\ x(0) = x \end{cases}$

(see Section II.6), it is evident that such results will be of utmost importance.

Among the many interesting types of asymptotic behavior, we first study *stability* of strongly continuous semigroups $(T(t))_{t \geq 0}$. By this we mean that the operators $T(t)$ should converge to *zero* as $t \to \infty$. However, as is to be expected in infinite-dimensional spaces, we have to distinguish different concepts of convergence.

a. Stability Concepts

For a strongly continuous semigroup $(T(t))_{t \geq 0}$ with generator $A : D(A) \subseteq X \to X$ we now make precise what we mean by

$$\text{“} \lim_{t \to \infty} T(t) = 0 \text{”}$$

and vary the topology and the "speed" of the convergence by proposing the following concepts.

3.1 Definition. *A strongly continuous semigroup $(T(t))_{t \geq 0}$ is called*

(a) *Uniformly exponentially stable if there exists $\varepsilon > 0$ such that*

$$(3.1) \qquad \lim_{t \to \infty} e^{\varepsilon t} \| T(t) \| = 0;$$

(b) *Uniformly stable if*

$$(3.2) \qquad \lim_{t \to \infty} \| T(t) \| = 0;$$

(c) *Strongly stable if*

$$(3.3) \qquad \lim_{t \to \infty} \| T(t) x \| = 0 \qquad \text{for all } x \in X;$$

(d) *Weakly stable if*

$$(3.4) \qquad \lim_{t \to \infty} \langle T(t) x, x' \rangle = 0 \qquad \text{for all } x \in X \text{ and } x' \in X'.$$

We start our discussion of these concepts by noting that the two "uniform" properties coincide and are even equivalent to a "pointwise" condition.

3.2 Proposition. *For a strongly continuous semigroup $(T(t))_{t \geq 0}$, the following assertions are equivalent.*

(a) $(T(t))_{t \geq 0}$ *is uniformly exponentially stable.*

(b) $(T(t))_{t \geq 0}$ *is uniformly stable.*

(c) *There exists $\varepsilon > 0$ such that $\lim_{t \to \infty} e^{\varepsilon t} \| T(t) x \| = 0$ for all $x \in X$.*

PROOF. Clearly, (a) implies (b) and (c). Because $e^{\omega_0 t} = r(T(t)) \leq \|T(t)\|$ for all $t \geq 0$ (see Proposition 1.22), (b) implies $\omega_0 < 0$, hence (a). If (c) holds, then $(e^{\varepsilon t} T(t))_{t \geq 0}$ is strongly, hence uniformly, bounded, which implies $\lim_{t \to \infty} e^{\varepsilon/2 t} \|T(t)\| = 0$. □

It is obvious from the definition that uniform (exponential) stability implies strong stability, which again implies weak stability. The following examples show that none of the converse implications holds.

3.3 Examples. (i) The (left) translation semigroup $(T(t))_{t \geq 0}$ on $X := L^p(\mathbb{R}_+)$, $1 \leq p < \infty$, is strongly stable, but one has

$$\|T(t)\| = 1$$

for all $t \geq 0$; hence it is not uniformly stable.

(ii) The (left) translation group $(T(t))_{t \in \mathbb{R}}$ on $X := L^p(\mathbb{R})$, $1 < p < \infty$, is a group of isometries, hence is not strongly stable. However, for functions $f \in X$, $g \in X' = L^q(\mathbb{R})$, $1/p + 1/q = 1$, with compact support and large t, one has that $T(t)f$ and g have disjoint supports, whence

$$\langle T(t)f, g \rangle = \int_{-\infty}^{\infty} f(s + t)g(s)\, ds = 0.$$

For arbitrary $f \in X$, $g \in X'$ and for each $n \in \mathbb{N}$, we choose $f_n \in X$ and $g_n \in X'$ with compact support such that $\|f - f_n\|_p \leq 1/n$ and $\|g - g_n\|_q \leq 1/n$. Then

$$|\langle T(t)f, g \rangle| \leq |\langle T(t)(f - f_n), g_n \rangle| + |\langle T(t)f, g - g_n \rangle| + |\langle T(t)f_n, g_n \rangle|$$
$$\leq \tfrac{1}{n}(\|g\|_q + 1 + \|f\|_p) + |\langle T(t)f_n, g_n \rangle|.$$

Because the last term is 0 for large t, we conclude that

$$\lim_{t \to \infty} \langle T(t)f, g \rangle = 0$$

for all $f \in X, g \in X'$; i.e., $(T(t))_{t \geq 0}$ is weakly stable.

It is now our goal to characterize the above stability concepts, it is hoped by properties of the generator. In the following subsection we try this for uniform exponential stability.

3.4 Exercises. (1) Discuss the above stability properties for multiplication semigroups on $L^p(\mathbb{R})$ and $C_0(\mathbb{R})$. (Hint: See [EN00, Expl. V.2.19.(ii) and (iii)].)

(2) Let μ be a probability measure on $\mathbb{R}$ that is absolutely continuous with respect to the Lebesgue measure. Use the Riemann–Lebesgue lemma (see Theorem A.20) to show that the multiplication semigroup $(T(t))_{t \geq 0}$ with

$$(T(t)f)(s) := e^{its} f(s), \qquad s \in \mathbb{R},$$

is weakly stable on $L^p(\mathbb{R}, \mu)$ for $1 \leq p < \infty$.

(3) Show that the adjoint semigroup of a strongly stable semigroup is *weak*-stable*; that is, $\lim_{t\to\infty} \langle T(t)x, x' \rangle = 0$ for all $x \in X$, $x' \in X'$, but not strongly stable in general.

(4) Show that a strongly continuous semigroup with compact resolvent which is weakly stable is necessarily uniformly exponentially stable. In particular, an immediately compact semigroup that is weakly stable is already uniformly exponentially stable.

b. Characterization of Uniform Exponential Stability

We start by recalling the definition of the *growth bound*

$$\omega_0 := \omega_0(\mathcal{T}) := \omega_0(A)$$
(3.5)
$$:= \inf \left\{ w \in \mathbb{R} : \exists\, M_w \geq 1 \text{ such that } \|T(t)\| \leq M_w e^{wt}\ \forall t \geq 0 \right\}$$
$$= \inf \left\{ w \in \mathbb{R} : \lim_{t\to\infty} e^{-wt} \|T(t)\| = 0 \right\}$$

of a semigroup $\mathcal{T} = \big(T(t)\big)_{t\geq 0}$ with generator A (compare Definition I.1.5). From this definition it is immediately clear that $\big(T(t)\big)_{t\geq 0}$ is uniformly exponentially stable if and only if

(3.6)
$$\omega_0 < 0.$$

Moreover, the identity

(3.7)
$$\omega_0 = \inf_{t>0} \frac{1}{t} \log \|T(t)\| = \lim_{t\to\infty} \frac{1}{t} \log \|T(t)\| = \frac{1}{t_0} \log r\big(T(t_0)\big)$$

for each $t_0 > 0$, proved in Proposition 1.22, yields the following characterizations of uniform exponential stability.

3.5 Proposition. *For a strongly continuous semigroup $\big(T(t)\big)_{t\geq 0}$, the following assertions are equivalent.*

   (a) $\omega_0 < 0$; *i.e.,* $\big(T(t)\big)_{t\geq 0}$ *is uniformly exponentially stable.*

   (b) $\lim_{t\to\infty} \|T(t)\| = 0$.

   (c) $\|T(t_0)\| < 1$ *for some* $t_0 > 0$.

   (d) $r\big(T(t_1)\big) < 1$ *for some* $t_1 > 0$.

All these stability criteria, as nice as they are, have the major disadvantage that they rely on the explicit knowledge of the semigroup $\big(T(t)\big)_{t\geq 0}$ and its orbits $t \mapsto T(t)x$. In most cases, however, only the generator (and its resolvent) is given. Therefore, direct characterizations of uniform exponential stability of the semigroup in terms of its generator are more desirable. Spectral theory provides the appropriate tool for this purpose, and the following classical Liapunov theorem for matrix semigroups serves as a prototype for the results for which we are looking.

3.6 Theorem. (LIAPUNOV 1892). *Let* $(e^{tA})_{t\geq 0}$ *be the one-parameter semigroup generated by* $A \in M_n(\mathbb{C})$. *Then the following assertions are equivalent.*

(a) *The semigroup is stable; i.e.,* $\lim_{t\to\infty} \|e^{tA}\| = 0$.

(b) *All eigenvalues of* A *have negative real part; i.e.,* $\operatorname{Re}\lambda < 0$ *for all* $\lambda \in \sigma(A)$.

In particular, one hopes that the inequality

$$(3.8) \qquad\qquad s(A) < 0$$

for the spectral bound $s(A) = \sup\{\operatorname{Re}\lambda : \lambda \in \sigma(A)\}$ of the generator A (see Definition II.1.12) characterizes uniform exponential stability. Counterexample 1.26 (see also Exercises 2.13.(2) and (3)) shows that this fails drastically. The reason is the failure of the spectral mapping theorem (SMT) as discussed in Section 2. On the other hand, if some (weak) spectral mapping theorem holds for the semigroup $(T(t))_{t\geq 0}$ and its generator A, then by Proposition 2.3 the growth bound ω_0 and the spectral bound $s(A)$ coincide, and hence the inequality (3.8) implies (3.6).

The coincidence of growth and spectral bounds clearly implies that uniform exponential stability is equivalent to the negativity of the spectral bound. So in this case the inequality $s(A) < 0$ characterizes uniform exponential stability of the semigroup $(T(t))_{t\geq 0}$ in terms of its generator A and its spectrum $\sigma(A)$. This is one reason for our thorough study of spectral mapping theorems in Section 2. The results obtained there, in particular Theorem 2.8 and its corollaries, pay off and yield the spectral bound equal growth bound condition (SBeGB) already stated in Corollary 2.9. We restate this as an infinite-dimensional version of Liapunov's stability theorem.

3.7 Theorem. *An eventually norm-continuous semigroup* $(T(t))_{t\geq 0}$ *is uniformly exponentially stable if and only if the spectral bound* $s(A)$ *of its generator* A *satisfies*

$$s(A) < 0.$$

Looking back at the stability results obtained so far, i.e., Proposition 3.5 and Theorem 3.7, we observe that in each case we needed information on the semigroup itself in order to conclude its stability. This can be avoided by restricting our attention to semigroups on Hilbert spaces only.

3.8 Theorem. (GEARHART 1978, PRÜSS 1984, GREINER 1985). *A strongly continuous semigroup* $(T(t))_{t\geq 0}$ *on a Hilbert space* H *is uniformly exponentially stable if and only if the half-plane* $\{\lambda \in \mathbb{C} : \operatorname{Re}\lambda > 0\}$ *is contained in the resolvent set* $\rho(A)$ *of the generator* A *with the resolvent satisfying*

$$(3.9) \qquad\qquad M := \sup_{\operatorname{Re}\lambda > 0} \|R(\lambda, A)\| < \infty.$$

This stability criterion is extremely useful for the stability analysis of concrete equations; see [BP05, Sects. 5.1 and 10.4], [CL03], [LZ99]. For a proof we refer to [EN00, Thm. V.1.11]. Its theoretical significance is emphasized by the following comments.

3.9 Comments. (i) The theorem does not hold without the boundedness assumption on the resolvent in the right half-plane. Take the semigroup $(T(t))_{t\geq 0}$ from Counterexample 2.4. Then $(e^{-t/2}T(t))_{t\geq 0}$ is a semigroup on a Hilbert space having spectral bound $s(A) = -1/2$, and hence we have $\{\lambda \in \mathbb{C} : \operatorname{Re}\lambda \geq 0\} \subset \rho(A)$, but its growth bound is $\omega_0 = 1/2$.

(ii) The theorem does not hold on arbitrary Banach spaces. In fact, for the semigroup in Counterexample 1.26 one has

$$\|R(\lambda + is, A)\| \leq \|R(\lambda, A)\|$$

for all $\lambda > s(A) = -1$ and $s \in \mathbb{R}$ (use the integral representation (1.20) of the resolvent in Section 1.b). Because $\|T(t)\| = 1$ for all $t \geq 0$, this semigroup is not uniformly exponentially stable, but the resolvent of its generator exists and is uniformly bounded in $\{\lambda \in \mathbb{C} : \operatorname{Re}\lambda \geq 0\}$.

3.10 Exercises. (1) Show that for a strongly continuous semigroup $\mathcal{T} = (T(t))_{t\geq 0}$ on a Hilbert space X with generator A its growth bound is given by

$$\omega_0 = \inf\Big\{\lambda > s(A) : \sup_{s\in\mathbb{R}} \|R(\lambda + is, A)\| < \infty\Big\}.$$

(2^*) Let $(T(t))_{t\geq 0}$ be a strongly continuous semigroup with generator A on a Hilbert space H.

 (i) Define $\mathcal{U}(t)T := T(t) \cdot T \cdot T(t)^*$ for $t \geq 0$ and $T \in \mathcal{L}(H)$ and show that $(\mathcal{U}(t))_{t\geq 0}$ is a semigroup on $\mathcal{L}(H)$ that is continuous for the weak operator topology on $\mathcal{L}(H)$.

 (ii) Define $R(\lambda)T := \int_0^\infty e^{-\lambda t}\mathcal{U}(t)T\,dt$, $T \in \mathcal{L}(H)$ and λ large, in the weak operator topology and show that $R(\lambda)$ is the resolvent of a Hille–Yosida operator $(G, D(G))$ on $\mathcal{L}(H)$.

 (iii) Formally, G is of the form $G(T) = AT - TA$ for $T \in D(G)$. Can you give a precise meaning to this statement? (Hint: See [Alb01].)

 (iv) Show that the following assertions are equivalent.

    (a) $(T(t))_{t\geq 0}$ is uniformly exponentially stable.

    (b) $(\mathcal{U}(t))_{t\geq 0}$ is uniformly exponentially stable.

    (c) $s(G) < 0$.

    (d) $\int_0^\infty \mathcal{U}(t)T\,dt$ exists for every $T \in \mathcal{L}(H)$.

    (e) There exists a positive definite $R \in \mathcal{L}(H)$ such that $GR = -I$.

(Hint: See [Nag86, D-IV, Sect. 2].)

c. Hyperbolic Decompositions

We now use the previous stability theorems in order to decompose a semigroup into a *stable* and an *unstable* part. More precisely, we try to decompose the Banach space into the direct sum of two closed subspaces such that the semigroup becomes "forward" exponentially stable on one subspace and "backward" exponentially stable on the other subspace.

3.11 Definition. *A semigroup* $(T(t))_{t \geq 0}$ *on a Banach space* X *is called hyperbolic if* X *can be written as a direct sum* $X = X_s \oplus X_u$ *of two* $(T(t))_{t \geq 0}$-*invariant, closed subspaces* X_s, X_u *such that the restricted semigroups* $(T_s(t))_{t \geq 0}$ *on* X_s *and* $(T_u(t))_{t \geq 0}$ *on* X_u *satisfy the following conditions.*

(i) *The semigroup* $(T_s(t))_{t \geq 0}$ *is uniformly exponentially stable on* X_s.

(ii) *The operators* $T_u(t)$ *are invertible on* X_u, *and* $(T_u(t)^{-1})_{t \geq 0}$ *is uniformly exponentially stable on* X_u.

It is easy to see that a strongly continuous semigroup $(T(t))_{t \geq 0}$ is hyperbolic if and only if there exists a projection P and constants $M, \varepsilon > 0$ such that each $T(t)$ commutes with P, satisfies $T(t) \ker P = \ker P$, and

$$(3.10) \qquad \|T(t)x\| \leq M e^{-\varepsilon t}\|x\| \qquad \text{for } t \geq 0 \text{ and } x \in \operatorname{rg} P,$$

$$(3.11) \qquad \|T(t)x\| \geq \frac{1}{M} e^{+\varepsilon t}\|x\| \qquad \text{for } t \geq 0 \text{ and } x \in \ker P.$$

As in the case of uniform exponential stability, we look for a spectral characterization of hyperbolicity. Using the spectra $\sigma(T(t))$ of the semigroup operators $T(t)$, this is easy.

3.12 Proposition. *For a strongly continuous semigroup* $(T(t))_{t \geq 0}$, *the following assertions are equivalent.*

(a) $(T(t))_{t \geq 0}$ *is hyperbolic.*

(b) $\sigma(T(t)) \cap \Gamma = \emptyset$ *for one/all* $t > 0$.

PROOF. The proof of the implication (a) $\Rightarrow$ (b) starts from the observation that $\sigma(T(t)) = \sigma(T_s(t)) \cup \sigma(T_u(t))$ because of the direct sum decomposition. By assumption, $(T_s(t))_{t \geq 0}$ is uniformly exponentially stable; hence $\operatorname{r}(T_s(t)) < 1$ for $t > 0$, and therefore $\sigma(T_s(t)) \cap \Gamma = \emptyset$.

By the same argument, we obtain that $\operatorname{r}(T_u(t)^{-1}) < 1$. Because

$$\sigma(T_u(t)) = \{\lambda^{-1} : \lambda \in \sigma(T_u(t)^{-1})\},$$

we conclude that $|\lambda| > 1$ for each $\lambda \in \sigma(T_u(t))$; hence $\sigma(T_u(t)) \cap \Gamma = \emptyset$.

To prove (b) $\Rightarrow$ (a), we fix $s > 0$ such that $\sigma(T(s)) \cap \Gamma = \emptyset$ and use the existence of a spectral projection P corresponding to the spectral set $\{\lambda \in \sigma(T(s)) : |\lambda| < 1\}$. Then the space X is the direct sum $X = X_s \oplus X_u$ of the $(T(t))_{t \geq 0}$-invariant subspaces $X_s := \operatorname{rg} P$ and $X_u := \ker P$. The restriction $T_s(s) \in \mathcal{L}(X_s)$ of $T(s)$ in X_s has spectrum

$$\sigma(T_s(s)) = \{\lambda \in \sigma(T(s)) : |\lambda| < 1\},$$

hence spectral radius $\mathrm{r}(T_s(s)) < 1$. From Proposition 3.5.(d), it follows that the semigroup $(T_s(t))_{t \geq 0} := (PT(t))_{t \geq 0}$ is uniformly exponentially stable on X_s. Similarly, the restriction $T_u(s) \in \mathcal{L}(X_u)$ of $T(s)$ in X_u has spectrum

$$\sigma(T_u(s)) = \{\lambda \in \sigma(T(s)) : |\lambda| > 1\},$$

hence is invertible on X_u. Clearly, this implies that $T_u(t)$ is invertible for $0 \leq t \leq s$, whereas for $t > s$ we choose $n \in \mathbb{N}$ such that $ns > t$. Then

$$T_u(s)^n = T_u(ns) = T(ns - t)T_u(t) = T_u(t)T_u(ns - t);$$

hence $T_u(t)$ is invertible, because $T_u(s)$ is bijective. Moreover, for the spectral radius we have $\mathrm{r}(T_u^{-1}(s)) < 1$, and again by Proposition 3.5.(d) this implies uniform exponential stability for the semigroup $(T_u(t)^{-1})_{t \geq 0}$. $\square$

The reader might be surprised by the extra condition in Definition 3.11.(ii) requiring the operators $T_u(t)$ to be invertible on X_u. However, this is necessary in order to obtain the spectral characterization in Proposition 3.12.

3.13 Example. Take the rescaled (left) shift semigroup $(T(t))_{t \geq 0}$ on $\mathrm{L}^1(\mathbb{R}_-)$ defined by

$$T(t)f(s) := \begin{cases} e^{\varepsilon t} f(s + t) & \text{for } s + t \leq 0, \\ 0 & \text{otherwise,} \end{cases}$$

for $f \in \mathrm{L}^1(\mathbb{R}_-)$, $s \leq 0$, and some fixed $\varepsilon > 0$. Then

$$\|T(t)f\| = e^{\varepsilon t}\|f\|$$

for all $f \in \mathrm{L}^1(\mathbb{R}_-)$; i.e., estimate (3.11) holds for all $f \in \mathrm{L}^1(\mathbb{R}_-)$. However, the operators $T(t)$ are not invertible and have spectrum

$$\sigma(T(t)) = \{\lambda \in \mathbb{C} : |\lambda| \leq e^{\varepsilon t}\}$$

for all $t > 0$.

This phenomenon is due to the fact that an injective operator on an infinite-dimensional Banach space need not be surjective. We can exclude this by assuming $\dim X_u < \infty$. See also Exercise 3.16.(2).

Up to now, our definition and characterization of hyperbolic semigroups use explicit knowledge of the semigroup itself. As in Section 3.b, we want to find a characterization in terms of the generator A and its spectrum $\sigma(A)$. As we should expect from Proposition 2.3, we need some extra relation between $\sigma(A)$ and $\sigma(T(t))$. Clearly, the spectral mapping theorem (SMT) or even the weak spectral mapping theorem (WSMT) from Section 2 is sufficient for this purpose. However, we show that an even weaker property does this job.

3.14 Definition. *We say that the strongly continuous semigroup $(T(t))_{t\geq0}$ with generator A satisfies the circular spectral mapping theorem if*

$$\text{(CSMT)} \qquad \Gamma \cdot \sigma(T(t)) \setminus \{0\} = \Gamma \cdot e^{t\sigma(A)} \qquad \text{for one/all } t > 0.$$

That "for one" implies "for all" in (CSMT) follows from Proposition 3.12 (and rescaling). Indeed, (CSMT) allows us to characterize hyperbolicity by a condition on the spectrum of the generator.

3.15 Theorem. *If (CSMT) holds for a strongly continuous semigroup $(T(t))_{t\geq0}$ with generator A, then the following assertions are equivalent.*

(a) $(T(t))_{t\geq0}$ *is hyperbolic.*

(b) $\sigma(T(t)) \cap \Gamma = \emptyset$ *for one/all $t > 0$.*

(c) $\sigma(A) \cap i\mathbb{R} = \emptyset$.

PROOF. The equivalence of (a) and (b) has been shown in Proposition 3.12. Property (b) always implies (c) (use Theorem 2.5), whereas (c) implies (b) if (CSMT) holds. □

We finally remark that

$$\text{(SMT)} \Rightarrow \text{(WSMT)} \Rightarrow \text{(CSMT)}$$

and refer to [GS91] and [KS05] where (CSMT) has been shown for interesting classes of generators and semigroups.

3.16 Exercises. (1) Show, by rescaling the semigroup and the estimates in (3.10) and (3.11), that a decomposition analogous to Definition 3.11 holds whenever

$$\sigma(T(t)) \cap \alpha\Gamma = \emptyset$$

for some $\alpha > 0$.

(2) Let $(T(t))_{t\geq0}$ satisfy (3.10) and (3.11) for a projection P commuting with $T(t)$ for all $t \geq 0$. Assume that for some $t_0 > 0$ the restriction $T_u(t_0)$ to $\ker P$ is compact. Show that $\dim \ker P < \infty$ and that $(T(t))_{t\geq0}$ is hyperbolic.

(3) Show that the generator A of a hyperbolic strongly continuous semi-group $(T(t))_{t \geq 0}$ is invertible and its inverse is given by

$$A^{-1}x = \int_0^\infty T_u(t)^{-1}(I - P)x \, dt - \int_0^\infty T_s(t)Px \, dt.$$

Derive an analogous representation of $R(\lambda, A)$ for $\operatorname{Re} \lambda < \varepsilon$, where ε is the constant in (3.10) and (3.11).

(4*) Given a hyperbolic semigroup $(T(t))_{t \geq 0}$ and a corresponding decomposition $X = X_s \oplus X_u$, prove that

$$X_s = \{x \in X : \lim_{t \to \infty} T(t)x = 0\}.$$

Conclude from this that X_s and X_u are uniquely determined.

4. Convergence to Equilibrium

In contrast to the previous section, we now suppose that 0 is an eigenvalue of the generator A of a strongly continuous semigroup $(T(t))_{t \geq 0}$ on the Banach space X. This means that *fixed space*

$$\operatorname{fix}(T(t))_{t \geq 0} := \{x \in X : T(t)x = x \text{ for all } t \geq 0\},$$

which coincides with $\ker A$ by Exercise 4.12.(1), is nontrivial. It is our goal to understand under which assumptions (and in which sense) each orbit

$$t \mapsto T(t)x$$

converges to such a fixed point (or, *equilibrium point*).

We first state some consequences if the semigroup converges for the weak operator topology.

4.1 Lemma. Let $(T(t))_{t \geq 0}$ be a strongly continuous semigroup with generator A on X and assume that there exists an operator $P \in \mathcal{L}(X)$ such that

$$\lim_{t \to \infty} \langle T(t)x, x' \rangle = \langle Px, x' \rangle \quad \text{for all } x \in X, \ x' \in X'.$$

Then $P = P^2$ is a projection onto the fixed space $\operatorname{fix}(T(t))_{t \geq 0}$ with $\ker P = \overline{\operatorname{rg} A}$ and commutes with every $T(t)$, $t \geq 0$.

PROOF. Because for convergence with respect to the weak operator topology we have

$$T(s) \cdot \lim_{t \to \infty} T(t) = \left(\lim_{t \to \infty} T(t)\right) \cdot T(s) = \lim_{t \to \infty} T(t + s) = P$$

for all $s \geq 0$, it follows that $\operatorname{rg} P = \operatorname{fix}(T(t))_{t \geq 0}$. By the same argument we conclude that

$$P^2 = \left(\lim_{t \to \infty} T(t)\right) P = \lim_{t \to \infty} (T(t)P) = P$$

is a projection which evidently commutes with each $T(t)$.

It only remains to show that $\ker P = \overline{\mathrm{rg}\, A}$. To that purpose we observe first that

$$\overline{\mathrm{rg}\, A} = \overline{\mathrm{lin}}\{x - T(t)x : x \in X,\ t \geq 0\}$$

by the definition of the generator A and formula (1.6) in Lemma II.1.3. This immediately shows that each $x - T(t)x$ belongs to $\ker P$. For the converse inclusion we show that each $x' \in X'$ vanishing on $\mathrm{lin}\{x - T(t)x : x \in X,\ t \geq 0\}$ also vanishes on $\ker P$. Indeed, for such x' we obtain $T(t)'x' = x'$, hence

$$\langle x, x' \rangle = \langle x, T(t)'x' \rangle = \langle T(t)x, x' \rangle \to \langle Px, x' \rangle \quad \text{as } t \to \infty.$$

For $x \in \ker P$ this yields $\langle x, x' \rangle = 0$ as claimed. $\qquad\square$

We observe that weak convergence implies, by the uniform boundedness principle, that the semigroup is uniformly bounded. Moreover, the space X splits into the direct sum

$$X = \mathrm{fix}\big(T(t)\big)_{t \geq 0} \oplus X_s$$

with $X_s := \overline{\mathrm{rg}\, A}$ such that the restricted semigroup on X_s is weakly stable (see Definition 3.1.(d)).

Up to now there are few sufficient conditions, and no satisfactory characterizations, for weakly converging (or weakly stable) semigroups. We refer to [EFNS05] for recent results in this direction. The case of strong convergence is much better understood and useful spectral criteria have been found by Arendt–Batty [AB88] and Lyubich–Vũ [LV88] (see also [EN00, Thm. V.2.21] and [CT06]). A systematic study of the asymptotic behavior of semigroups can be found in [Nee96]. We restrict our considerations to the case of uniform convergence and again start with a necessary condition.

4.2 Lemma. *Let $\big(T(t)\big)_{t \geq 0}$ be a strongly continuous semigroup with generator A on X and assume that*

$$P := \| \cdot \|\text{-}\lim_{t \to \infty} T(t) \neq 0$$

exists. Then 0 is a dominant eigenvalue, i.e., $\mathrm{Re}\,\lambda \leq \varepsilon < 0$ for all $0 \neq \lambda \in \sigma(A)$. In addition, this eigenvalue is a simple pole of the resolvent $R(\cdot, A)$.

PROOF. By Lemma 4.1, $P = P^2$ is a projection onto $\mathrm{fix}\big(T(t)\big)_{t \geq 0}$ commuting with $\big(T(t)\big)_{t \geq 0}$. Hence we can decompose the space X into the direct sum of the two $\big(T(t)\big)_{t \geq 0}$-invariant subspaces

$$X = \mathrm{rg}\, P \oplus \mathrm{rg}(I - P) = \mathrm{fix}\big(T(t)\big)_{t \geq 0} \oplus \mathrm{rg}(I - P) = \ker A \oplus \mathrm{rg}(I - P).$$

Now $T(t)|_{\mathrm{rg}\, P} = I_{\mathrm{rg}\, P}$ and from $\lim_{t \to \infty} T(t)(I - P) = P - P^2 = 0$ and Proposition 3.2 we conclude that $\big(T_s(t)\big)_{t \geq 0} := \big(T(t)|_{\mathrm{rg}(I-P)}\big)_{t \geq 0}$ is uniformly exponentially stable. Hence, $\big(T(t)\big)_{t \geq 0}$ can be decomposed as

$$T(t) = I_{\mathrm{rg}\, P} \oplus T_s(t), \quad t \geq 0.$$

The proposition in Paragraph II.2.3 yields the corresponding decomposition

$$A = 0_{\mathrm{rg}\,P} \oplus A_s$$

for the generator A, where $A_s := A|_{\mathrm{rg}(I-P)}$. Because the growth bound of $(T_s(t))_{t\geq 0}$, and hence also the spectral bound of A_s is negative, all claims follow easily. $\square$

If the spectral mapping theorem (SMT) from Section 2 holds on every closed subspace $Y \subseteq X$ (e.g., if $(T(t))_{t\geq 0}$ is eventually norm-continuous), then these spectral conditions are even sufficient for uniform convergence.

4.3 Proposition. *Let $(T(t))_{t\geq 0}$ be a strongly continuous semigroup with generator A on X such that* (SMT) *holds for every subspace semigroup $(T(t)_{|Y})_{t\geq 0}$ and any $(T(t))_{t\geq 0}$-invariant, closed subspace $Y \subseteq X$. Then the following are equivalent.*

(a) $P := \lim_{t\to\infty} T(t)$ *exists in the operator norm with $P \neq 0$.*

(b) 0 *is a dominant eigenvalue of A and a first-order pole of $R(\cdot, A)$.*

PROOF. By the previous lemma it suffices to show that (b) implies (a).

Let P_0 be the residue of the resolvent $R(\cdot, A)$ in $\lambda = 0$. Then by Paragraph 1.18 the operator P_0 is the spectral projection of A with respect to the decomposition $\sigma(A) = \{0\} \cup (\sigma(A) \setminus \{0\}) =: \sigma_c \cup \sigma_u$ of $\sigma(A)$; cf. Proposition 1.17. Let $X_c := \mathrm{rg}\,P_0 = \ker(I - P_0)$ and $X_u := \ker P_0 = \mathrm{rg}(I - P_0)$. Then X_c and X_u are $(T(t))_{t\geq 0}$-invariant and $X = X_c \oplus X_u$. Because by assumption $\lambda = 0$ is a first-order pole of $R(\cdot, A)$, by (1.13) in Paragraph 1.18 we conclude $AP = 0$ and hence $A_c := A|_{X_c} = 0_{X_c}$. This implies that $T_c(t) := T(t)|_{X_c} = I_{X_c}$. Next we define $T_u(t) := T(t)|_{X_u}$ which by Paragraph II.2.3 defines a strongly continuous semigroup $(T_u(t))_{t\geq 0}$ on X_u with generator $A_u := A|_{X_u}$. Because 0 is dominant in $\sigma(A) = \{0\} \cup \sigma(A_u)$ and $\sigma(A_u) = \sigma_u$, we obtain $\mathrm{s}(A_u) < 0$. Hence (SMT) applied to $(T_u(t))_{t\geq 0}$ and Proposition 2.3 imply $\omega_0(A_u) = \mathrm{s}(A_u) < 0$ and therefore $\lim_{t\to\infty} T_u(t) = 0$. Summarizing these facts we conclude that

$$\lim_{t\to\infty} T(t) = \lim_{t\to\infty} T_c(t) \oplus \lim_{t\to\infty} T_u(t) = I_{X_c} \oplus 0_{X_u} = P_0 = P.$$

                                                                        $\square$

From Theorem 2.8 we know that eventually norm-continuous semigroups are covered by this result. However, many semigroups arising naturally do not satisfy (SMT), hence Proposition 4.3 does not apply. In addition, it is clear from the proof above that we do not need a spectral mapping theorem for the entire spectrum.

In order to handle these aspects we introduce a new class of semigroups.

4.4 Definition. *A strongly continuous semigroup $(T(t))_{t\geq 0}$ on a Banach space X is called quasi-compact if there exists $t_0 > 0$ such that the essential spectral radius $\mathrm{r}_{\mathrm{ess}}(T(t_0)) < 1$.*

The operators in a quasi-compact semigroup $\big(T(t)\big)_{t\geq 0}$ need not be compact, but only have to approach the subspace $\mathcal{K}(X)$ of all compact operators on X. More precisely, the following holds.

4.5 Proposition. *For a strongly continuous semigroup $\big(T(t)\big)_{t\geq 0}$ on a Banach space X the following assertions are equivalent.*

(a) $\big(T(t)\big)_{t\geq 0}$ *is quasi-compact.*

(b) $r_{\mathrm{ess}}\big(T(t)\big) < 1$ *for all $t > 0$.*

(c) $\lim\limits_{t\to\infty} \mathrm{dist}\big(T(t),\mathcal{K}(X)\big) := \lim\limits_{t\to\infty} \inf\big\{\, \|T(t) - K\| : K \in \mathcal{K}(X) \big\} = 0.$

(d) $\|T(t_0) - K\| < 1$ *for some $t_0 > 0$ and $K \in \mathcal{K}(X)$.*

PROOF. To show that (a) implies (b) we observe that for the essential spectral radius $r_{\mathrm{ess}}(\cdot)$ from Paragraph 1.19 we have

$$r_{\mathrm{ess}}\big(T(t)\big) = e^{\omega_{\mathrm{ess}}\, t},$$

where

$$\omega_{\mathrm{ess}} := \inf_{t>0} \frac{1}{t} \log \|T(t)\|_{\mathrm{ess}}$$

denotes the *essential growth bound* of $\big(T(t)\big)_{t\geq 0}$. This can be proved exactly as the corresponding formula for the spectral radius $r(\cdot)$ given in (1.18) from Proposition 1.22. By assumption (a) there exists $t_0 > 0$ such that $r_{\mathrm{ess}}\big(T(t_0)\big) = e^{t_0\,\omega_{\mathrm{ess}}} < 1$ which implies $\omega_{\mathrm{ess}} < 0$. Hence $r_{\mathrm{ess}}\big(T(t)\big) = e^{t\,\omega_{\mathrm{ess}}} < 1$ for all $t > 0$ as claimed.

To prove that (b) implies (c) we note that

$$r_{\mathrm{ess}}\big(T(1)\big) = \lim_{n\to\infty} \|T(1)^n\|_{\mathrm{ess}}^{1/n} = \lim_{n\to\infty} \|T(n)\|_{\mathrm{ess}}^{1/n} < 1$$

for $\|S\|_{\mathrm{ess}} := \mathrm{dist}\big(S, \mathcal{K}(X)\big)$. Thus, we find $n_0 \in \mathbb{N}$ and $a < 1$ such that

$$\|T(n)\|_{\mathrm{ess}} < a^n \qquad \text{for all } n \geq n_0.$$

Now choose compact operators $K_n \in \mathcal{K}(X)$ such that $\|T(n) - K_n\| < a^n$ for $n \geq n_0$ and define $M := \sup_{0\leq s\leq 1} \|T(s)\|$. We then obtain

$$\|T(t) - T(t-n)K_n\| \leq \|T(t-n)\| \cdot \|T(n) - K_n\| \leq Ma^n$$

for $t \in [n, n+1]$ and $n \geq n_0$. Because $T(t-n)K_n$ is compact for all $n \geq n_0$ this implies $\lim_{t\to\infty} \mathrm{dist}\big(T(t),\mathcal{K}(X)\big) = 0$ as claimed.

Clearly, (c) implies (d), and (d) $\Rightarrow$ (a) follows from

$$r_{\mathrm{ess}}\big(T(t_0)\big) \leq \|T(t_0)\|_{\mathrm{ess}} = \|T(t_0) - K\|_{\mathrm{ess}} \leq \|T(t_0) - K\| < 1.$$

This completes the proof. $\square$

The simplest examples of quasi-compact semigroups are eventually compact semigroups on one side and uniformly exponentially stable semigroups on the other side.

In the next theorem we show that any quasi-compact semigroup can be decomposed into the direct sum of a semigroup on a finite-dimensional space and a uniformly exponentially stable semigroup.

4.6 Theorem. *Let $(T(t))_{t \geq 0}$ be a quasi-compact strongly continuous semigroup with generator A on a Banach space X. Then the following holds.*

(i) *The set $\{\lambda \in \sigma(A) : \operatorname{Re} \lambda \geq 0\}$ is finite (or empty) and consists of poles of $R(\cdot, A)$ of finite algebraic multiplicity.*

If we denote these poles by $\lambda_1, \ldots, \lambda_m$ with corresponding orders $k_1, \ldots, k_m$ and spectral projections $P_1, \ldots, P_m$, we have

(ii) *$T(t) = T_1(t) + T_2(t) + \cdots + T_m(t) + R(t)$, where*

(4.1) $$T_i(t) = e^{\lambda_i t} \sum_{j=0}^{k_i - 1} \frac{t^j}{j!} (A - \lambda_i)^j P_i, \quad t \geq 0 \text{ and } 1 \leq i \leq m,$$
and

(4.2) $\|R(t)\| \leq M e^{-\varepsilon t}$ *for some $\varepsilon > 0$, $M \geq 1$ and all $t \geq 0$.*

PROOF. Let $T := T(t_0)$ where $t_0 > 0$ such that $r_{\mathrm{ess}}(T(t_0)) < 1$. Because every $\mu \in \sigma(T)$ satisfying $|\mu| > r_{\mathrm{ess}}(T)$ is isolated, the set $\sigma(T) \cap \{z \in \mathbb{C} : |z| \geq 1\}$ is finite. Hence we can write

$$\sigma_c := \sigma(T) \cap \{z \in \mathbb{C} : |z| \geq 1\} = \{\mu_1, \ldots, \mu_l\}.$$

Now let $\sigma_u := \sigma(T) \setminus \sigma_c$. Then $\sigma(T)$ is the disjoint union of the closed sets σ_c and σ_u and hence we can define the associated spectral projection P_c as in (1.7). This projection yields the spectral decomposition

$$X = \operatorname{rg}(P_c) \oplus \ker(P_c) =: X_c \oplus X_u.$$

Observing that σ_c is finite and any of its elements is a pole of $R(\cdot, T)$ of finite algebraic multiplicity we conclude that X_c is finite-dimensional. Moreover, because for all $\lambda \in \rho(T)$ the resolvent $R(\lambda, T) = R(\lambda, T(t_0))$ commutes with every $T(t)$, $t \geq 0$, the spaces X_c and $X_u = \operatorname{rg}(I - P_c)$ are $(T(t))_{t \geq 0}$-invariant. Hence we can consider the subspace semigroups $\mathcal{T}_c := (T_c(t))_{t \geq 0}$ and $\mathcal{T}_u := (T_u(t))_{t \geq 0}$ on X_c and X_u, respectively, defined by $T_c(t) := T(t)|_{X_c}$ and $T_u(t) := T(t)|_{X_u}$. By Paragraph II.2.3 the corresponding generators are given by the parts $A_c := A|_{X_c} \in \mathcal{L}(X_c)$ and $A_u := A|_{X_u}$. Because X_c is finite-dimensional, $\sigma(A_c)$ is finite. Moreover, for $\mathcal{T}_c$ the Spectral Mapping Theorem 2.8 holds and hence for all $t \geq 0$ we can write

$$\sigma(A_c) = \{\lambda_1, \ldots, \lambda_m\} \quad \text{and} \quad \sigma(T_c(t)) = \{e^{\lambda t} : \lambda \in \sigma(A_c)\}.$$

In particular, for $t = t_0$ we obtain

$$\sigma_c = \sigma(T_c(t_0)) = \{e^{\lambda t_0} : \lambda \in \sigma(A_c)\} \subset \{z \in \mathbb{C} : |z| \geq 1\}$$

and hence $\operatorname{Re} \lambda \geq 0$ for all $\lambda \in \sigma(A_c)$.

Next we show that $\mathcal{T}_u$ is uniformly exponentially stable. By contradiction assume that $\omega_0(\mathcal{T}_u) \geq 0$. Then (1.18) in Proposition 1.22 implies that $\mathrm{r}\big(T_u(t_0)\big) \geq 1$; i.e., there exists $\widetilde{\mu} \in \sigma\big(T_u(t_0)\big)$ satisfying $|\widetilde{\mu}| \geq 1$. Because by (1.8), $\sigma_u = \sigma\big(T_u(t_0)\big)$ we obtain $\widetilde{\mu} \in \dot{\sigma}_u$. However, by construction, $\sigma_u \subset \{z \in \mathbb{C} : |z| < 1\}$. Hence $|\widetilde{\mu}| < 1$ which is a contradiction. Therefore $\omega_0(\mathcal{T}_u) < 0$ which also implies that $\mathrm{s}(A_u) < 0$. Hence we conclude from the disjoint decomposition $\sigma(A) = \sigma(A_c) \cup \sigma(A_u)$ that $\{\lambda \in \sigma(A) : \mathrm{Re}\,\lambda \geq 0\} = \sigma_c$ is finite. Moreover, because X_c is finite-dimensional and $A = A_c \oplus A_u$, every element of σ_c is a pole of finite algebraic multiplicity of $R(\cdot, A) = R(\cdot, A_c) \oplus R(\cdot, A_u)$. This proves (i).

In order to verify (ii) we define the spectral projection $P := \sum_{i=1}^{m} P_i$ of A corresponding to the spectral set $\{\lambda_1, \ldots, \lambda_m\}$; cf. Proposition 1.17. Then $P = P_c$ by Proposition 1.17.(vi). Next we decompose $T(t) = T(t)P_1 + \cdots + T(t)P_m + T(t)(I - P)$ where, by Paragraph II.2.3, the restricted semigroup $\big(T(t)_{|\,\mathrm{rg}\,P_i}\big)_{t \geq 0}$ has generator $A_{|\,\mathrm{rg}\,P_i}$. Because $\mathrm{rg}\,P_i$ is finite-dimensional and $\big((A - \lambda_i)_{|\,\mathrm{rg}\,P_i}\big)^{k_i} = 0$ we obtain as in the proof of Proposition I.2.6,

$$T_i(t) = T(t)P_i = \mathrm{e}^{\lambda_i t} \sum_{j=0}^{k_i - 1} \frac{t^j}{j!}\,(A - \lambda_i)^j P_i \quad \text{for all } t \geq 0.$$

This proves (4.1). In order to verify (4.2) it suffices to note that $R(t) = T(t)(I - P) = T_u(t)(I - P_c)$ and $\omega_0(\mathcal{T}_u) < 0$. $\square$

Because the spectral mapping theorem (SMT) holds for the above finite-dimensional semigroups $\big(T_i(t)\big)_{t \geq 0}$, $1 \leq i \leq m$, we obtain the following stability criterion.

4.7 Corollary. *A quasi-compact strongly continuous semigroup with generator A is uniformly exponentially stable if and only if*

$$\mathrm{s}(A) < 0.$$

From (4.1) and (4.2) it is now clear which additional hypotheses imply norm convergence of $T(t)$ to an equilibrium as $t \to \infty$. First, we assume the existence of a *dominant eigenvalue* λ_0; i.e.,

$$(4.3) \qquad \mathrm{Re}\,\lambda_0 > \sup\{\mathrm{Re}\,\lambda : \lambda_0 \neq \lambda \in \sigma(A)\}.$$

Moreover, λ_0 has to be a pole of order 1; hence $T_0(t)$ simply becomes $\mathrm{e}^{\lambda_0 t} P_0$. Considering the rescaled semigroup $\big(\mathrm{e}^{-\lambda_0 t} T(t)\big)_{t \geq 0}$ we obtain by estimate (4.2),

$$\big\|\mathrm{e}^{-\lambda_0 t} T(t) - P_0\big\| \leq \mathrm{e}^{-\mathrm{Re}\,\lambda_0 t}\,\|T(t) - T_0(t)\| = \mathrm{e}^{-\mathrm{Re}\,\lambda_0 t}\,\|R(t)\| \leq M \mathrm{e}^{-\varepsilon t}$$

for some $\varepsilon > 0$ and $M \geq 1$.

4.8 Corollary. *Let* $(T(t))_{t\geq 0}$ *be a quasi-compact strongly continuous semigroup. If* λ_0 *is a dominant eigenvalue of the generator and a first-order pole of the resolvent with residue* P_0, *then there exist constants* $\varepsilon > 0$ *and* $M \geq 1$ *such that*

$$\left\|e^{-\lambda_0 t}T(t) - P_0\right\| \leq Me^{-\varepsilon t}$$

for all $t \geq 0$.

It should be evident that the most interesting case occurs if $\lambda_0 = 0$ in the above corollary, and we refer to [Nag86, B-IV, Thm. 2.5 and Expl. 2.6] for an important class of examples.

Generators of quasi-compact semigroups can now be perturbed by an arbitrary compact operator destroying the uniform exponential stability but not the quasi-compactness.

4.9 Proposition. *Let* $(T(t))_{t\geq 0}$ *be a quasi-compact strongly continuous semigroup with generator* A *on the Banach space* X *and take a compact operator* $K \in \mathcal{L}(X)$. *Then* $A + K$ *generates a quasi-compact semigroup.*

PROOF. By (IE) in Corollary III.1.7 we know that the semigroup $(S(t))_{t\geq 0}$ generated by $A + K$ can be represented as

$$S(t) = T(t) + \int_0^t T(t-s)KS(s)\,ds$$

where the integral is understood in the strong sense. In view of Proposition 4.5 it is now enough to show that the operator $\int_0^t T(t-s)KS(s)\,ds$ is compact.

Because the mapping $(t,x) \mapsto T(t)x$ is jointly continuous on $\mathbb{R}_+ \times X$ and because K is compact, the set $M := \{T(s)Kx : 0 \leq s \leq t, \|x\| \leq 1\}$ is relatively compact in X. Having in mind that $\int_0^t T(t-s)KS(s)x\,ds$, $x \in X$, is the norm limit of Riemann sums, we observe that

$$\frac{1}{ct}\int_0^t T(t-s)KS(s)x\,ds$$

is an element of the closed convex hull $\overline{\mathrm{co}}\,M$, provided that $c := \sup\{\|S(s)\| : 0 \leq s \leq t\}$ and $\|x\| \leq 1$. Because $\overline{\mathrm{co}}\,M$ is compact by Proposition A.1, the assertion follows. $\square$

We have noted above that every exponentially stable semigroup is quasi-compact. Therefore we obtain from Proposition 4.9 the following important class of quasi-compact semigroups.

4.10 Example. If $(T(t))_{t\geq 0}$ generates an exponentially stable semigroup with generator A and $K \in \mathcal{L}(X)$ is compact, then $A + K$ generates a quasi-compact semigroup.

We now discuss quasi-compactness and its consequences in a concrete example.

4.11 Example. On the Banach space $X := C(\mathbb{R}_- \cup \{-\infty\})$ we consider the first-order differential operator

$$(4.4) \qquad\qquad Af := f' + mf$$

with domain

$$(4.5) \quad D(A) := \{f \in X : f \text{ is differentiable, } f' \in X \text{ and } f'(0) = Lf\},$$

where $m \in X$ is real-valued and L is a continuous linear form on X. As in Paragraph II.3.29 we can show that the operator $(A, D(A))$ generates a strongly continuous semigroup $(T(t))_{t \geq 0}$.

Lemma 1. *The semigroup* $(T(t))_{t \geq 0}$ *satisfies*

$$
T(t)f(s) = \begin{cases}
e^{\int_s^0 m(\sigma)\,d\sigma}\left[e^{(s+t)m(0)}f(0)\right.\\
\qquad\left.+ \displaystyle\int_0^{s+t} e^{\tau m(0)} L\,T(s+t-\tau)f\,d\tau\right] & \text{for } s+t > 0,\\[2mm]
e^{\int_s^{s+t} m(\sigma)\,d\sigma} f(s+t) & \text{for } s+t \leq 0.
\end{cases}
$$

PROOF. For $f \in D(A)$ and $0 \leq r \leq t$ we have

$$\frac{d}{dr}\left(e^{rm(0)}\big(T(t-r)f\big)(0) + \int_0^r e^{\tau m(0)} L\,T(t-\tau)f\,d\tau\right) = 0.$$

This implies

$$\big(T(t)f\big)(0) = e^{tm(0)}f(0) + \int_0^t e^{\tau m(0)} L\,T(t-\tau)f\,d\tau.$$

On the other hand, we have

$$\frac{d}{dr}\left(e^{\int_s^{s+r} m(\sigma)\,d\sigma}\big(T(t-r)f\big)(s+r)\right) = 0.$$

Therefore, we obtain

$$
\big(T(t)f\big)(s) = \begin{cases}
e^{\int_s^0 m(\sigma)\,d\sigma}\big(T(s+t)f\big)(0) & \text{for } s+t > 0,\\[2mm]
e^{\int_s^{s+t} m(\sigma)\,d\sigma} f(s+t) & \text{for } s+t \leq 0.
\end{cases}
$$

$\square$

This lemma allows us to give a condition that forces the semigroup $(T(t))_{t \geq 0}$ to be quasi-compact.

Lemma 2. *If* $m(-\infty) < 0$, *then the semigroup* $(T(t))_{t \geq 0}$ *is quasi-compact.*

PROOF. We define operators $K(t) \in \mathcal{L}(X)$ by

$$
K(t)f(s) := \begin{cases} e^{\int_s^0 m(\sigma)\,d\sigma} \Big[e^{(s+t)\,m(0)} f(0) \\[2mm] \quad + \displaystyle\int_0^{s+t} e^{(s+t-\tau)\,m(0)} L\,T(\tau)f\,d\tau \Big] & \text{for } 0 < s+t, \\[4mm] (t+s+1) \cdot e^{\int_s^0 m(\sigma)\,d\sigma} f(0) & \text{for } -1 < s+t \leq 0, \\[2mm] 0 & \text{for } s+t \leq -1. \end{cases}
$$

These operators are compact by the Arzelà–Ascoli theorem. On the other hand, because $m(-\infty) < 0$, we have

$$
\lim_{t \to \infty} \|T(t) - K(t)\| = 0.
$$

Therefore, the semigroup $(T(t))_{t \geq 0}$ is quasi-compact. $\square$

Assume in the following that $m(-\infty) < 0$. In order to apply Theorem 4.6 and Corollary 4.8, we have to find the eigenvalues λ of A with $\operatorname{Re}\lambda \geq 0$. An eigenfunction $f \in D(A)$ with eigenvalue λ satisfies

$$
f' = \lambda f - mf,
$$

hence is of the form $f = cg_\lambda$, where

$$
g_\lambda(s) := e^{\int_s^0 m(\sigma)\,d\sigma} e^{\lambda s}
$$

for all $s \in \mathbb{R}_-$. Because $\operatorname{Re}\lambda \geq 0$, the functions g_λ and g_λ' vanish at $-\infty$, hence belong to X. Consequently, $g_\lambda \in D(A)$ if and only if

$$
\lambda - Lg_\lambda - m(0) = 0.
$$

This shows that λ is an eigenvalue of A if and only if the *characteristic equation*

(4.6) $$\xi(\lambda) := \lambda - Lg_\lambda - m(0) = 0$$

holds.

Now, suppose that λ with $\operatorname{Re}\lambda \geq 0$ is not an eigenvalue of A. For each $g \in X$ we want to find a function $f \in D(A)$ such that

$$
f' = \lambda f - mf - g.
$$

This equation is solved by

$$f = cg_\lambda + h_\lambda,$$

where

$$h_\lambda(s) := \int_s^0 e^{\int_s^\tau m(\sigma)\,d\sigma} e^{\lambda(s-\tau)} g(\tau)\,d\tau$$

for all $s \in \mathbb{R}_-$. If the constant c is chosen as

(4.7)
$$c := \frac{g(0) + Lh_\lambda}{\lambda - Lg_\lambda - m(0)},$$

we then obtain the unique $f \in D(A)$ satisfying $(\lambda - A)f = g$. This even yields an explicit representation of the resolvent of A in λ.

In the remaining part of this section we look for conditions implying the existence of a dominant eigenvalue and convergence to an equilibrium. In Chapter VI we show that positivity of the semigroup is the key to such results. Here it suffices to assume that L is of the form

(4.8)
$$L = L_0 + a\delta_0,$$

where a is a real number and L_0 is a *positive* linear form on X. We then have the following lemma proving the existence of a dominant eigenvalue.

Lemma 3. *Suppose that $m(-\infty) < 0$. If $\xi(0) \leq 0$, i.e., $Lg_0 \geq -m(0)$, then the characteristic function ξ has a unique zero $\lambda_0 \geq 0$ that is a dominant eigenvalue of the operator A.*

PROOF. *The function $\xi : \mathbb{R}_+ \ni \lambda \mapsto \lambda - L_0 g_\lambda - a - m(0)$ is strictly increasing from $\xi(0)$ to ∞. Consequently, if $\xi(0) \leq 0$, it has a unique zero λ_0 that is an eigenvalue of A. Now take an arbitrary eigenvalue λ of A with $\operatorname{Re}\lambda \geq \lambda_0$. Then, we have*

$$|\lambda - a - m(0)| = |L_0 g_\lambda| \leq L_0 g_{\lambda_0} = \lambda_0 - a - m(0).$$

This implies $\lambda = \lambda_0$, and therefore λ_0 is a dominant eigenvalue of A.   □

The eigenspace corresponding to the dominant eigenvalue λ_0 is spanned by the function g_{λ_0}, hence is one-dimensional. Moreover, it is a first-order pole, as can be seen from (4.7).

After these preparations, we can give a precise description of the asymptotic behavior of the semigroup $(T(t))_{t\geq 0}$. In particular, it follows that the rescaled semigroup $(e^{-\lambda_0 t} T(t))_{t\geq 0}$ converges in norm to a one-dimensional projection.

Proposition 4. *Assume that* $m(-\infty) < 0$, $L = L_0 + a\delta_0$ *as in* (4.8), *and* $L_0 g_0 + a \geq -m(0)$. *Then there is a dominant eigenvalue* $\lambda_0 \geq 0$ *of* A, *a continuous linear form* φ *on* X, *and constants* ε, $M > 0$ *such that*

$$\|e^{-\lambda_0 t} T(t)f - (g_{\lambda_0} \otimes \varphi)f\| \leq Me^{-\varepsilon t}\|f\| \qquad \text{for all } f \in X, t \geq 0,$$

where $(g_{\lambda_0} \otimes \varphi)f := \varphi(f) \cdot g_{\lambda_0}$.

4.12 Exercises. (1) Show that for a strongly continuous semigroup $\mathcal{T} = (T(t))_{t \geq 0}$ with generator A on a Banach space X the fixed space $\text{fix}(T(t))_{t \geq 0}$ and the kernel $\ker A$ coincide. (Hint: Use Lemma II.1.3.(iv).)

(2) Let $(T(t))_{t \geq 0}$ be an eventually compact semigroup such that the spectrum $\sigma(A)$ of its generator A is infinite. Show that there exists a sequence $(\mu_n)_{n \in \mathbb{N}}$ in $\mathbb{C}$ such that $\sigma(A) = P\sigma(A) = \{\mu_n : n \in \mathbb{N}\}$ and $\lim_{n \to \infty} \text{Re}\, \mu_n = -\infty$. (Hint: Use Theorem 4.6 and Theorem II.5.3.)

Chapter VI

Positive Semigroups

In many concrete problems solvable by semigroups, there is a natural notion of "positivity," and only "positive" solutions make sense. In terms of the corresponding semigroup $(T(t))_{t\geq 0}$ this means that the operators $T(t)$ should be "positive" on some ordered Banach space.

The complete theory of such "one-parameter semigroups of positive operators" on Banach lattices and other ordered vector spaces can be found in [Nag86]. In the following we present the basic ideas and some typical results from this theory.

1. Basic Properties

For our purposes it suffices to restrict our attention to Banach spaces of type $X := L^p(\Omega, \mu)$ or $C_0(\Omega)$. On these spaces we call a function $f \in X$ *positive* (in symbols: $0 \leq f$) if

$$0 \leq f(s) \qquad \text{for (almost) all } s \in \Omega.$$

For real-valued functions $f, g \in X$ we then write $f \leq g$ if $0 \leq g - f$ and obtain an ordering making (the real part of) X into a *vector lattice*; cf. [Sch74, Sect. II.1]. To indicate that $0 \leq f$ and $0 \neq f$ we use the notation $0 < f$. Moreover, for an arbitrary (complex-valued) function $f \in X$ we define its absolute value $|f|$ as

$$|f|(s) := |f(s)| \qquad \text{for } s \in \Omega.$$

Recalling the definition of the norm on X, we see that

(1.1) $|f| \leq |g|$ implies $\|f\| \leq \|g\|$ for all $f, g \in X$.

These properties make the space X a *Banach lattice*, and we refer to [Sch74], [MN91], or [AB85] for the abstract definitions. It is convenient to use this general terminology and to state the results for general Banach lattices. However, the reader not accustomed to this terminology may always think of the space X as one of the concrete function spaces $L^p(\Omega, \mu)$ or $C_0(\Omega)$ with the canonical ordering.

This is why in this chapter we use the symbol f to denote an element in a Banach lattice X.

1.1 Definition. *A strongly continuous semigroup $\big(T(t)\big)_{t \geq 0}$ on a Banach lattice X is called* positive *if each operator $T(t)$ is positive, i.e., if*

$$0 \leq f \in X \quad \text{implies} \quad 0 \leq T(t)f \qquad \text{for each } t \geq 0,$$

or equivalently, if

$$|T(t)f| \leq T(t)|f| \quad \text{holds for each } f \in X, t \geq 0.$$

As in the preceding chapters, it is important to characterize positivity of the semigroup through a property of its generator. At least in the finite-dimensional case this is simple.

1.2 Proposition. *A matrix $A = (a_{ij})_{n \times n} \in M_n(\mathbb{C})$ generates a positive semigroup $\big(T(t)\big)_{t \geq 0}$ if and only if it is real and positive off-diagonal; i.e., $a_{ii} \in \mathbb{R}$ and $a_{ij} \geq 0$ for all $1 \leq i, j \leq n$, $i \neq j$.*

PROOF. By Proposition I.2.7 we know that $T(t) = e^{tA}$ and

$$A = \lim_{t \downarrow 0} \frac{e^{tA} - I}{t},$$

which means

(1.2) $$a_{ij} = \lim_{t \downarrow 0} \left\langle \frac{e^{tA} e_j - e_j}{t}, e_i \right\rangle$$

for $i, j = 1, \ldots, n$ and e_i the ith unit vector in $\mathbb{C}^n$. If we denote the (i, j)th entry of e^{tA} by $\tau_{ij}(t)$, then (1.2) implies

(1.3) $$a_{ij} = \begin{cases} \lim_{t \downarrow 0} \frac{\tau_{ij}(t)}{t} & \text{for } i \neq j, \\ \lim_{t \downarrow 0} \frac{\tau_{ij}(t) - 1}{t} & \text{for } i = j. \end{cases}$$

For $\big(e^{tA}\big)_{t \geq 0}$ positive, i.e., $\tau_{ij}(t) \geq 0$ for all t, i, and j, this implies

$$a_{ij} \geq 0 \qquad \text{for } i \neq j$$

and

$$a_{ij} \in \mathbb{R} \qquad \text{for } i = j.$$

To prove the converse implication we suppose that A is real and positive off-diagonal. Thus we can find $\rho \in \mathbb{R}$ such that

$$(1.4) \qquad\qquad B_\rho := A + \rho I \geq 0$$

(e.g., take $\rho := \max_{1 \leq i \leq n} |a_{ii}|$). Hence we obtain

$$e^{tA} = e^{[t(A+\rho I) - t\rho I]}$$

$$= e^{-t\rho} \cdot e^{tB_\rho} \geq 0$$

for all $t \geq 0$. □

Various characterizations of generators of positive semigroups on infinite-dimensional Banach spaces can be found in [Nag86, C-II]. We give only an elementary characterization in terms of the resolvent.

1.3 Characterization Theorem. *A strongly continuous semigroup* $\mathcal{T} := (T(t))_{t \geq 0}$ *on a Banach lattice* X *is positive if and only if the resolvent* $R(\lambda, A)$ *of its generator* A *is positive for all sufficiently large* λ.

PROOF. The positivity of $\mathcal{T}$ implies the positivity of $R(\lambda, A)$ by the integral representation (1.13) in Section II.1. Conversely, the positivity of $T(t) = \lim_{n \to \infty} [n/t R(n/t, A)]^n$ (see Corollary IV.2.5) follows from that of $R(\lambda, A)$ for λ large. □

In the next two sections we show the special features of a positive semigroup with respect to its spectrum and its asymptotic behavior.

2. Spectral Theory for Positive Semigroups

In the years 1907–1912, O. Perron and G. Frobenius discovered very beautiful symmetry properties of the spectrum of positive matrices. Many of these properties still hold for the spectra of positive operators on arbitrary Banach lattices (cf. [Sch74, Sects. V.4 and 5]), and even carry over to generators of positive semigroups (cf. [Nag86]).

In order to prove the basic results of this theory, we need the following lemma. It shows that for positive semigroups the integral representation of the resolvent holds even for $\operatorname{Re} \lambda > s(A)$ and not only for $\operatorname{Re} \lambda > \omega_0(A)$ as shown in Theorem II.1.10.

2.1 Lemma. *For a positive strongly continuous semigroup* $(T(t))_{t \geq 0}$ *with generator* A *on a Banach lattice* X *we have*

$$(2.1) \qquad\qquad R(\lambda, A)f = \int_0^\infty e^{-\lambda s} T(s)f \, ds, \qquad f \in X,$$

for all $\operatorname{Re} \lambda > s(A)$. *Moreover, the following properties are equivalent for* $\lambda_0 \in \rho(A)$.

(a) $0 \leq R(\lambda_0, A)$.

(b) $s(A) < \lambda_0$.

PROOF. Using the rescaling techniques from Paragraph I.1.10 it suffices to prove the representation (2.1) for $\operatorname{Re}\lambda > 0$ whenever $s(A) < 0$.

Because the integral representation (2.1) certainly holds for $\operatorname{Re}\lambda > \omega_0(A)$, we obtain from the positivity of $(T(t))_{t\geq 0}$ the positivity of $R(\lambda, A)$ for $\lambda > \omega_0(A)$. The power series expansion (1.3) in Proposition V.1.3 of the resolvent yields $0 \leq R(\lambda, A)$ for all $\lambda > s(A)$.

The assumption $s(A) < 0$ and Lemma II.1.3.(iv) then imply

$$0 \leq V(t) := \int_0^t T(s)\,ds = R(0, A) - R(0, A)T(t) \leq R(0, A),$$

hence $\|V(t)\| \leq M$ for all $t \geq 0$ and some constant M. From this estimate we deduce that

$$\int_0^\infty e^{-\lambda s}V(s)\,ds, \qquad \operatorname{Re}\lambda > 0,$$

exists in operator norm. An integration by parts yields

$$\int_0^t e^{-\lambda s}T(s)\,ds = e^{-\lambda t}V(t) + \lambda \int_0^t e^{-\lambda s}V(s)\,ds,$$

which converges to $\lambda \int_0^\infty e^{-\lambda s}V(s)\,ds$ as $t \to \infty$. This first proves (2.1) by Theorem II.1.10.(i) and then the implication (b) $\Rightarrow$ (a).

Moreover, as shown in Theorem 2.2 below, the integral representation (2.1) implies that $s(A) \in \sigma(A)$. Therefore, by Corollary V.1.14, we obtain for the spectral radius of the resolvent

$$(2.2) \qquad\qquad r\big(R(\lambda, A)\big) = \frac{1}{\lambda - s(A)}$$

for all $\lambda > s(A)$.

In order to prove (a) $\Rightarrow$ (b) we now assume that $R(\lambda_0, A) \geq 0$ and observe that this can be true only for λ_0 real. As we have shown above, $R(\lambda, A)$ is positive for $\lambda > \max\{\lambda_0, s(A)\}$. Hence, an application of the resolvent equation yields

$$R(\lambda_0, A) = R(\lambda, A) + (\lambda - \lambda_0)R(\lambda, A)R(\lambda_0, A) \geq R(\lambda, A) \geq 0$$

for $\lambda > \max\{\lambda_0, s(A)\}$. It follows from (2.2) and (1.1) that

$$\frac{1}{\lambda - s(A)} = r\big(R(\lambda, A)\big) \leq \|R(\lambda, A)\| \leq \|R(\lambda_0, A)\|$$

for all $\lambda > \max\{\lambda_0, s(A)\}$. This implies that λ_0 is greater than $s(A)$. $\qquad\square$

O. Perron proved in 1907 that the spectral radius of a positive matrix is always an eigenvalue. The semigroup version of this result assures that the spectral bound of the generator of a positive semigroup is always a spectral value.

2.2 Theorem. *Let* $(T(t))_{t \geq 0}$ *be a positive strongly continuous semigroup with generator A on a Banach lattice X. If* $s(A) > -\infty$, *then*

$$s(A) \in \sigma(A).$$

PROOF. The positivity of the operators $T(t)$ means that

$$|T(t)f| \leq T(t)|f| \qquad \text{for all } f \in X, t \geq 0.$$

We therefore obtain from the integral representation (2.1) that

$$|R(\lambda, A)f| \leq \int_0^\infty e^{-\operatorname{Re}\lambda \cdot s} T(s)|f|\, ds$$

for all $\operatorname{Re}\lambda > s(A)$ and $f \in X$. Using the inequality in (1.1) we deduce that

$$(2.3) \qquad \|R(\lambda, A)\| \leq \|R(\operatorname{Re}\lambda, A)\| \qquad \text{for all } \operatorname{Re}\lambda > s(A).$$

By Corollary V.1.14, there exist $\lambda_n \in \rho(A)$ such that $\operatorname{Re}\lambda_n \downarrow s(A)$ and $\|R(\lambda_n, A)\| \uparrow \infty$. The estimate (2.3) then implies $\|R(\operatorname{Re}\lambda_n, A)\| \uparrow \infty$ and therefore $s(A) \in \sigma(A)$ by Proposition V.1.3.(iii). $\qquad\square$

For positive matrix semigroups much more can be said on the spectral value $s(A)$.

2.3 Proposition. *If the matrix $A = (a_{ij})_{n \times n}$ is real and positive off-diagonal, then $s(A)$ is a dominant eigenvalue of A.*

PROOF. Take the matrix $B_\rho = A + \rho I$ from (1.4) above which is positive by assumption for $\rho \in \mathbb{R}$ sufficiently large. Therefore, Perron's Theorem (see [Sch74, Chap. I, Prop. 2.3]) implies that the spectral radius $\operatorname{r}(B_\rho)$ is an eigenvalue of B_ρ. Evidently, $\operatorname{r}(B_\rho)$ is dominant in $\sigma(B_\rho)$. Because

$$\sigma(B_\rho) = \sigma(A) + \rho \qquad \text{and} \qquad s(B_\rho) = s(A) + \rho,$$

we obtain that $s(A)$ is dominant in $\sigma(A)$. $\qquad\square$

Another useful property of positive semigroups is the monotonicity of the spectral bound under positive perturbations.

2.4 Corollary. *Let A be the generator of a positive strongly continuous semigroup $(T(t))_{t \geq 0}$ and let $B \in \mathcal{L}(X)$ be a positive operator on the Banach lattice X. Then the following hold.*

(i) *$A+B$ generates a positive semigroup $(S(t))_{t \geq 0}$ satisfying $0 \leq T(t) \leq S(t)$ for all $t \geq 0$.*

(ii) *$s(A) \leq s(A + B)$ and $R(\lambda, A) \leq R(\lambda, A + B)$ for all $\lambda > s(A + B)$.*

PROOF. Because B is bounded, we obtain the generation property of $A+B$ from Theorem III.1.3. Moreover, the perturbed resolvent is

$$R(\lambda, A + B) = R(\lambda, A) + R(\lambda, A) \sum_{n=1}^{\infty} (BR(\lambda, A))^n \qquad \text{for } \lambda \text{ large}$$

(see Section III.1, (1.3)). Because B and $R(\lambda, A)$ are positive for $\lambda > s(A)$, this implies

(2.4) $$0 \leq R(\lambda, A) \leq R(\lambda, A + B)$$

for λ large. The inequality in (i) then follows from the Post–Widder inversion formula in Corollary IV.2.5. Next, we use the representation (2.1) for the resolvents of A and $A + B$, respectively, and infer that (2.4) and hence

$$\|R(\lambda, A)\| \leq \|R(\lambda, A + B)\|$$

hold for all $\lambda > \max\{s(A), s(A+B)\}$. The inequality in (ii) for the spectral bounds then follows, because $s(A) \in \sigma(A)$ by Theorem 2.2 and therefore $\overline{\lim}_{\lambda \downarrow s(A)} \|R(\lambda, A)\| = \infty$. $\qquad \square$

Due to these results, the spectral bound becomes the supremum of all *real* spectral values only, hence is much easier to compute. Moreover, Lemma 2.1 says that

$$s(A) < 0$$

if and only if $0 \in \rho(A)$ with $0 \leq R(0, A) = -A^{-1}$. So in order to have $s(A) < 0$ for a positive semigroup it suffices to show that A is invertible with negative inverse. This is behind many maximum principles for partial differential operators.

On the other hand, we know from the example in Section V.2.a that the spectral bound and the growth bound do not coincide in general, hence $s(A) < 0$ does not imply uniform exponential stability. Counterexample V.1.26 and Exercise V.2.13.(2) show that this even happens for positive semigroups on Banach lattices. However, on special Banach lattices positivity, as eventual norm continuity in Corollary V.2.9, makes the spectral bound and the growth bound coincide.

2.5 Theorem. Let $(T(t))_{t\geq 0}$ be a positive strongly continuous semigroup with generator A on a Banach lattice $L^p(\Omega, \mu)$, $1 \leq p < \infty$. Then

$$s(A) = \omega_0$$

holds.

PROOF. We only prove the case $p = 2$. By the usual rescaling technique and because $s(A) \leq \omega_0(A)$ it suffices to show that $(T(t))_{t\geq 0}$ is uniformly exponentially stable if $s(A) < 0$.

Because $L^2(\Omega, \mu)$ is a Hilbert space, we can use Theorem V.3.8 and therefore have to show that

$$\sup_{\operatorname{Re}\lambda>0} \|R(\lambda, A)\| < \infty.$$

However this follows from the estimate $\|R(\lambda, A)\| \leq \|R(\operatorname{Re}\lambda, A)\|$, $\operatorname{Re}\lambda > 0$, proved in (2.3) above.

For the proof for $p = 1$ we refer to [ABHN01, Thm. 5.3.7] whereas the general case, due to Weis [Wei95], [Wei98], can be found in [ABHN01, Thm. 5.3.6]. □

3. Convergence to Equilibrium, Revisited

In this section we return to the question of when the semigroup $(T(t))_{t\geq0}$ converges to a nontrivial projection

$$P = \lim_{t\to\infty} T(t).$$

For a strongly continuous semigroup $(T(t))_{t\geq0}$ with generator A satisfying (SMT) we showed in Proposition V.4.3 that this is true if and only if 0 is a dominant eigenvalue of A and a first-order pole of the resolvent $R(\cdot, A)$. We see now that *positivity* and quasi-compactness of the semigroup combined with *irreducibility* imply these conditions, hence convergence.

Although all the following results hold in any Banach lattice, we again restrict our considerations to concrete function spaces. Hence we assume that $(T(t))_{t\geq0}$ is a positive strongly continuous semigroup on a Banach lattice

$$X = L^p(\Omega, \mu)$$

for some σ-finite measure space (Ω, μ) and $1 \leq p < \infty$. We call a function $f \in X$ *strictly positive* (in symbols, $0 \ll f$), if

$$0 < f(s) \text{ for almost all } s \in \Omega.$$

Similarly, a positive linear form $\varphi \in X'$ is called *strictly positive* (in symbols, $0 \ll \varphi$), if

$$0 \leq f \in X \quad \text{and} \quad \langle f, \varphi \rangle = 0 \quad \text{imply} \quad f = 0.$$

Observe that if we identify X' with $L^q(\Omega, \mu)$, then a strictly positive linear form corresponds to a strictly positive function. With this terminology we can now introduce the concept of an irreducible semigroup in the following way.

3.1 Definition. *A positive semigroup with generator A on the Banach lattice $X = \mathrm{L}^p(\Omega, \mu)$ is irreducible, if for some $\lambda > s(A)$ and all $0 < f \in \mathrm{L}^p(\Omega, \mu)$ the resolvent satisfies*

$$0 \ll R(\lambda, A)f.$$

This notion is fundamental for the theory of positive semigroups and we refer to [Nag86, C-III] for a list of different characterizations and many nice properties. In particular, it is shown there that a strongly continuous semigroup is irreducible if and only if $0 \ll R(\lambda, A)f$ for all $\lambda > s(A)$ and all $0 < f \in X$.

For our purposes we need that for an irreducible positive semigroup every positive fixed element is strictly positive.

3.2 Lemma. *Let $(T(t))_{t \geq 0}$ be an irreducible positive semigroup. Then*

$$0 < f \in \mathrm{fix}\big(T(t)\big)_{t \geq 0} \qquad implies \qquad 0 \ll f$$

and

$$0 < \varphi \in \mathrm{fix}\big(T(t)'\big)_{t \geq 0} \qquad implies \qquad 0 \ll \varphi.$$

PROOF. Assume that $0 < f \in \mathrm{fix}\big(T(t)\big)_{t \geq 0} = \ker A$ is not strictly positive. Then $R(\lambda, A)f = 1/\lambda f$ is also not strictly positive for $\lambda > s(A)$, hence $(T(t))_{t \geq 0}$ is not irreducible.

For the second statement we take $0 < \varphi \in \mathrm{fix}\big(T(t)'\big)_{t \geq 0} = \ker A'$. Again we obtain $R(\lambda, A')\varphi = 1/\lambda\, \varphi > 0$ for $\lambda > s(A)$. If φ is not strictly positive there exists $0 < f \in X$ such that $\langle f, \varphi \rangle = 0$. This implies

$$0 = \langle f, 1/\lambda\, \varphi \rangle = \langle f, R(\lambda, A')\varphi \rangle = \langle R(\lambda, A)f, \varphi \rangle,$$

hence $R(\lambda, A)f$ is not strictly positive and $(T(t))_{t \geq 0}$ not irreducible. □

We now give some typical examples of irreducible semigroups.

3.3 Examples. (i) On $X = \mathbb{C}^n$, the semigroup $\big(e^{tA}\big)_{t \geq 0}$ generated by a real, positive off-diagonal matrix A is irreducible if and only if there is no permutation matrix Q such that

$$QAQ^{-1} = \left(\begin{array}{c|c} * & * \\ \hline 0 & * \end{array} \right).$$

For more details, see [Sch74, Chap. I] and [BP79].

(ii) If $(T(t))_{t \geq 0}$ is the semigroup induced by a measure-preserving flow $(\varphi_t)_{t \geq 0}$ on Ω, i.e.,

$$T(t)f = f \circ \varphi_t \quad \text{for } f \in \mathrm{L}^p(\Omega, \mu),\ t \geq 0,$$

then $(T(t))_{t \geq 0}$ is irreducible if and only if $(\varphi_t)_{t \geq 0}$ is ergodic. See [Kre85].

(iii) The diffusion semigroup on $\mathrm{L}^p(\mathbb{R}^n)$ defined in Paragraph II.2.12 is irreducible. This follows because each operator $T(t)$, hence each resolvent operator $R(\lambda, A)$, λ large, is an integral operator with strictly positive kernel (see (2.8) in Paragraph II.2.12).

From these and many other examples one sees that irreducibility occurs naturally. In the next result we show that it has strong consequences on the spectrum of the generator.

3.4 Proposition. *Let* $(T(t))_{t\geq 0}$ *be an irreducible, positive, strongly continuous semigroup with generator A on the Banach lattice X and assume that* s$(A) = 0$. *If 0 is a pole of the resolvent $R(\cdot, A)$, then the following properties hold.*

(i) fix$(T(t))_{t\geq 0}$ = ker A = lin$\{h\}$ *for some strictly positive function* $h \in X$.

(ii) fix$(T(t)')_{t\geq 0}$ = ker A' = lin$\{\varphi\}$ *for some strictly positive linear form* $\varphi \in X'$.

(iii) 0 *is a first-order pole with residua* $P = \varphi \otimes h$, *where $h \in$ fix$(T(t))_{t\geq 0}$,* $\varphi \in$ fix$(T(t)')_{t\geq 0}$, *and* $\langle h, \varphi \rangle = 1$.

PROOF. (i) We first show that fix$(T(t))_{t\geq 0}$ contains a strictly positive element. If s$(A) = 0$ is a pole of order k, it follows from the positivity of $(T(t))_{t\geq 0}$, hence of $R(\lambda, A)$ for $\lambda > 0$, and from Paragraph V.1.18 that

$$U_{-k} = \lim_{\lambda \downarrow 0} \lambda^k R(\lambda, A) > 0$$

and

$$AU_{-k} = U_{-(k+1)} = 0.$$

Hence there exists $0 \neq f \in X$ such that $U_{-k}f \neq 0$. Again by positivity we obtain

$$h := U_{-k}|f| \geq |U_{-k}f| > 0.$$

Because $Ah = 0$, the positive function h belongs to fix$(T(t))_{t\geq 0}$. The irreducibility of $(T(t))_{t\geq 0}$ implies $0 \ll h$ by Lemma 3.2.

(ii) By analogous arguments and by taking adjoints we obtain that fix$(T(t)')_{t\geq 0}$ contains strictly positive elements.

We now continue the proof of (i) and show that fix$(T(t))_{t\geq 0}$ is one-dimensional. Because $f \in$ fix$(T(t))_{t\geq 0}$ if and only if $\bar{f} \in$ fix$(T(t))_{t\geq 0}$ it suffices to prove that the real vector space

$$\{f \in \text{fix}(T(t))_{t\geq 0} : f = \bar{f}\}$$

is one-dimensional. Take $0 \neq f \in$ fix$(T(t))_{t\geq 0}$. The positivity of the operators $T(t)$ yields

$$|f| = |T(t)f| \leq T(t)|f|.$$

Moreover,

$$\langle T(t)|f| - |f|, \varphi \rangle = \langle |f|, T(t)'\varphi \rangle - \langle |f|, \varphi \rangle = 0$$

for every $0 < \varphi \in \text{fix}\big(T(t)'\big)_{t \geq 0}$. Because we have shown that such φ exists and is strictly positive, we conclude that

$$T(t)|f| = |f|, \qquad \text{i.e.,} \qquad |f| \in \text{fix}\big(T(t)\big)_{t \geq 0}.$$

Then also

$$f^+ := \tfrac{1}{2}(f + |f|) \qquad \text{and} \qquad f^- := \tfrac{1}{2}(-f + |f|)$$

belong to $\text{fix}\big(T(t)\big)_{t \geq 0}$. Again the strict positivity of the elements in the fixed space $\text{fix}\big(T(t)\big)_{t \geq 0}$ implies

$$f^+ = 0 \qquad \text{or} \qquad f^- = 0;$$

i.e., for each $f \in \text{fix}\big(T(t)\big)_{t \geq 0}$ we have

$$f \geq 0 \qquad \text{or} \qquad f \leq 0.$$

This means that $\text{fix}\big(T(t)\big)_{t \geq 0}$ is a totally ordered Banach lattice, hence one-dimensional by [Sch74, Prop. II.3.4].

(iii) It remains to show that 0 is a first-order pole of $R(\cdot, A)$. Assume that the pole order is $k > 1$ and take some $0 \ll g \in \ker A$. With the notation from Paragraph V.1.18 and P the residua of $R(\cdot, A)$ in 0 we have

$$PA^{k-1} = A^{k-1}P = U_{-k},$$

which is a positive operator as already seen above. This operator vanishes on g, hence on all functions $f \in X$ such that

$$|f| \leq n \cdot g \qquad \text{for some } n \in \mathbb{N}.$$

These functions form a dense subspace in X because g is strictly positive. Therefore $U_{-k} = 0$, a contradiction. $\qquad\qquad\qquad\qquad\qquad \square$

If we now add quasi-compactness to the above properties, we obtain not only convergence but convergence to a (up to scalars) unique equilibrium.

3.5 Theorem. *Let $\big(T(t)\big)_{t \geq 0}$ be a quasi-compact, irreducible, positive strongly continuous semigroup with generator A and assume that $s(A) = 0$. Then 0 is a dominant eigenvalue of A and a first-order pole of $R(\cdot, A)$. Moreover, there exist strictly positive elements $0 \ll h \in X$, $0 \ll \varphi \in X'$, and constants $M \geq 1$, $\varepsilon > 0$ such that*

$$\big\| T(t)f - \langle f, \varphi \rangle \cdot h \big\| \leq M e^{-\varepsilon t} \|f\| \quad \text{for all } t \geq 0, \ f \in X.$$

PROOF. The quasi-compactness of $\big(T(t)\big)_{t \geq 0}$ implies that $\sigma(A) \cap i\mathbb{R}$ consists of finitely many poles only (see Theorem V.4.6). Assume that there exists $0 \neq i\alpha \in \sigma(A) \cap i\mathbb{R}$. Then $i\alpha$ is an eigenvalue and we have

$$Af = i\alpha f \qquad \text{and} \qquad T(t)f = e^{i\alpha t} f \quad \text{for all } t \geq 0$$

and some $0 \neq f \in X$. This means that

$$f \in \ker(I - T) =: Y,$$

where we set $T := T\left(\frac{2\pi}{|\alpha|}\right)$. The characterization of quasi-compactness in Proposition V.4.5 implies that

$$r_{\mathrm{ess}}(T) < 1.$$

Therefore (use the characterization of the essential spectral radius in (1.16) in Paragraph V.1.19) 1 is a pole of finite algebraic multiplicity such that

$$\dim Y < \infty.$$

Clearly this subspace is $(T(t))_{t\geq 0}$-invariant and the generator of the restricted semigroup has 0 as spectral bound and $i\alpha$ in its spectrum. In addition, Y is a (closed) sublattice of X as can be seen as follows.

For $y \in Y$ we have $|y| = |Ty| \leq T|y|$ by the positivity of T. Applying a strictly positive linear form $\varphi \in \mathrm{fix}(T(t)')_{t\geq 0}$ yields

$$\langle T|y| - |y|, \varphi \rangle = \langle |y|, T'\varphi \rangle - \langle |y|, \varphi \rangle = 0,$$

hence $T|y| = |y|$ and $|y| \in Y$.

We showed that Y is a finite-dimensional (complex) Banach lattice, hence isomorphic to $\mathbb{C}^n$. The restricted semigroup is still positive, hence has dominant spectral bound by Proposition 2.3. This contradicts our assumption that $0 \neq i\alpha \in \sigma(A)$.

The remaining assertions now follow from Theorem V.4.6 and Proposition 3.4. □

3.6 Remark. If the semigroup $(T(t))_{t\geq 0}$ satisfies all the assumptions above but has spectral bound $\mathrm{s}(A) > 0$, one obtains, via rescaling, that

$$\lim_{t\to\infty} \left\| e^{-\mathrm{s}(A)t}T(t) - P \right\| = 0$$

for a projection $P := \varphi \otimes h$ as above. Such a behavior is called *balanced exponential growth* and occurs frequently in models on population growth (see [Web85] and [Web87]).

3.7 Outlook. The above theorem is a typical but not the most general example for what positivity can do for the spectral theory and asymptotic behavior of semigroups. We mention two generalizations.

(i) If the semigroup is only quasi-compact and positive, then the spectral bound is still a dominant eigenvalue of the generator, but, in general, no longer a pole of first order. The asymptotic behavior is again described by a positive matrix semigroup; see [Nag86, C-IV, Thm. 2.1 and Rems. 2.2] for more details.

(ii) If the semigroup is only positive and $s(A)$ a pole of the resolvent, then the boundary spectrum

$$\sigma(A) \cap \{s(A) + i\mathbb{R}\}$$

is cyclic; i.e., if $s(A) + i\alpha \in \sigma(A)$ then also $s(A) + ik\alpha \in \sigma(A)$ for all $k \in \mathbb{Z}$. Recall that if, in addition, the semigroup is eventually norm-continuous, then this boundary spectrum must be bounded by Theorem II.5.3. Therefore, $s(A)$ again becomes a dominant eigenvalue of A. See [Nag86, C-III. Cors. 2.12 and 2.13].

4. Semigroups for Age-Dependent Population Equations

In this section we show how the previous results on the asymptotic behavior of positive quasi-compact semigroups can be used to study a model for an age-dependent population described by the Cauchy problem

(APE)
$$\begin{cases} \dfrac{\partial f}{\partial t}(a,t) + \dfrac{\partial f}{\partial a}(a,t) + \mu(a)f(a,t) = 0 & \text{for } a,\, t \geq 0, \\[2mm] f(0,t) = \displaystyle\int_0^\infty \beta(a)f(a,t)\,da & \text{for } t \geq 0, \\[2mm] f(a,0) = f_0(a) & \text{for } a \geq 0. \end{cases}$$

Here t and a are nonnegative real variables representing time and age, respectively, $f(\cdot,t)$ describes the age structure of a population at time t and f_0 is the initial age structure at time $t = 0$. Moreover, μ and β are supposed to be bounded, measurable, and positive functions describing the *mortality rate* and *birth rate*, respectively. For further details as well as for nonlinear and vector-valued generalizations of this model we refer to [Gre84] and [Web85].

In order to rewrite (APE) as an abstract Cauchy problem we take the Banach space $X := \mathrm{L}^1(\mathbb{R}_+)$ and define on it the closed and densely defined operator A_m by

$$A_m f := -f' - \mu f, \quad f \in D(A_m) := \mathrm{W}^{1,1}(\mathbb{R}_+).$$

In the sequel we will always assume that

(4.1) $$\mu_\infty := \lim_{a \to \infty} \mu(a) > 0 \quad \text{exists.}$$

Then for $\operatorname{Re}\lambda > -\mu_\infty$ it's an exercise in calculus to show that

$$(4.2) \qquad \operatorname{rg}(\lambda - A_m) = X \qquad \text{and} \qquad \ker(\lambda - A_m) = \operatorname{lin}\{\varepsilon_\lambda\},$$

where

$$(4.3) \qquad \varepsilon_\lambda(a) := e^{-\int_0^a (\lambda + \mu(s))\, ds}.$$

Next we define the restriction

$$(4.4) \qquad Af := A_m f, \quad D(A) := \left\{ f \in D(A_m) : f(0) = \int_0^\infty \beta(a) f(a)\, da \right\}$$

of A_m which incorporates the birth process given by the second equation of (APE) into the domain $D(A)$. Then (APE) is equivalent to the abstract Cauchy problem

$$\text{(ACP)} \qquad \begin{cases} \dot{u}(t) = Au(t) & \text{for } t \geq 0, \\ u(0) = f_0 \end{cases}$$

for $u(t) := f(\cdot, t)$.

So instead of studying (APE) we solve (ACP) by semigroup methods. To this end, by Theorem II.6.6, we have to prove that A generates a strongly continuous semigroup $(T(t))_{t \geq 0}$ on X. In this case the unique solution of (APE) is given by $f(a, t) := (T(t) f_0)(a)$.

As a preparatory step we discuss the case $\beta = 0$; i.e., we consider the operator

$$A_0 f := A_m f, \quad D(A_0) := \{ f \in D(A_m) : f(0) = 0 \}.$$

Then it is not difficult to verify that A_0 generates a positive strongly continuous semigroup $(T_0(t))_{t \geq 0}$ given by

$$(4.5) \qquad \left[T_0(t) f \right](a) = \begin{cases} 0 & \text{for } 0 \leq a < t, \\ e^{-\int_{a-t}^a \mu(s)\, ds} \cdot f(a - t) & \text{for } t \leq a. \end{cases}$$

Moreover, the following holds.

4.1 Proposition. *The spectra of A_0 and $T_0(t)$, $t > 0$, are given by*

$$\sigma(A_0) = \{ \lambda \in \mathbb{C} : \operatorname{Re}\lambda \leq -\mu_\infty \}, \qquad \sigma(T_0(t)) = \{ \lambda \in \mathbb{C} : |\lambda| \leq e^{-\mu_\infty t} \}$$

PROOF. For $\lambda \in \mathbb{C}$ we define

$$h_\lambda(a) := e^{\int_0^a (\lambda + \mu(s))\, ds}.$$

Then $h_\lambda \in X' = L^\infty(\mathbb{R}_+)$ for $\mathrm{Re}\,\lambda < -\mu_\infty$ and $\langle(\lambda - A_0)f, h_\lambda\rangle = 0$ for all $f \in D(A_0)$. This shows that $\lambda - A_0$ is not surjective for $\mathrm{Re}\,\lambda < -\mu_\infty$ and hence $\{\lambda \in \mathbb{C} : \mathrm{Re}\,\lambda \leq -\mu_\infty\} \subseteq \sigma(A_0)$. On the other hand (4.1) implies that

$$\mu(a) = \lim_{t \to 0} \int_a^{a+t} \mu(s)\,ds$$

converges uniformly in $a \in \mathbb{R}_+$. Using this fact together with the explicit representation of $T_0(t)$ in (4.5) we conclude from Proposition V.1.22 that $\omega_0(A_0) \leq -\mu_\infty$. Now the assertion follows because $e^{t\sigma(A_0)} \subseteq \sigma(T_0(t))$ by the Spectral Inclusion Theorem V.2.5 and $\mathrm{r}(T(t_0)) = e^{t\,\omega_0(A_0)}$ by Proposition V.1.22.   □

We now consider the case $\beta \neq 0$, i.e., the operator A given by (4.4). The functions ε_λ defined by (4.3) are positive and satisfy $\|\varepsilon_\lambda\| \leq {}^1\!/\!_\lambda$ for all $\lambda > 0$. Thus for $\lambda > \|\beta\|_\infty$ the operator

$$\Phi_\lambda := \frac{1}{1 - \langle\varepsilon_\lambda, \beta\rangle} \cdot \varepsilon_\lambda \otimes \beta \in \mathcal{L}(X)$$

is well-defined and positive.

4.2 Lemma. *For* $\lambda > \|\beta\|_\infty$ *the operator* $I + \Phi_\lambda$ *is invertible with inverse*

$$(4.6) \qquad\qquad (I + \Phi_\lambda)^{-1} = I - \varepsilon_\lambda \otimes \beta.$$

Moreover, $(I + \Phi_\lambda)D(A_0) = D(A)$ *and*

$$(4.7) \qquad\qquad (\lambda - A)(I + \Phi_\lambda) = \lambda - A_0.$$

PROOF. Formula (4.6) can be verified by inspection. Moreover, from $\varepsilon_\lambda \in D(A_m)$ and $D(A) = \{f \in D(A_m) : f(0) = \langle f, \beta\rangle\}$ it follows easily that

$$(I + \Phi_\lambda)D(A_0) \subseteq D(A) \qquad \text{and} \qquad (I + \Phi_\lambda)^{-1}D(A) \subseteq D(A_0);$$

i.e., $(I + \Phi_\lambda)D(A_0) = D(A)$. Finally, for $f \in D(A_0)$ we have

$$(\lambda - A)(I + \Phi_\lambda)f = (\lambda - A_m)f + (\lambda - A_m)\Phi_\lambda f = (\lambda - A_0)f$$

because $\mathrm{rg}\,\Phi_\lambda = \mathrm{lin}\{\varepsilon_\lambda\} = \ker(\lambda - A_m)$.   □

The following proposition opens the door for the application of the results from the previous section. Here and in the sequel we always assume that

$$(4.8) \qquad\qquad \beta \in D(A_0)' = \mathrm{W}^{1,\infty}(\mathbb{R}_+).$$

4.3 Proposition. *The operator* A *is the generator of a positive strongly continuous semigroup* $(T(t))_{t \geq 0}$ *on* X. *Moreover,* $T(t) - T_0(t)$ *is compact and positive for every* $t \geq 0$.

PROOF. By (4.7) the operators $A - \lambda$ and $(I + \Phi_\lambda)^{-1}(A_0 - \lambda) = A_0 - \lambda + \lambda\varepsilon_\lambda \otimes \beta - (\varepsilon_\lambda \otimes \beta)A_0$ are similar for $\lambda > \|\beta\|_\infty$. From the assumption (4.8) it follows that $(\varepsilon_\lambda \otimes \beta)A_0$ has the bounded extension $\varepsilon_\lambda \otimes A_0'\beta$ to X. Thus by the Bounded Perturbation Theorem III.1.3 and by similarity (see Paragraph I.1.9) we conclude that A generates a strongly continuous semigroup $(T(t))_{t\geq0}$ on X. Next we observe that (4.7) implies

$$R(\lambda, A) = (I + \Phi_\lambda)R(\lambda, A_0) \geq R(\lambda, A_0) \geq 0 \quad \text{for } \lambda > \|\beta\|_\infty.$$

From the Post–Widder inversion formula in Corollary IV.2.5 we then obtain

$$T(t) \geq T_0(t) \geq 0; \quad \text{i.e., } \cdot T(t) - T_0(t) \geq 0 \quad \text{for all } t \geq 0.$$

It only remains to show that $T(t) - T_0(t)$ is compact. To this end we recall that by the above considerations

$$A - \lambda = (I + \Phi_\lambda) \cdot (A_0 - \lambda + \lambda\varepsilon_\lambda \otimes \beta - \varepsilon_\lambda \otimes A_0'\beta) \cdot (I + \Phi_\lambda)^{-1}$$

for $\lambda > \|\beta\|_\infty$. If $(S(t))_{t\geq0}$ denotes the semigroup generated by the operator $A_0 + \lambda\varepsilon_\lambda \otimes \beta - \varepsilon_\lambda \otimes A_0'\beta$, this implies

$$T(t) = (I + \Phi_\lambda) \cdot S(t) \cdot (I + \Phi_\lambda)^{-1}.$$

Because the operator $\lambda\varepsilon_\lambda \otimes \beta - \varepsilon_\lambda \otimes A_0'\beta$ is compact, by (the proof of) Proposition V.4.9 the perturbed semigroup $(S(t))_{t\geq0}$ has the form

$$S(t) = T_0(t) + K(t),$$

where $K(t)$ is compact for all $t \geq 0$. Combining these facts and using (4.6) we finally obtain

$$T(t) = (I + \Phi_\lambda) \cdot (T_0(t) + K(t)) \cdot (I - \varepsilon_\lambda \otimes \beta).$$

Because Φ_λ and $\varepsilon_\lambda \otimes \beta$ are both compact, this implies the compactness of $T(t) - T_0(t)$ for all $t \geq 0$. $\qquad\square$

Summarizing the above results we obtain the following where, in particular, we make the assumptions (4.1) and (4.8).

4.4 Theorem. *The operator A generates a positive quasi-compact semigroup $(T(t))_{t\geq0}$ on $\mathrm{L}^1(\mathbb{R}_+)$. This semigroup is irreducible if and only if*

(4.9)   *there exists no $a_0 \geq 0$ such that $\beta|_{[a_0,\infty)} = 0$ almost everywhere.*

PROOF. We already showed above that $(T(t))_{t\geq0}$ is positive. The essential spectral radius is invariant under compact perturbations (see Paragraph V.1.19), hence its quasi-compactness follows from Propositions 4.1 and 4.3 because

$$\mathrm{r_{ess}}(T(1)) = \mathrm{r_{ess}}(T_0(1) + [T(1) - T_0(1)]) = \mathrm{r_{ess}}(T_0(1))$$
$$\leq \mathrm{r}(T_0(1)) = \mathrm{e}^{-\mu_\infty} < 1.$$

In order to prove the claim concerning irreducibility we first observe that it is a simple exercise in linear ODEs to show that for $\lambda > \|\beta\|_\infty$ the resolvent of A_0 is given by

$$(4.10) \quad [R(\lambda, A_0)f](a) = \int_0^a e^{-\int_s^a (\lambda + \mu(r)) \, dr} f(s) \, ds, \quad f \in X, \ a \geq 0.$$

Now assume that (4.9) does not hold, i.e., that there exists a_0 such that $\beta|_{[a_0, \infty)} = 0$. Then for each $f > 0$ such that $f|_{[0, a_0]} = 0$ we obtain from (4.10) that $[R(\lambda, A_0)f](a) = 0$ for all $a \in [0, a_0]$. This implies

$$R(\lambda, A)f = (I + \Phi_\lambda)R(\lambda, A_0)f$$
$$= R(\lambda, A_0)f + \frac{1}{1 - \langle \varepsilon_\lambda, \beta \rangle} \cdot \langle R(\lambda, A_0)f, \beta \rangle \cdot \varepsilon_\lambda$$
$$= R(\lambda, A_0)f.$$

Hence, $R(\lambda, A)f$ is not strictly positive in general for $0 < f$ and $\lambda > \|\beta\|_\infty$ and hence $(T(t))_{t \geq 0}$ not irreducible.

Conversely, assume that (4.9) holds and take some $\lambda > \|\beta\|_\infty$ and $0 < f$. Then by (4.10) we conclude that $[R(\lambda, A_0)f](a) > 0$ for $a \geq 0$ sufficiently large and therefore $\langle R(\lambda, A_0)f, \beta \rangle > 0$. Now we obtain as above

$$R(\lambda, A)f = R(\lambda, A_0)f + \frac{1}{1 - \langle \varepsilon_\lambda, \beta \rangle} \cdot \langle R(\lambda, A_0)f, \beta \rangle \cdot \varepsilon_\lambda$$
$$\geq \frac{1}{1 - \langle \varepsilon_\lambda, \beta \rangle} \cdot \langle R(\lambda, A_0)f, \beta \rangle \cdot \varepsilon_\lambda.$$

Because $\langle \varepsilon_\lambda, \beta \rangle < 1$ and ε_λ is strictly positive this implies that $R(\lambda, A)f$ is strictly positive; i.e., $(T(t))_{t \geq 0}$ is irreducible. $\square$

After these preparations we are in the position to analyze the stability and convergence of the semigroup $(T(t))_{t \geq 0}$ solving the age-dependent population equation (APE).

4.5 Corollary. *The following assertions are equivalent.*

(a) *$(T(t))_{t \geq 0}$ is uniformly exponentially stable.*

(b) *$s(A) < 0$.*

(c) *$\langle \varepsilon_0, \beta \rangle = \int_0^\infty \beta(a) \cdot e^{-\int_0^a \mu(s) \, ds} \, da < 1$.*

PROOF. Because by Theorem 4.4 we know that $(T(t))_{t \geq 0}$ is positive and quasi-compact, the equivalence of (a) and (b) follows from Corollary V.4.7 or Theorem 2.5.

To prove the remaining implications we observe that by the same arguments as in the proof of Lemma 4.2 it follows that

$$(4.11) \qquad \lambda - A = (\lambda - A_0)(I - \varepsilon_\lambda \otimes \beta) \quad \text{for all } \lambda \in \mathbb{R}.$$

Next we define the function

$$r(\lambda) := \langle \varepsilon_\lambda, \beta \rangle = \int_0^\infty \beta(a) \cdot e^{-\int_0^a (\lambda + \mu(s)) \, ds} \, da, \quad \lambda \in \mathbb{R},$$

which is continuous and strictly decreasing. Moreover,

$$(4.12) \qquad \lim_{\lambda \to -\infty} r(\lambda) = \infty, \qquad \lim_{\lambda \to \infty} r(\lambda) = 0$$

and $\sigma(\varepsilon_\lambda \otimes \beta) = \{r(\lambda)\}$.

Now assume (b). Because $s(A_0) = -\mu_\infty < 0$, the assumption $s(A) < 0$ together with (4.11) implies that $I - \varepsilon_\lambda \otimes \beta$ is invertible for all $\lambda \geq 0$; i.e., $r(\lambda) \neq 1$ for all $\lambda \geq 0$. By the continuity and monotonicity of $r(\cdot)$ and (4.12) this is possible only if $r(0) = \langle \varepsilon_0, \beta \rangle < 1$. This proves (b) $\Rightarrow$ (c).

Conversely, if $\langle \varepsilon_0, \beta \rangle < 1$, then $(I - \varepsilon_0 \otimes \beta)^{-1} = \Phi_0$ exists and (4.11) implies that

$$-A^{-1} = -(I + \Phi_0)A_0^{-1}.$$

Here, because $s(A_0) = -\mu_\infty < 0$, the inverse A_0^{-1} exists and is negative by Lemma 2.1. Now $\Phi_0 \geq 0$ shows $R(0, A) = -A^{-1} \geq 0$ and again by Lemma 2.1 we conclude that $s(A) < 0$. This proves (c) $\Rightarrow$ (b) and completes the proof. $\qquad\square$

Our final result deals with convergence to a one-dimensional projection in case $s(A) = 0$.

4.6 Corollary. *Assume that* (4.9) *holds and*

$$\langle \varepsilon_0, \beta \rangle = \int_0^\infty \beta(a) \cdot e^{-\int_0^a \mu(s) \, ds} \, da = 1.$$

Then there exists a strictly positive linear form $\varphi \in \mathrm{fix}(T(t)')_{t \geq 0}$ and constants $M \geq 1$, $\varepsilon > 0$ such that

$$\|T(t)f - \langle f, \varphi \rangle \cdot \varepsilon_0\| \leq M e^{-\varepsilon t} \|f\| \quad \text{for all } t \geq 0 \; f \in X.$$

PROOF. First we observe that by Theorem 4.4 the condition (4.9) implies that $(T(t))_{t \geq 0}$ is irreducible. Moreover, the assumption $\langle \varepsilon_0, \beta \rangle = r(0) = 1$ implies by the same reasoning as in the proof of Corollary 4.5 that $s(A) = 0$. The assertion then follows immediately from Theorem 3.5. $\qquad\square$

4.7 Remark. The conditions imposed above can be relaxed without changing the conclusions. In particular, one can eliminate the assumptions (4.1), (4.8), and (4.9), still getting stability and convergence of the semigroup $(T(t))_{t \geq 0}$ as in Corollaries 4.5 and 4.6, respectively. Moreover, these results can be generalized to higher dimensions. For further details see [Gre84] and [Web85].

Appendix

A Reminder of Some Functional Analysis and Operator Theory

This book is written in a functional-analytic spirit. Its main objects are operators on Banach spaces, and we use many, sometimes quite sophisticated, results and techniques from functional analysis and operator theory. As a rule, we refer to textbooks such as [Con85], [DS58], [Lan93], [RS72], [Rud73], [TL80], or [Yos65]. However, for the convenience of the reader we add this appendix, where we

- Introduce our notation,
- List some basic results, and
- Prove a few of them.

To start with, we introduce the following classical sequence and function spaces. Here, J is a real interval and Ω, depending on the context, is a domain in $\mathbb{R}^n$, a locally compact metric space, or a measure space.

$$\ell^\infty := \left\{ (x_n)_{n\in\mathbb{N}} \subset \mathbb{C} : \sup_{n\in\mathbb{N}} \|x_n\| < \infty \right\}, \quad \|(x_n)_{n\in\mathbb{N}}\| := \sup_{n\in\mathbb{N}} \|x_n\|,$$

$$c := \left\{ (x_n)_{n\in\mathbb{N}} \subset \mathbb{C} : \lim_{n\to\infty} x_n \text{ exists} \right\} \subset \ell^\infty,$$

$$c_0 := \left\{ (x_n)_{n\in\mathbb{N}} \subset \mathbb{C} : \lim_{n\to\infty} x_n = 0 \right\} \subset c,$$

$$\ell^p := \left\{ (x_n)_{n\in\mathbb{N}} \subset \mathbb{C} : \sum_{n\in\mathbb{N}} |x_n|^p < \infty \right\}, \ p \geq 1, \ \|(x_n)_{n\in\mathbb{N}}\| := \left(\sum_{n\in\mathbb{N}} |x_n|^p \right)^{1/p},$$

$C(\Omega) := \{f : \Omega \to \mathbb{K} \mid f \text{ is continuous}\}$,

$\quad \|f\|_\infty := \sup\limits_{s \in \Omega} |f(s)| \quad \text{(if } \Omega \text{ is compact)}$,

$C_0(\Omega) := \{f \in C(\Omega) : f \text{ vanishes at infinity}\}$;   cf. p. 20,

$C_b(\Omega) := \{f \in C(\Omega) : f \text{ is bounded}\}$,

$C_c(\Omega) := \{f \in C(\Omega) : f \text{ has compact support}\}$;   cf. p. 21,

$C_{ub}(\Omega) := \{f \in C(\Omega) : f \text{ is bounded and uniformly continuous}\}$,

$AC(J) := \{f : J \to \mathbb{K} \mid f \text{ is absolutely continuous}\}$,

$C^k(J) := \{f \in C(J) : f \text{ is } k\text{-times continuously differentiable}\}$,

$C^\infty(J) := \{f \in C(J) : f \text{ is infinitely many times differentiable}\}$,

$L^p(\Omega, \mu) := \{f : \Omega \to \mathbb{K} \mid f \text{ is } p\text{-integrable on } \Omega\}$,

$$\|f\|_p := \left(\int_\Omega |f|^p(s) \, d\mu(s) \right)^{1/p},$$

$L^\infty(\Omega, \mu) := \{f : \Omega \to \mathbb{K} \mid f \text{ is measurable and } \mu\text{-essentially bounded}\}$,

$\quad \|f\|_\infty := \operatorname{ess\,sup} |f|$;   cf. p. 28,

$W^{k,p}(\Omega) := \left\{ f \in L^p(\Omega) : \begin{array}{l} f \text{ is } k\text{-times distributionally differentiable} \\ \text{with } D^\alpha f \in L^p(\Omega) \text{ for all } |\alpha| \le k \end{array} \right\}$,

$H^k(\Omega) := W^{k,2}(\Omega)$,

$H_0^k(J) := \{f \in H^k(J) : f(s) = 0 \text{ for } s \in \partial J\}$,

$\mathscr{S}(\mathbb{R}^n) :=$ Schwartz space of rapidly decreasing functions;   cf. p. 55.

Clearly, we may combine the various sub- and superscripts for the spaces of continuous functions and obtain, e.g., $C_c^1(J) = C^1(J) \cap C_c(J)$.

For an abstract complex Banach space X we denote its dual by X' and the *canonical bilinear form* by

$$\langle x, x' \rangle \quad \text{for } x \in X, \quad x' \in X'.$$

As usual, we also write $x'(x)$ for $\langle x, x' \rangle$ and denote by $\sigma(X, X')$ the *weak topology* on X and by $\sigma(X', X)$ the weak* topology on X'. Then the following properties hold.

A.1 Proposition.

(i) *For convex sets in X (in particular, for subspaces) the weak and norm closure coincide.*

(ii) *The closed, convex hull $\overline{\text{co}}\, K$ of a weakly compact set K in X is weakly compact (Kreĭn's theorem).*

(iii) *The dual unit ball $U^0 := \{x' \in X' : \|x'\| \le 1\}$ is weak* compact (Banach–Alaoglu's theorem).*

The space of all bounded, linear operators on X is denoted[1] by $\mathcal{L}(X)$ and becomes a Banach space for the norm

$$\|T\| := \sup\{\|Tx\| : \|x\| \le 1\}, \quad T \in \mathcal{L}(X).$$

The operators $T \in \mathcal{L}(X)$ satisfying

$$\|Tx\| \le \|x\| \qquad \text{for all } x \in X$$

are called *contractions*, whereas *isometries* are defined by

$$\|Tx\| = \|x\| \qquad \text{for all } x \in X.$$

Besides the *uniform operator topology* on $\mathcal{L}(X)$, which is the one induced by the above operator norm, we frequently consider two more topologies on $\mathcal{L}(X)$.

We write $\mathcal{L}_s(X)$ if we endow $\mathcal{L}(X)$ with the *strong operator topology*, which is the topology of pointwise convergence on $(X, \|\cdot\|)$.

Finally, $\mathcal{L}_\sigma(X)$ denotes $\mathcal{L}(X)$ with the *weak operator topology*, which is the topology of pointwise convergence on $(X, \sigma(X, X'))$.

A net $(T_\alpha)_{\alpha \in \mathcal{A}} \subset \mathcal{L}(X)$ converges to $T \in \mathcal{L}(X)$ if and only if

(A.1) $\|T_\alpha - T\| \to 0$ (uniform operator topology),

(A.2) $\|T_\alpha x - Tx\| \to 0 \, \forall \, x \in X$ (strong operator topology),

(A.3) $|\langle T_\alpha x - Tx, x' \rangle| \to 0 \, \forall \, x \in X, \, x' \in X'$ (weak operator topology).

With these notions, the *principle of uniform boundedness* can be stated as follows.

A.2 Proposition. *For a subset $K \subset \mathcal{L}(X)$ the following properties are equivalent.*

(a) *K is bounded for the weak operator topology.*

(b) *K is bounded for the strong operator topology.*

(c) *K is uniformly bounded; i.e., $\|T\| \le c$ for all $T \in K$.*

Continuity with respect to the strong operator topology is shown frequently by using the following property (b) (see [Sch80, Sect. III.4.5]).

A.3 Proposition. *On bounded subsets of $\mathcal{L}(X)$, the following topologies coincide.*

(a) *The strong operator topology.*

(b) *The topology of pointwise convergence on a dense subset of X.*

(c) *The topology of uniform convergence on relatively compact subsets of X.*

---

[1] For the space of all bounded, linear operators between two normed spaces X and Y we use the notation $\mathcal{L}(X, Y)$.

The advantage of using the strong or weak operator topology instead of the norm topology on $\mathcal{L}(X)$ is that the former yield more continuity and more compactness. This becomes evident already from the definition of a strongly continuous semigroup in Section I.1.

As an example for the functional-analytic constructions made throughout the text, we consider the following setting.

Let $\mathcal{X}_{t_0} := \mathrm{C}\big([0, t_0], \mathcal{L}_s(X)\big)$ be the space of all functions on $[0, t_0]$ into $\mathcal{L}(X)$ that are continuous for the strong operator topology. For each $F \in \mathcal{X}_{t_0}$ and $x \in X$, the functions $t \mapsto F(t)x$ are continuous, hence bounded, on $[0, t_0]$. The uniform boundedness principle then implies

$$\|F\|_\infty := \sup_{s \in [0, t_0]} \|F(s)\| < \infty.$$

Clearly, this defines a norm making $\mathcal{X}_{t_0}$ a complete space.

A.4 Proposition. *The space*

$$\mathcal{X}_{t_0} := \Big(\mathrm{C}\big([0, t_0], \mathcal{L}_s(X)\big), \|\cdot\|_\infty\Big)$$

is a Banach space.

PROOF. Let $(F_n)_{n \in \mathbb{N}}$ be a Cauchy sequence in $\mathcal{X}_{t_0}$. Then, by the definition of the norm in $\mathcal{X}_{t_0}$, $\big(F_n(\cdot)x\big)_{n \in \mathbb{N}}$ is a Cauchy sequence in $\mathrm{C}([0, t_0], X)$ for all $x \in X$. Because $\mathrm{C}([0, t_0], X)$ is complete, the limit $\lim_{n \to \infty} F_n(\cdot)x =: F(\cdot)x \in \mathrm{C}([0, t_0], X)$ exists, and we obtain $\lim_{n \to \infty} F_n = F$ in $\mathcal{X}_{t_0}$.   □

Familiarity with linear operators, in particular unbounded operators, is essential for an understanding of our semigroups and their generators. The best introduction is still Kato's monograph [Kat80] (see also [DS58], [GGK90], [Gol66], [TL80], [Wei80]), but we briefly restate some of the basic definitions and properties.[2]

A.5 Definition. *A linear operator A with domain $D(A)$ in a Banach space X, i.e., $D(A) \subset X \to X$, is closed if it satisfies one of the following equivalent properties.*

(a) *If for the sequence $(x_n)_{n \in \mathbb{N}} \subset D(A)$ the limits $\lim_{n \to \infty} x_n = x \in X$ and $\lim_{n \to \infty} Ax_n = y \in X$ exist, then $x \in D(A)$ and $Ax = y$.*

(b) *The graph $\mathcal{G}(A) := \{(x, Ax) : x \in D(A)\}$ is closed in $X \times X$.*

(c) *$X_1 := \big(D(A), \|\cdot\|_A\big)$ is a Banach space[3] for the graph norm*

$$\|x\|_A := \|x\| + \|Ax\|, \quad x \in D(A).$$

---

[2] Most of the following concepts also make sense for operators acting between different Banach spaces. However, for simplicity we state them for a single Banach space only and leave the straightforward generalization to the reader.

[3] This definition of X_1 also makes sense if A has an empty resolvent set. Because if $\rho(A) \neq \emptyset$, the graph norm and the norms $\|\cdot\|_{1,\lambda}$ from Exercise II.2.22.(1) are all equivalent, this definition of X_1 will not conflict with Definition II.2.14 for $n = 1$.

(d) *A is weakly closed; i.e., property (a) (or property (b)) holds for the $\sigma(X, X')$-topology on X.*

If $\lambda - A$ is injective for some $\lambda \in \mathbb{C}$, then the above properties are also equivalent to

(e) *$(\lambda - A)^{-1}$ is closed.*

Next we consider perturbations of closed operators. Whereas the additive perturbation of a closed operator A by a bounded operator $B \in \mathcal{L}(X)$ yields again a closed operator, the situation is slightly more complicated for multiplicative perturbations.

A.6 Proposition. *Let $(A, D(A))$ be a closed operator and take $B \in \mathcal{L}(X)$. Then the following hold.*

(i) *AB with domain $D(AB) := \{x \in X : Bx \in D(A)\}$ is closed.*

(ii) *BA with domain $D(BA) := D(A)$ is closed if $B^{-1} \in \mathcal{L}(X)$.*

PROOF. (i) is easy to check and implies (ii) after the similarity transformation $BA = B(AB)B^{-1}$. □

It will be important to find closed extensions of not necessarily closed operators. Here are the relevant notions.

A.7 Definition. *An operator $(B, D(B))$ is an extension of $(A, D(A))$, in symbols $A \subset B$, if $D(A) \subset D(B)$ and $Bx = Ax$ for $x \in D(A)$. The smallest closed extension of A, if it exists, is called the closure of A and is denoted by $\overline{A}$. Operators having a closure are called closable.*

A.8 Proposition. *An operator $(A, D(A))$ is closable if and only if for every sequence $(x_n)_{n \in \mathbb{N}} \subset D(A)$ with $x_n \to 0$ and $Ax_n \to z$ one has $z = 0$. In that case, the graph of the closure is given by*

$$\mathcal{G}(\overline{A}) = \overline{\mathcal{G}(A)}.$$

A simple operator that is not closable is

$$Af := f'(0) \cdot \mathbb{1} \quad \text{with domain} \quad D(A) := \mathrm{C}^1[0,1]$$

in the Banach space $X := \mathrm{C}[0,1]$. This follows, e.g., from the following characterization of bounded linear forms and the fact that the kernel of a closed operator is always closed.[4]

A.9 Proposition. *Let X be a normed vector space and take a linear functional $x' : X \to \mathbb{C}$. Then x' is bounded if and only if its kernel $\ker x'$ is closed in X. Hence, x' is unbounded if and only if $\ker x'$ is dense in X.*

---

[4] Here, for a linear map $\Phi : X \to Y$ between two vector spaces X and Y its kernel is defined by $\ker \Phi := \{x \in X : \Phi x = 0\}$.

PROOF. If x' is bounded, then clearly $\ker(x')$ is closed. On the other hand, if $\ker x'$ is closed, then the quotient $X/_{\ker x'}$ is a normed vector space of dimension 1. Moreover, we can decompose $x' = i\,\widehat{x'}$ by the canonical maps $i: X/_{\ker x'} \to \mathbb{C}$ and $\widehat{x'}: X \to X/_{\ker x'}$. Because $\|\widehat{x'}\| \leq 1$, this proves that x' is bounded. The remaining assertions follow from the fact that for each linear form $x' \neq 0$ the codimension of $\ker x'$ in X is 1. $\qquad\square$

A subspace D of $D(A)$ that is dense in $D(A)$ for the graph norm is called a *core* for A. If $(A, D(A))$ is closed, one can recover A from its restriction to a core D; i.e., the closure of (A, D) becomes $(A, D(A))$. See Exercise II.1.15.(2).

The closed graph theorem states that everywhere defined closed operators are already bounded. It can be phrased as follows.

A.10 Theorem. *For a closed operator $A: D(A) \subset X \to X$ the following properties are equivalent.*

(a) $(A, D(A))$ *is a bounded operator; i.e., there exists $c \geq 0$ such that*

$$\|Ax\| \leq c\,\|x\| \qquad \text{for all } x \in D(A).$$

(b) $D(A)$ *is a closed subspace of X.*

By the closed graph theorem, one obtains the following surprising result.

A.11 Corollary. *Let $A: D(A) \subset X \to X$ be closed and assume that a Banach space Y is continuously embedded in X such that the range $\operatorname{rg} A := A(D(A))$ is contained in Y. Then A is bounded from $(D(A), \|\cdot\|_A)$ into Y.*

If an operator A has dense domain $D(A)$ in X, we can define its adjoint operator on the dual space X'.[5]

A.12 Definition. *For a densely defined operator $(A, D(A))$ on X, we define the adjoint operator $(A', D(A'))$ on X' by*

$$D(A') := \{x' \in X' : \exists\, y' \in X' \text{ such that } \langle Ax, x'\rangle = \langle x, y'\rangle \ \forall\, x \in D(A)\},$$
$$A'x' := y' \text{ for } x' \in D(A').$$

A.13 Example. Take $A_p := d/ds$ on $X_p := \mathrm{L}^p(\mathbb{R})$, $1 \leq p < \infty$, with domain $D(A_p) := \mathrm{W}^{1,p}(\mathbb{R}) := \{f \in X_p : f$ absolutely continuous, $f' \in X_p\}$. Then $A_p' = -A_q$ on X_q, where $1/p + 1/q = 1$. For a proof and many more examples we refer to [Gol66, Sect. II.2 and Chap. VI] and [Kat80, Sect. III.5]. Compare also Exercise II.4.14.(11).

Although the adjoint operator is always closed, it may happen that $D(A') = \{0\}$ (e.g., take the nonclosable operator following Proposition A.8).

---

[5] Similarly, one can define the Hilbert space adjoint A^* by replacing the canonical bilinear form $\langle \cdot, \cdot \rangle$ by the *inner product* $(\cdot\,|\,\cdot)$.

On reflexive Banach spaces there is a nice duality between densely defined and closable operators.

A.14 Proposition. *Let $(A, D(A))$ be a densely defined operator on a reflexive Banach space X. Then the adjoint A' is densely defined if and only if A is closable. In that case, one has*

$$(A')' = \overline{A}.$$

We now prove a close relationship between inverses and adjoints.

A.15 Proposition. *Let $(A, D(A))$ be a densely defined closed operator on X. Then the inverse $A^{-1} \in \mathcal{L}(X)$ exists if and only if the inverse $(A')^{-1} \in \mathcal{L}(X')$ exists. In that case, one has*

$$(A')^{-1} = (A^{-1})'.$$

PROOF. Assume $A^{-1} \in \mathcal{L}(X)$. Because $(A^{-1})' \in \mathcal{L}(X')$, one has

$$\langle x, (A^{-1})'A'x' \rangle = \langle A^{-1}x, A'x' \rangle = \langle AA^{-1}x, x' \rangle = \langle x, x' \rangle$$

for all $x \in X$, $x' \in D(A')$; i.e., A' has a left inverse. Similarly,

$$\langle Ax, (A^{-1})'x' \rangle = \langle A^{-1}Ax, x' \rangle = \langle x, x' \rangle$$

holds for all $x \in D(A)$, $x' \in X'$; i.e., $(A^{-1})'x' \in D(A')$ and $A'(A^{-1})'x' = x'$.

On the other hand, assume $(A')^{-1} \in \mathcal{L}(X')$. Then

$$\langle Ax, (A')^{-1}x' \rangle = \langle x, A'(A')^{-1}x' \rangle = \langle x, x' \rangle$$

for all $x \in D(A)$ and $x' \in X'$. For every $x \in D(A)$, choose $x' \in X'$ such that $\|x'\| = 1$ and $|\langle x, x' \rangle| = \|x\|$ and obtain

$$\|x\| = |\langle Ax, (A')^{-1}x' \rangle| \le \|Ax\| \cdot \|(A')^{-1}\|.$$

This shows that A is injective and its inverse satisfies

$$\|A^{-1}\| \le \|(A')^{-1}\|,$$

hence is bounded. By Theorem A.10, $D(A^{-1}) = \operatorname{rg} A$ must be closed. A simple Hahn–Banach argument shows that $\operatorname{rg} A = X$, hence $A^{-1} \in \mathcal{L}(X)$. □

A.16 Corollary. *For a densely defined closed operator $(A, D(A))$ the spectra of A and of A' coincide; i.e.,*

$$\sigma(A) = \sigma(A')$$

and $R(\lambda, A)' = R(\lambda, A')$ for all $\lambda \in \rho(A)$.

Now we turn again to the unbounded situation and define iterates of unbounded operators.

A.17 Definition. *The nth power A^n of an operator $A : D(A) \subset X \to X$ is defined successively as*

$$A^n x := A(A^{n-1}x),$$
$$D(A^n) := \{x \in D(A) : A^{n-1}x \in D(A)\}.$$

In general, it may happen that $D(A^2) = \{0\}$ even if A is densely defined and closed. However, if $A^{-1} \in \mathcal{L}(X)$ exists (or if $\rho(A) \neq \emptyset$), the infinite intersection

$$D(A^\infty) := \bigcap_{n=1}^{\infty} D(A^n)$$

is still dense. This is proved in Proposition II.1.8 for semigroup generators and in [Len94] or [AEMK94, Prop. 6.2] for the general case.

Next, we give some results concerning the continuity and differentiability of products of operator-valued functions.

A.18 Lemma. *Let J be some real interval and $P, Q : I \to \mathcal{L}(X)$ be two strongly continuous operator-valued functions defined on J. Then the product $(PQ)(\cdot) : J \to \mathcal{L}(X)$, defined by $(PQ)(t) := P(t)Q(t)$, is strongly continuous as well.*

PROOF. We fix $x \in X$ and $t \in J$ and take a sequence $(t_n)_{n\in\mathbb{N}} \subset J$ with $\lim_{n\to\infty} t_n = t$. Then, by the uniform boundedness principle, the set $\{P(t_n) : n \in \mathbb{N}\} \subset \mathcal{L}(X)$ is bounded, and therefore

$$\|P(t_n)Q(t_n)x - P(t)Q(t)x\| \leq \|P(t_n)\| \cdot \|Q(t_n)x - Q(t)x\|$$
$$+ \|(P(t_n) - P(t))Q(t)x\|,$$

where the right-hand side converges to zero as $n \to \infty$. $\square$

A.19 Lemma. *Let J be some real interval and $P, Q : J \to \mathcal{L}(X)$ be two strongly continuous operator-valued functions defined on J. Moreover, assume that $P(\cdot)x : J \to X$ and $Q(\cdot)x : J \to X$ are differentiable for all $x \in D$ for some subspace D of X, which is invariant under Q. Then $(PQ)(\cdot)x : J \to X$, defined by $(PQ)(t)x := P(t)Q(t)x$, is differentiable for every $x \in D$ and*

$$\tfrac{d}{dt}\Big(P(\cdot)Q(\cdot)x\Big)(t_0) = \tfrac{d}{dt}\Big(P(\cdot)Q(t_0)x\Big)(t_0) + P(t_0)\Big(\tfrac{d}{dt}Q(\cdot)x\Big)(t_0).$$

PROOF. Let $t_0 \in J$ and $(h_n)_{n\in\mathbb{N}} \subset \mathbb{R}$ be a sequence such that $\lim_{n\to\infty} h_n = 0$ and $t_0 + h_n \in J$ for all $n \in \mathbb{N}$. Then, for $x \in D$, we have

$$\frac{P(t_0 + h_n)Q(t_0 + h_n)x - P(t_0)Q(t_0)x}{h_n}$$

$$= P(t_0 + h_n)\frac{Q(t_0 + h_n)x - Q(t_0)x}{h_n} + \frac{P(t_0 + h_n) - P(t_0)}{h_n}Q(t_0)x$$

$$=: L_1(n, x) + L_2(n, x).$$

Clearly, the sequence $(L_2(n, x))_{n \in \mathbb{N}}$ converges for all $x \in D$ and its limit is $\lim_{n \to \infty} L_2(n, x) = P'(t_0)Q(t_0)x$. In order to show that $(L_1(n, x))_{n \in \mathbb{N}}$ converges for $x \in D$, note that

$$\left\{ \frac{Q(t_0 + h_n)x - Q(t_0)x}{h_n} : n \in \mathbb{N} \right\}$$

is relatively compact in X and that $\{P(t_0 + h_n) : n \in \mathbb{N}\}$ is bounded. Because by Proposition A.3 the topologies of pointwise convergence and of uniform convergence on relatively compact sets coincide, we conclude that $(L_1(n, x))_{n \in \mathbb{N}}$ converges for $x \in D$ and

$$\lim_{n \to \infty} L_1(n, x) = P(t_0)Q'(t_0)x.$$

This completes the proof. □

In the context of operators on spaces of vector-valued functions it is convenient to use the following *tensor product* notation.

Assume that X, Y are Banach spaces, $\mathrm{F}(J, Y)$ is a Banach space of Y-valued functions defined on an interval $J \subseteq \mathbb{R}$, $T \in \mathcal{L}(X, Y)$ is a bounded linear operator, and $f : J \to \mathbb{C}$ is a complex-valued function. If the map $f \otimes y : J \ni s \mapsto f(s) \cdot y \in Y$ belongs to $\mathrm{F}(J, Y)$ for all $y \in Y$, then we define the linear operator $f \otimes T : X \to \mathrm{F}(J, Y)$ by

$$((f \otimes T)x)(s) := (f \otimes Tx)(s) = f(s) \cdot Tx$$

for all $x \in X$, $s \in J$.

Independently, for a Banach space X and elements $x \in X$, $x' \in X'$, we frequently use the tensor product notation $x \otimes x'$ for the rank-one operator on X defined by

$$(x \otimes x')\, v := x'(v) \cdot x, \qquad v \in X.$$

We conclude this appendix with the following vector-valued version of the Riemann–Lebesgue lemma.

A.20 Theorem. *If $f \in \mathrm{L}^1(\mathbb{R}, X)$, then $\widehat{f} \in \mathrm{C}_0(\mathbb{R}, X)$; i.e., we have* $\lim_{s \to \pm\infty} \widehat{f}(s) = 0$.

For the proof it suffices to consider step functions, for which, as in the scalar case, the assertion follows by integration by parts.

References

[AB85]   C.D. Aliprantis and O. Burkinshaw, *Positive Operators*, Academic Press, 1985.

[AB88]   W. Arendt and C.J.K. Batty, *Tauberian theorems for one-parameter semigroups*, Trans. Amer. Math. Soc. **306** (1988), 837–852.

[ABHN01]   W. Arendt, C.J.K. Batty, M. Hieber, and F. Neubrander, *Vector-valued Laplace Transforms and Cauchy Problems*, Monographs Math., vol. 96, Birkhäuser Verlag, 2001.

[AEMK94]   W. Arendt, O. El-Mennaoui, and V. Keyantuo, *Local integrated semigroups: Evolution with jumps of regularity*, J. Math. Anal. Appl. **186** (1994), 572–595.

[Alb01]   J. Alber, *On implemented semigroups*, Semigroup Forum **63** (2001), 371–386.

[Ama90]   H. Amann, *Ordinary Differential Equations. An Introduction to Nonlinear Analysis*, de Gruyter Stud. Math., vol. 13, de Gruyter, 1990.

[Are87]   W. Arendt, *Resolvent positive operators*, Proc. London Math. Soc. **54** (1987), 321–349.

[Bea82]   B. Beauzamy, *Introduction to Banach Spaces and Their Geometry*, North-Holland Math. Stud., vol. 68, North-Holland, 1982.

[Bla01]   M.D. Blake, *A spectral bound for asymptotically norm-continuous semigroups*, J. Operator Theory **45** (2001), 111–130.

[BLX05]   C.J.K. Batty, J. Liang, and T.-J. Xiao, *On the spectral and growth bound of semigroups associated with hyperbolic equations*, Adv. Math. **191** (2005), 1–10.

[BP79]   A. Berman and R.J. Plemmons, *Nonnegative Matrices in the Mathematical Sciences*, Academic Press, 1979.

[BP05]   A. Bàtkai and S. Piazzera, *Semigroups for Delay Equations*, AK Peters, 2005.

[CL03]   D. Cramer and Y. Latushkin, *Gearhart–Prüss Theorem in stability for wave equations: a survey*, Evolution Equations (G. Ruiz Goldstein, R. Nagel, and S. Romanelli, eds.), Lect. Notes in Pure and Appl. Math., vol. 234, Marcel Dekker, 2003, pp. 105–119.

[Con85]   J.B. Conway, *A Course in Functional Analysis*, Graduate Texts in Math., vol. 96, Springer-Verlag, 1985.

[CPY74]   S.R. Caradus, W.E. Pfaffenberger, and B. Yood, *Calkin Algebras and Algebras of Operators on Banach Spaces*, Lect. Notes in Pure and Appl. Math., vol. 9, Marcel Dekker, 1974.

[CT06]   R. Chill and Y. Tomilov, *Stability of operator semigroups: ideas and results*, Banach Center Publ. (2006), (to appear).

[deL94]   R. deLaubenfels, *Existence Families, Functional Calculi and Evolution Equations*, Lect. Notes in Math., vol. 1570, Springer-Verlag, 1994.

[DS58]   N. Dunford and J.T. Schwartz, *Linear Operators I. General Theory*, Interscience Publishers, 1958.

[DS88]   W. Desch and W. Schappacher, *Some perturbation results for analytic semigroups*, Math. Ann. **281** (1988), 157–162.

[Dug66]   J. Dugundji, *Topology*, Allyn and Bacon, 1966.

[Dys49]   F.J. Dyson, *The radiation theories of Tomonaga, Schwinger, and Feynman*, Phys. Rev. **75** (1949), 486–502.

[EFNS05]   T. Eisner, B. Farkas, R. Nagel, and A. Serény, *Almost weak stability of C_0-semigroups*, Tübinger Berichte zur Funktionalanalysis **14** (2005), 66–76.

[EN00]   K.-J. Engel and R. Nagel, *One-Parameter Semigroups for Linear Evolution Equations*, Graduate Texts in Math., vol. 194, Springer-Verlag, 2000.

[Gea78]   L. Gearhart, *Spectral theory for contraction semigroups on Hilbert spaces*, Trans. Amer. Math. Soc. **236** (1978), 385–394.

[Gel39]   I.M. Gelfand, *On one-parameter groups of operators in a normed space*, Dokl. Akad. Nauk SSSR **25** (1939), 713–718.

[GGK90]   I. Gohberg, S. Goldberg, and M.A. Kaashoek, *Classes of Linear Operators I*, Oper. Theory Adv. Appl., vol. 49, Birkhäuser Verlag, 1990.

[Gol66]   S. Goldberg, *Unbounded Linear Operators*, McGraw-Hill, 1966.

[Gol85]   J.A. Goldstein, *Semigroups of Operators and Applications*, Oxford University Press, 1985.

[Gre84]   G. Greiner, *A typical Perron–Frobenius theorem with applications to an age-dependent population equation*, Infinite-Dimensional Systems (F. Kappel and W. Schappacher, eds.), Lect. Notes in Math., vol. 1076, Springer-Verlag, 1984, pp. 86–100.

[Gre85]   G. Greiner, *Some applications of Fejers theorem to one-parameter semigroups*, Semesterbericht Funktionalanalysis Tübingen **7** (Wintersemester 1984/85), 33–50, see also [Nag86, Sect. A-III.7].

[GS91]   G. Greiner and M. Schwarz, *Weak spectral mapping theorems for functional differential equations*, J. Differential Equations **94** (1991), 205–216.

[Haa06]   M. Haase, *The Functional Calculus for Sectorial Operators*, Oper. Theory Adv. Appl., Birkhäuser Verlag, 2006, (to appear).

[Hal63]   P.R. Halmos, *What does the spectral theorem say?*, Amer. Math. Monthly **70** (1963), 241–247.

[Hal74]   P.R. Halmos, *Measure Theory*, Graduate Texts in Math., vol. 18, Springer-Verlag, 1974.

[Hes70]   P. Hess, *Zur Störungstheorie linearer Operatoren in Banach-räumen*, Comment. Math. Helv. **45** (1970), 229–235.

[HHK06]   B. Haak, M. Haase, and P. Kunstmann, *Perturbation, interpolation, and maximal regularity*, Adv. Diff. Equations **11** (2006), 201–240.

[Hil48]   E. Hille, *Functional Analysis and Semigroups*, Amer. Math. Soc. Coll. Publ., vol. 31, Amer. Math. Soc., 1948.

[HP57]   E. Hille and R.S. Phillips, *Functional Analysis and Semigroups*, Amer. Math. Soc. Coll. Publ., vol. 31, Amer. Math. Soc., 1957.

[HW03]   M. Hieber and I. Wood, *Asymptotics of perturbations to the wave equation*, Evolution Equations (G. Ruiz Goldstein, R. Nagel, and S. Romanelli, eds.), Lecture Notes in Pure and Appl. Math., vol. 234, Marcel Dekker, 2003, pp. 243–252.

[Kat59]   T. Kato, *Remarks on pseudo-resolvents and infinitesimal generators of semigroups*, Proc. Japan Acad. **35** (1959), 467–468.

[Kat80]   T. Kato, *Perturbation Theory for Linear Operators*, 2nd ed., Grundlehren Math. Wiss., vol. 132, Springer-Verlag, 1980.

[Kel75]   J.L. Kelley, *General Topology*, Graduate Texts in Math., vol. 27, Springer-Verlag, 1975.

[Kre85]   U. Krengel, *Ergodic Theorems*, de Gruyter, 1985.

234   References

[KS05]   M. Kramar and E. Sikolya, *Spectral properties and asymptotic periodicity of flows in networks*, Math. Z. **249** (2005), 139–162.

[Lan93]   S. Lang, *Real and Functional Analysis*, Graduate Texts in Math., vol. 142, Springer-Verlag, 1993.

[Len94]   C. Lennard, *A Baire category theorem for the domains of iterates of a linear operator*, Rocky Mountain J. Math. **24** (1994), 615–627.

[Lia92]   A.M. Liapunov, *Stability of Motion*, Ph.D. thesis, Kharkov, 1892, English translation, Academic Press, 1966.

[Lot85]   H.P. Lotz, *Uniform convergence of operators on L^∞ and similar spaces*, Math. Z. **190** (1985), 207–220.

[LP61]   G. Lumer and R.S. Phillips, *Dissipative operators in a Banach space*, Pacific J. Math. **11** (1961), 679–698.

[Lun95]   A. Lunardi, *Analytic Semigroups and Optimal Regularity in Parabolic Problems*, Birkhäuser Verlag, 1995.

[LV88]   Yu.I. Lyubich and Quôc Phóng Vũ, *Asymptotic stability of linear differential equations on Banach spaces*, Studia Math. **88** (1988), 37–42.

[LZ99]   Z. Liu and S. Zheng, *Semigroups Associated with Dissipative Systems*, Res. Notes Math., vol. 398, Chapman & Hall/CRC, 1999.

[MM96]   J. Martinez and J.M. Mazón, *C_0-semigroups norm continuous at infinity*, Semigroup Forum **52** (1996), 213–224.

[MN91]   P. Meyer-Nieberg, *Banach Lattices*, Springer-Verlag, 1991.

[Nag86]   R. Nagel (ed.), *One-Parameter Semigroups of Positive Operators*, Lect. Notes in Math., vol. 1184, Springer-Verlag, 1986.

[Nee96]   J.M.A.A. van Neerven, *The Asymptotic Behaviour of Semigroups of Linear Operators*, Oper. Theory Adv. Appl., vol. 88, Birkhäuser Verlag, 1996.

[Neu88]   F. Neubrander, *Integrated semigroups and their application to the abstract Cauchy problem*, Pacific J. Math. **135** (1988), 111–157.

[NP94]   J.M.A.A. van Neerven and Ben de Pagter, *The adjoint of a positive semigroup*, Compositio Math. **90** (1994), 99–118.

[NP00]   R. Nagel and J. Poland, *The critical spectrum of a strongly continuous semigroup*, Adv. Math. **152** (2000), 120–133.

[Ped89]   G.K. Pedersen, *Analysis Now*, Graduate Texts in Math., vol. 118, Springer-Verlag, 1989.

[Phi53]   R.S. Phillips, *Perturbation theory for semi-groups of linear operators*, Trans. Amer. Math. Soc. **74** (1953), 199–221.

[Prü84]   J. Prüss, *On the spectrum of C_0-semigroups*, Trans. Amer. Math. Soc. **284** (1984), 847–857.

[Rao87]   M.M. Rao, *Measure Theory and Integration*, John Wiley & Sons, 1987.

[Ren94]   M. Renardy, *On the linear stability of hyperbolic PDEs and viscoelastic flows*, Z. Angew. Math. Phys. **45** (1994), 854–865.

[RR93]   M. Renardy and R.C. Rogers, *An Introduction to Partial Differential Equations*, Texts in Appl. Math., vol. 13, Springer-Verlag, 1993.

[RS72]   M. Reed and B. Simon, *Methods of Modern Mathematical Physics I. Functional Analysis*, Academic Press, New York, 1972.

[RS75]   M. Reed and B. Simon, *Methods of Modern Mathematical Physics II. Fourier Analysis and Self-Adjointness*, Academic Press, New York, 1975.

[Rud73]   W. Rudin, *Functional Analysis*, McGraw-Hill, 1973.

[Rud86]   W. Rudin, *Real and Complex Analysis*, 3rd ed., McGraw-Hill, 1986.

[Sch74]   H.H. Schaefer, *Banach Lattices and Positive Operators*, Grundlehren Math. Wiss., vol. 215, Springer-Verlag, 1974.

[Sch80]   H.H. Schaefer, *Topological Vector Spaces*, Graduate Texts in Math., vol. 3, Springer-Verlag, 1980.

[SD80]   W. Schempp and B. Dreseler, *Einführung in die Harmonische Analyse*, Teubner Verlag, 1980.

[Sin05]   E. Sinestrari, *Product semigroups and extrapolation spaces*, Semigroup Forum **71** (2005), 102–118.

[Sto32]   M.H. Stone, *On one-parameter unitary groups in Hilbert space*, Ann. of Math. **33** (1932), 643–648.

[Tay85]   A.E. Taylor, *General Theory of Functions and Integration*, Dover Publications, 1985.

[TL80]   A.E. Taylor and D.C. Lay, *Introduction to Functional Analysis*, 2nd ed., John Wiley & Sons, 1980.

[Tro58]   H. Trotter, *Approximation of semigroups of operators*, Pacific J. Math. **8** (1958), 887–919.

[Ulm92]   M.G. Ulmet, *Properties of semigroups generated by first order differential operators*, Results Math. **22** (1992), 821–832.

[Web85]   G.F. Webb, *Theory of Non-Linear Age-Dependent Population Dynamics*, Marcel Dekker, 1985.

[Web87]   G.F. Webb, *An operator-theoretic formulation of asynchronous exponential growth*, Trans. Amer. Math. Soc. **303** (1987), 751–763.

[Wei80]   J. Weidmann, *Linear Operators in Hilbert Space*, Graduate Texts in Math., vol. 68, Springer-Verlag, 1980.

[Wei95]   L. Weis, *The stability of positive semigroups on L_p-spaces*, Proc. Amer. Math. Soc. **123** (1995), 3089–3094.

[Wei98]   L. Weis, *A short proof for the stability theorem for positive semigroups on $L_p(\mu)$*, Proc. Amer. Math. Soc. **126** (1998), 3253–3256.

[Wol81]   M. Wolff, *A remark on the spectral bound of the generator of semigroups of positive operators with applications to stability theory*, Functional Analysis and Approximation (P.L. Butzer, B. Sz.-Nagy, and E. Görlich, eds.), Birkhäuser Verlag, 1981, pp. 39–50.

[Yos48]   K. Yosida, *On the differentiability and the representation of one-parameter semigroups of linear operators*, J. Math. Soc. Japan **1** (1948), 15–21.

[Yos65]   K. Yosida, *Functional Analysis*, Grundlehren Math. Wiss., vol. 123, Springer-Verlag, 1965.

Selected References to Recent Research

Semigroups for Functional Partial Differential Equations

[BP05]   A. Bàtkai and S. Piazzera, *Semigroups for Delay Equations*, AK Peters, 2005.

[BS04]   A. Bátkai and R. Schnaubelt, *Asymptotic behaviour of parabolic problems with delays in the highest order derivatives*, Semigroup Forum **69** (2004), 369–399.

Semigroups for Stochastic Partial Differential Equations

[DP04]   G. Da Prato, *An introduction to Markov semigroups*, Functional Analytic Methods for Evolution Equations (M. Iannelli, R. Nagel, and S. Piazzera, eds.), Lecture Notes in Math., vol. 1855, Springer, 2004, pp. 1–63.

[DP04b]   G. Da Prato, *Kolmogorov Equations for Stochastic PDEs*, Advanced Courses in Mathematics. CRM Barcelona, Birkhäuser Verlag, 2004.

[Jac05]   N. Jacob, *Pseudo Differential Operators and Markov Processes. Vol. I–III*, Imperial College Press, 2001–2005.

[MR04]   S. Magnilia, A. Rhandi, *Gaussian Measures on Separable Hilbert Spaces and Applications*, Quaderno dell' Università di Lecce, 1/2004.

[Tai03]   K. Taira, *Semigroups, Boundary Value Problems and Markov Processes*, Springer-Verlag, 2003.

Semigroups for Dynamical Boundary Conditions

[CENN03] V. Casarino, K.-J. Engel, R. Nagel, and G. Nickel, *A semigroup approach to boundary feedback systems*, Integral Equations Operator Theory **47** (2003), 289–306.

[FGGR02] A. Favini, G. Ruiz Goldstein, J. Goldstein, S. Romanelli, *The heat equation with generalized Wentzell boundary condition*, J. Evol. Equ. **2** (2002) 1–19.

[Nic04] G. Nickel, *A semigroup approach to dynamic boundary value problems*, Semigroup Forum **69** (2004), 159–183.

Semigroups for Maximal Regularity

[KW04] P.C. Kunstmann and L. Weis, *Maximal L^p-regularity for parabolic equations, Fourier multiplier theorems and H^∞-functional calculus*, Functional Analytic Methods for Evolution Equations (M. Iannelli, R. Nagel, and S. Piazzera, eds.), Lect. Notes in Math., vol. 1855, Springer, 2004, pp. 65–311.

Semigroups for Dynamical Networks

[KMS06] M. Kramar Fijavž, D. Mugnolo, and E. Sikolya, *Variational and semigroups methods for waves and diffusion in networks*, Appl. Math. Optim. (2006).

[KS05] M. Kramar and E. Sikolya, *Spectral properties and asymptotic periodicity of flows in networks*, Math. Z. **249** (2005), 139–162.

[MS06] T. Mátrai and E. Sikolya, *Asymptotic behavior of flows in networks*, Forum Math. (2006) (to appear).

[Sik05] E. Sikolya, *Flows in networks with dynamic ramification nodes*, J. Evol. Equ. **5** (2005), 441–463.

Semigroups for Numerical Analysis

[HV03] W. Hundsdorfer, J. Verwer, *Numerical Solution of Time-Dependent Advection-Diffusion-Reaction Equations*, Springer Ser. Comput. Math., vol 33, Springer-Verlag, 2003.

[IK02] K. Ito, F. Kappel, *Evolution Equations and Approximations*, Ser. Adv. Math. Appl. Sci., vol. 61, World Sci. Publishing, 2002.

[JL00] T. Jahnke, C. Lubich, *Error bounds for exponential operator splittings*, BIT **40** (2000), 735–744.

Semigroups for Boundary Control

[Las04]   I. Lasiecka, *Optimal control problems and Riccati equations for systems with unbounded controls and partially analytic generators—applications to boundary and point control problems*, Functional Analytic Methods for Evolution Equations (M. Iannelli, R. Nagel, and S. Piazzera, eds.), Lecture Notes in Math., vol. 1855, Springer, 2004, pp. 313–369.

Semigroups for Diffusion on Manifolds

[Ouh04]   E.-M. Ouhabaz, *Analysis of Heat Equations on Domains*, London Math. Sob. Monogr., vol. 31, Oxford University Press, 2004.

[Pre04]   M. Preunkert, *A semigroup version of the isoperimetric inequality*, Semigroup Forum **68** (2004), 233–245.

List of Symbols and Abbreviations

242   *Symbols and Abbreviations*

Index